D0850770

FLORIDA STATE
UNIVERSITY LIBRARIES

JUL 27 1995

TALLAHASSEE, FLORIDA

Encyclopaedia of
Mathematical Sciences

Volume 22

Editor-in-Chief: R. V. Gamkrelidze

A.A. Kirillov (Ed.)

Representation Theory and Noncommutative Harmonic Analysis I

Fundamental Concepts. Representations of Virasoro
and Affine Algebras

With 11 Figures

Springer-Verlag
Berlin Heidelberg New York
London Paris Tokyo
Hong Kong Barcelona
Budapest

Consulting Editors of the Series:
A. A. Agrachev, A. A. Gonchar, E. F. Mishchenko,
N. M. Ostianu, V. P. Sakharova, A. B. Zhishchenko

Sci
QA
176
T4613
1994

Title of the Russian edition:
Itogi nauki i tekhniki, Sovremennye problemy matematiki,
Fundamental'nye napravleniya, Vol. 22, Teoriya predstavlenij i nekommutativnyj
garmonicheskij analiz 1
Publisher VINITI, Moscow 1988

Mathematics Subject Classification (1991):
16Gxx, 17B68

ISBN 3-540-18698-0 Springer-Verlag Berlin Heidelberg New York
ISBN 0-387-18698-0 Springer-Verlag New York Berlin Heidelberg

Library of Congress Cataloging-in-Publication Data
Teoriia predstavleniĭ i nekommutativnyĭ garmonicheskiĭ analiz I.
English. Representation theory and noncommutative harmonic analysis I:
fundamental concepts, representations of Virasoro and affine algebras / A. A. Kirillov (ed.) p. cm. --
(Encyclopaedia of mathematical sciences; v. 22)
Includes bibliographical references and indexes.
ISBN 0-387-18698-0
1. Representations of groups. 2. Kac-Moody algebras. 3. Harmonic analysis.
I. Kirillov, A. A. (Aleksandr Aleksandrovich), 1936- . II. Title. III. Series.
QA176.T4613 1994 512'.55—dc20 94-11597

This work is subject to copyright. All rights are reserved, whether the whole or part of the material is concerned, specifically the rights of translation, reprinting, reuse of illustrations, recitation, broadcasting, reproduction on microfilm or in any other way, and storage in data banks. Duplication of this publication or parts thereof is permitted only under the provisions of the German Copyright Law of September 9, 1965, in its current version, and permission for use must always be obtained from Springer-Verlag. Violations are liable for prosecution under the German Copyright Law.
© Springer-Verlag Berlin Heidelberg 1994
Printed in Germany

Typesetting: Camera ready copy from the translator using a Springer T_EX macro package
SPIN 10008369   41/3140 - 5 4 3 2 1 0 - Printed on acid-free paper

List of Editors, Authors and Translators

Editor-in-Chief

R. V. Gamkrelidze, Russian Academy of Sciences, Steklov Mathematical Institute,
ul. Vavilova 42, 117966 Moscow, Institute for Scientific Information (VINITI),
ul. Usievicha 20a, 125219 Moscow

Consulting Editor

A. A. Kirillov, Moscow University, Mehmat, 117234 Moscow, Russia

Authors

A. A. Kirillov, Moscow University, Mehmat, 117234 Moscow, Russia
Yu. A. Neretin, Moscow Institute of Electronic Engineering, MIEM,
Bol'shoj Vuzovskij Per. 3/12, 109028 Moscow, Russia

Translator

V. Souček, Charles University, Mathematical Institute, Sokolovská 83,
18600 Prague, Czech Republic

Contents

I. Introduction to the Theory of Representations and Noncommutative Harmonic Analysis

A. A. Kirillov

Translated from the Russian
by V. Souček

Contents

Chapter 1
A Historical Sketch

§1. Foreword

Noncommutative harmonic analysis and its basic tool – the theory of group representations – has existed as an independent domain of mathematics for about 100 years.

Harmonic analysis in the most general sense can be defined as the mathematical apparatus applicable to the study and use of symmetry in the surrounding world and in its mathematical models. As a rule, the symmetry is described in terms of transformation groups and the subject of harmonic analysis is the study of the corresponding representations of groups. Taking such a point of view, harmonic analysis covers a broad field and includes a considerable part of the whole of classical and contemporary mathematics. Besides this, the idea of a symmetry plays quite a substantial role in theoretical physics and in other natural sciences.

The full account of the history of harmonic analysis has not yet been written (see the review papers quoted in the bibliography) and to write it would need more space, time and qualification than the author has at his disposal. My task is much more modest – to make the reader familiar with the history of evolution of the theory of group representations up to the contemporary level. Understandably, priority is inevitably given to the fields that are the closest to the author's interests.

§2. Finite-Dimensional Representations

The first ideas of harmonic analysis – the notions of the generating function of a sequence and of "group character" of a finite abelian group – had appeared in number theory and in probability theory at the beginning of the eighteenth century, even before the creation of the notion of a group itself. With the appearance of Fourier series and the Fourier transform in mathematical physics in the nineteenth century, commutative harmonic analysis became the most powerful tool for the solution of various translation-invariant problems. It is sufficient to recall the theory of differential equations with constant coefficients, Fourier series and Fourier transform.

The birth of noncommutative harmonic analysis is usually connected with a series of papers by F.G.Frobenius, published in the period $1896-1901$. The original aim of Frobenius was the solution of a problem posed by his teacher R.I.W.Dedekind. The problem was the following. Denote by $x_1, ..., x_n$ elements of a finite group G of order n. The multiplication table (also called the Cayley table) of the group G can be interpreted as an $n \times n$-matrix; the determinant of the matrix is a homogeneous polynomial of order n, depending on n variables $x_1, ..., x_n$. Dedekind called it the group determinant and found that the determinant can be split, for a commutative group G, into linear factors, the coefficients of which coincide with the group characters. For example, for the cyclic group of order 3 we have the identity

$$\begin{vmatrix} x & y & z \\ z & x & y \\ y & z & x \end{vmatrix} = (x + y + z)(x + \varepsilon y + \varepsilon^2 z)(x + \varepsilon^2 y + \varepsilon z),$$

where $\varepsilon = e^{2\pi i/3}$.

Dedekind was able to decompose the group determinant into simple (not necessarily linear) factors for some noncommutative groups, including S_3 and the group of quaternions, but he did not find a general pattern underlying it. Frobenius discovered that in the general case the group determinant is a product of the form $P_1^{n_1} \ldots P_k^{n_k}$, where the P_i are irreducible polynomials of degree n_i. He also introduced expressions

$$\chi_j^{(i)} = \frac{\partial P_i}{\partial x_j}(1, 0, \ldots, 0),$$

calling them group characters, and showed that they give (in the modern terminology) multiplicative functionals on the centre of the group algebra of the group G. It was only in his next paper that he found an interpretation of the group characters as traces of matrix representations and from that moment he began the study of the theory of representations in the contemporary sense of the word. Frobenius gave the full classification of irreducible representations of the most important finite groups – the group of symmetries of regular polygons, the permutation group S_n and alternating group A_n. Moreover, he proved the complete reducibility of representations of finite groups. The result was at the same time (1897) and independently found by Fedor Eduardovich Molin. The result is usually called Maschke's theorem, after the mathematician who proved it one year later for representations over a field of any characteristic p such that the order of the group is not divisible by p.

Another important theoretical result of Frobenius was a theorem involving a relation between representations of a group and subgroup, which shows a duality between the functors of restriction and induction. The result is now called the Frobenius reciprocity principle.

I.Schur, who was a student of Frobenius, introduced the notion of averaging over the group, proved the orthogonality relations for matrix elements and studied the behaviour of representations under extension of the base field. It was Schur who began a systematic study of projective representations and along the way created the theory of central extensions of a group.

W.Burnside was the first to apply the representation theory of finite groups to the study of the structure of groups. The theorem on the solvability of groups of order $p^\alpha q^\beta$ for prime numbers p and q was the starting point for a gigantic work of classification of all finite simple groups that was finished only recently. It is interesting that the role of representation theory in the solution of the classification problem was considered in turn negligible Burnside (1901); Gorenstein (1984), or very substantial Burnside (1917); Curtis (1979).

An excellent achievement in the representation theory and harmonic analysis was the duality principle, found by L.S.Pontryagin in 1934 for discrete and compact abelian groups in connection with his topological studies, and generalized by E.R. van Kampen in 1935 to locally compact commutative groups. Although the mentioned result belongs to commutative harmonic analysis, it has played and is still playing a very important role in the noncommutative harmonic analysis.

A fundamental series of papers by A.Young, devoted to the representations of the permutation group S_n, brought the first period of the evolution of representation theory to an end. Even if the subject is far from being exhausted (and also difficult papers by Young cannot be considered to be completely understood), the main interest in representation theory moved into the field of compact topological groups. The most important results of the second period are the theorem by A.Haar and J. von Neumann on the existence of a finite invariant measure and the Peter-Weyl theorems on the completness of the system of finite-dimensional irreducible unitary representations of compact groups. At the same time H. Weyl and É. Cartan created the theory of finite-dimensional representations of semisimple Lie groups. These results were not only exceptionally beautiful, but they have also found many applications (the theory of symmetric spaces in geometry, the theory of momentum in quantum mechanics, spinors in field theory). Noncommutative harmonic analysis has changed from a purely algebraic and a very special theory into an applied science, which was becoming more and more popular.

The methods of the theory of groups became, thanks to papers by H.Weyl, J. von Neumann, E.P.Wigner and P.A.M.Dirac, a principal working tool in the new quantum physics.

§3. Infinite-Dimensional Representations

The necessity to consider noncompact groups and their infinite-dimensional representations very soon became clear. The exact formulation of the canonical commutation relations

$$PQ - QP = i\hbar$$

looks very awkward because of the necessity of considering the unbounded operators P and Q on their common domain of definition. The suitable form of the canonical commutation relation was suggested by H. Weyl in 1927; by introducing the operators $U_t = \exp(itP), V_s = \exp(isQ)$:

$$U_t V_s = e^{i\hbar st} V_s U_t.$$

The loosely formulated problem of describing realizations of the canonical commutation relations was in this way transformed into the precisely formulated question of classifying all unitary representations of a special 3-dimensional Lie group (which was later called the Heisenberg group). The problem was solved independently by M.H.Stone in 1930 and J. von Neumann in 1931. Thus one of the basic principles of quantum mechanics was shown to be a variant of the manifestation of "symmetry" in nature.

Another attempt to use infinite-dimensional representations in physics was made by E.P.Wigner in 1939. It was realised that the classification of irreducible representations of the nonhomogeneous Lorentz group (the group of motions in Minkowski space) coincides with the classification of elementary particles. Namely, both representations and elementary particles are characterised by two parameters – one is continuous (mass) and one is discrete (spin). Hence also in this case symmetry plays a decisive role.

The systematic study of infinite-dimensional representations of groups, which forms the basic content of the third period, began in the 1940s.

The work Gel'fand, Rajkov (1943) on the completeness of the system of unitary irreducible representations of locally compact groups can be naturally identified with the beginning of the third period.

The transition from compact groups to locally compact ones, from finite-dimensional representations to infinite-dimensional ones, needed a substantial reworking of the theory of representations. Instead of purely algebraic constructions, the basic role is played by topology, measure theory and functional analysis. The problem of how to decompose representations into irreducible components has appeared to be much more difficult in the infinite-dimensional case than in the finite-dimensional one. Besides the necessity of employing continuous sums (integrals) instead of ordinary (discrete) sums, new features appeared here, connected with the fact that the decomposition is not unique and that there exist so-called factor representations of type II and III. The multiplicity of an irreducible component in a discrete sum is equal to a positive integer or to infinity; the multiplicity for factor-representations of type II can be any real number or infinity and it is always infinity for type III.

The theory of decompositions of unitary representations was created independently by I.Segal and F.I.Mautner in 1950 using the well-known results of F.Murray and J. von Neumann (1936–1943) on the theory of factors (for details see Chap. 6). At the same time papers by G.M.Adel'son-Vel'skij and R.Godement appeared.

The first classification theorems for infinite-dimensional representations were published for the Lorentz group and for the group of affine transformations on the real line in 1947 by I.M.Gel'fand and M.A.Najmark and for the restricted Lorentz group $SL(2, \mathbb{R}) \simeq SO(2, 1)$ by V.Bargmann.

In 1950 the monograph "Unitary representations of classical groups" by I.M.Gel'fand and M.A.Najmark appeared, in which infinite-dimensional representations of the groups $SL(n, \mathbb{C}), SO(n, \mathbb{C})$ and $Sp(n, \mathbb{C})$ were constructed for any n. The book has become quite famous and the flow of papers on the theory of infinite-dimensional representations has been growing ever since. We shall follow here its basic lines. Before this it is necessary to remark that research in the theory of representations of finite and compact groups did not cease. Quite to the contrary, some ideas and methods of the theory of infinite-dimensional representations have also proved to be useful in the finite-dimensional situation. For example, the theory of representations of finite simple groups of Lie type was constructed in recent papers by P.Deligne and G.Lusztig following the pattern of the theory of (infinite-dimensional) representations of semisimple Lie groups.

§4. The General Theory of Infinite-Dimensional Representations

After the discovery of factors of type II and III by F.Murray and J. von Neumann the problem of the possible types of factor representations of a given group arose. Following deep investigations by G.Mackey, J.M.G.Fell, J.Dixmier and J.Glimm a special class of groups was found, which are termed groups of type I. A summary of the activity in this direction can be found in the well-known monograph on C^*-algebras and their representations by J.Dixmier.

Harish-Chandra proved in 1953 that all semisimple Lie groups are of type I. A few years later a simpler and more beautiful proof was found by R.Godement. J.Bernstein has recently succeeded in transfering Godement's method to the case of semisimple p-adic groups. E.Takenouchi has shown that all exponential groups are of type I and J.Dixmier has proved the same for real algebraic groups. The first example of a wild group (that is not of type I) was shown already in papers by F.Murray and J. von Neumann. The first wild Lie group was found, it seems, by F.I.Mautner. His example was rediscovered many times. E.Thoma noticed that a countable discrete group is always a wild group with the only exception being when the group is an extension of a finite group by an Abelian group.

L.Auslander and B.Kostant have found for solvable groups a criterion for a group to be of type I in terms of orbits (for the method of orbits see below).

Recently, mainly due to the efforts of A.Connes, the theory of factors II and III was substantially advanced. This will clearly increase the "applicability" of wild groups. There are already papers in which factors of types II and III are used in mathematical physics (the structure of the algebra of observables), in topology (the classification of knots and links), and in geometry and analysis (noncommutative integral geometry).

§5. Induced Representations

The method of induced representations was found already by G.Frobenius in the case of finite groups. The analogue of the Frobenius reciprocity principle was discovered for compact groups by É.Cartan and H.Weyl. Different versions of the principle for infinite-dimensional representations were created by I.M.Gel'fand, I.I.Pyatetskij-Shapiro and G.I.Ol'shanskij.

It was noted already in papers by E.P.Wigner, I.M.Gel'fand and M.A. Najmark that their method of the construction of infinite-dimensional representations is a generalization of the notion of an induced representation. A systematic study of the construction was made by G.W.Mackey in the case of locally compact groups; he gave a suitable criterion for a representation to be induced and contributed much to the popularity of the method. Similarly to the case of finite groups, the class of monomial locally compact groups (i.e. groups all of whose irreducible unitary representations are unitarily induced by a one-dimensional representation of a subgroup) is contained in the class of solvable groups and contains the class of connected nilpotent Lie groups. A further generalization has led to the notion of a holomorphically induced representation (V.Bargmann, I.M.Gel'fand-M.I.Graev, R.J.Blattner) and to the notion of a representation in cohomology (A.Borel, A.Weil, R.Bott, R.P.Langlands). The latter method has allowed the unified construction of irreducible representations of semisimple algebraic groups over the complex, real, p-adic and finite fields to be carried out.

The most important results in this respect are due to P.Deligne, D.Kazhdan and G.Lusztig.

The method of induced representations is not directly applicable to infinite-dimensional groups (or more precisely to pairs $G \supset H$ with an infinite-dimensional factor space G/H). This is why other approaches have been used for the construction of representations of such groups (see Chap. 7 and Ol'shanskij (1984)).

§6. Representations of Semisimple Groups

This is, apparently, the most beautiful, the deepest and the most important (for applications) part of the theory of infinite-dimensional representations. Its founders were on the one hand I.M.Gel'fand and M.A.Najmark, who constructed already in 1950 the principal, complementary and degenerated series of representations of the classical complex groups, computed their characters and studied their reducibility (at the same time I.M.Gel'fand introduced the notion of infinitesimal character), and on the other hand Harish-Chandra.

The important series of papers by Harish-Chandra on the theory of infinite-dimensional representations of semisimple groups began in 1951. He independently discovered infinitesimal characters, carried out a deep study of universal enveloping algebras and proved a remarkable theorem showing that semisimple groups belong to type I and that generalized characters are regular. Papers by I.M.Gel'fand and M.I.Graev and independently by Harish-Chandra on representations of real semisimple groups appeared in the mid-1950s.

The work by F.A.Berezin on the Laplace operator on semisimple Lie groups appeared in 1957; the complete classification of irreducible infinite-dimensional representations of semisimple groups in Banach spaces was given there for the first time. The proof of this important result contained a gap in the first version, which was removed in the complementary paper published in 1963. Another approach to the classification problem was outlined in the series of papers by M.A.Najmark in the 1950s and elaborated upon in the work of D.P.Zhelobenko in the 1960s and 1970s. It is based on the fact that irreducible representations of a semisimple group G have a spectrum with finite multiplicities after the restriction to the maximal compact subgroup $K \subset G$ and that the component with minimal highest weight appears with multiplicity one.

The problem of distinguishing unitary irreducible representations inside the known list of irreducible representations in Banach spaces suprisingly appeared to be quite difficult. In 1967 the work by E.M.Stein appeared in which singular complementary representations of $\mathrm{GL}(n, \mathbb{C})$ (missing in the list of Gel'fand-Najmark) were discovered. And only recently the complete description of $\hat{G}$ [1] for $G = \mathrm{GL}(n, \mathbb{C})$ has been proved in a paper by M.Tadić. It was shown that the representations found by E.M.Stein, together with those described by I.M.Gel'fand and M.A.Najmark, exhaust $\hat{G}$. The same problem for the other complex simple groups has not yet been solved. Already at the end of the 1950s it became clear that representations of semisimple groups are closely connected with a new branch of functional analysis, namely integral geometry. The main problem in this field is the study of special integral operators of geometrical origin. An important role is played in the representation theory by the operator defined for any function f on the group G as the integral over the space of conjugacy classes of elements. Integral geometry is nowadays quite a developed independent field of mathematics. A review of its results will be contained in one of the future volumes of the Encyclopaedia.

At the beginning of the 1960s the study of representations of semisimple groups over the field of p-adic numbers and over the ring of adèles began. The interest in such groups and representations was inspired by work of A.Weil and A.Selberg in number theory and the theory of automorphic forms. A connection of their work with representation theory was discovered and made known by I.M.Gel'fand and R.P.Langlands. The Langlands conjecture formulating the noncommutative analogue of class field theory triggered a whole stream of papers on representation theory. The review of this deep direction of study will be given in the algebraic volumes of the Encyclopaedia. It

---

[1] The symbol $\hat{G}$ denotes the set of equivalence classes of unitary irreducible representations of the group G.

will only be noted here that the theory of representations of real and p-adic semisimple groups is nowadays substantially based on Langlands' classification of tempered representations. Important contributions to this field were given by R.E.Howe, T.J.Enright, W.Schmid, G.Zuckerman, A.Knapp, D.Vogan, J. Wolf and N.Wallach.

The theory of infinite-dimensional representations of semisimple groups also led to new results in classical questions of finite-dimensional theory. I.M.Gel'fand and M.L.Tsetlin have found an explicit basis for finite-dimensional representations of the linear and orthogonal groups. D.P.Zhelobenko constructed a similar basis for representations of the symplectic groups.

Recently a series of papers has appeared in which "models of representations" (i.e. the infinite-dimensional representations of semisimple groups which are the direct sum of all finite-dimensional representations of the group) are constructed.

§7. The Method of Orbits

This method appeared for the first time in a paper by the author in 1962 in which representations of nilpotent Lie groups were considered. It allowed one to give simple and intuitive answers to all basic questions concerning infinite-dimensional unitary representations of a nilpotent Lie group in terms of a finite-dimensional object – the orbits of the group G in its coadjoint representation. It was also observed that the method can be generalized to other classes of groups. The method was applied to exponential, solvable, semisimple and general Lie groups in papers by P.Bernat, B.Kostant, L.Auslander, L.Pukanszky, M.Duflo and the author. A connection of the method with mechanics was noticed by B.Kostant and J.M.Souriau. The classification of homogeneous symplectic manifolds by orbits in the coadjoint representation was discovered independently by B. Kostant, J.-M.Souriau and the author in 1970. As has been recently clarified, the Poisson structure on the space dual to the Lie algebra was already known to Sophus Lie. The use of the structure for the construction of systems with symmetries in quantum mechanics became the subject of the new field in mathematical physics called geometrical quantization. For a review of the field see Kirillov (1980). Another direction in mathematical physics which is connected with the method of orbits is the theory of completely integrable Hamiltonian systems. It turned out that almost all known examples of such systems can be arranged into a relatively simple general scheme (the so-called Adler-Kostant scheme). For more details see volumes 4,5,15 of the Encyclopaedia.

§8. Infinite-Dimensional Groups

Representations of infinite-dimensional Lie groups and Lie algebras arise in different fields of mathematics and physics. Their systematic study began only recently, but it continues to attract more and more attention. There are three types of infinite-dimensional groups which are the most interesting and the richest in applications.

Firstly, groups of invertible operators on infinite-dimensional spaces (for example, the group $U(H)$ of unitary operators in a Hilbert space H). The study of representations of groups of this type began at the beginning of the 1970s. Generalized characters of the group $U(\infty)$ corresponding to its factor representations of type II_1 were computed in papers by D.Voiculescu and S.Stratila. The present author showed that among the different variants of infinite-dimensional unitary groups there was one belonging to type I. Then G.I.Ol'shanskij created the theory of (G, K)-pairs and their representations, which is in many respects similar to the classical theory of finite-dimensional representations of semisimple groups. More details can be found in Ol'shanskij (1984) and in further volumes of the Encyclopaedia.

Secondly, "continuous products" of finite-dimensional groups, which are called current groups in physics, are quite interesting. They are defined to be groups G^M consisting of smooth maps from the manifold M into the finite-dimensional Lie group G. The study of representations of the group G^M started with papers by H.Araki and E.Woods and was substantially advanced by I.M.Gel'fand, M.I.Graev and A.M.Vershik, S.Albeverio - R.Høegh-Krohn and P.Delorme. The case of $M = S^1$ is especially interesting. It turned out that the corresponding "loop groups" are connected with infinite-dimensional analogues of semisimple Lie algebras – the so-called Kac-Moody algebras. The theory of representations of Kac-Moody algebras is already well-developed not only due to the efforts of mathematicians, but also by the effort of specialists in mathematical physics. An account of the research in this field can be found in the V.Kac' book and in two volumes of Publications of Mathematical Sciences Research Institute in Berkeley (vols. 3 and 4).

Thirdly, important examples of infinite-dimensional Lie groups and algebras are groups of diffeomorphisms of smooth manifolds and Lie algebras of vector fields. Here again the most simple and the most important examples are those connected with the circle. I.M.Gel'fand and D.B.Fuchs noted that Lie algebra of vector fields on the circle had nontrivial central extensions. The corresponding central extensions of the group of diffeomorphisms were constructed by R.Bott. The extended Lie algebra is called the Virasoro algebra by physicists. The theory of representations of the Virasoro algebra is one of mathematical tools in the "string approach" to quantum field theory. This fact contributed a lot to the popularity of the Virasoro algebra and during the past five or ten years more than a hundred papers in mathematics and physics journals, dedicated to the representations of the Virasoro algebra, have been published.

The Virasoro algebra and Kac-Moody algebras can be naturally unified into one infinite-dimensional Lie algebra. The corresponding group plays the role of the symmetry group of (the gauge group) of so-called conformal field theory.

Groups of diffeomorphisms of other manifolds were studied by R.S.Ismagilov, who presented new original constructions and found many new phenomena.

§9. Representations of Lie Supergroups and Superalgebras

During the past ten years the ideas of supersymmetry, under the influence of theoretical physics, have penetrated more and more into mathematics. F.A.Berezin was one of the mathematicians who initiated the field and made it popular. He discovered that there is a surprising analogy in the theory of the second quantization between particles of two sorts – fermions and bosons. Later it became clear that the analogy is far-reaching: almost all notions in "ordinary" mathematics have their "superanalogues". Thus, Grassmann algebra corresponds to the ring of polynomials, Clifford algebra to the ring of differential operators, supermanifolds to manifolds, Lie supergroups and superalgebras to Lie groups and algebras. The typical feature of "superobjects" is the existence of particular "odd" coordinates and "sign rules" for computations with them. At first "supermathematics" was considered only as a new, sometimes useful, way to express old facts (for example, the alternating sum in the Lefschetz formula was interpreted as the "supertrace" of a linear operator). At present, however, a new generation of mathematicians and physicists has already grown up and is thinking naturally in the new language. An important role in the development of the new theory was played by papers by E.Witten.

The theory of representations of Lie superalgebras and supergroups is nowadays an independent field of mathematics, but it is still far from having the clarity and completeness which is typical for the classical theory. At present the classification of simple finite-dimensional Lie superalgebras is known (V.G.Kac, I.Kaplansky) and a systematic study of their finite-dimensional representations has begun. The main difficulty is the fact that such representations are not completely reducible, and thus the study of general representations cannot be reduced to the study of irreducible ones. Nevertheless, many results of the ordinary theory (the description of irreducible representations by their characters, the structure of the centre of the universal enveloping algebra, invariant integration) have been successfully transferred to Lie supergroups and superalgebras.

The superanalogues of the Virasoro algebra and Kac-Moody algebras play a very important role in field theory. The study of the coadjoint representations of the corresponding supergroups has proved to be very interesting. In particular, it is possible to find in this direction natural generalizations (in the super sense) of the classical integrable systems of the type of the Korteweg-de Vries equations.

Chapter 2
Basic Notions of the Theory of Representations

§1. Group Actions

1.1. Actions. Any realization of an (abstract) group G in the form of a specific group of transformations of a set will be called a *representation of the group G in the broad sense*. As a rule, the set is equipped with various additional structures and it is understood that the transformations of the considered group respect all these additional structures. In particular, representations with the property that the elements of the group are realized by continuous transformations (homeomorphisms) of topological spaces, smooth transformations (diffeomorphisms) of manifolds, isometries of Riemannian spaces, linear operators on Banach, Hilbert or other spaces, projective maps etc., are widely used.

The term "representation" in such a broad sense means the same as "*group action*". Let us recall that the map

$$G \times X \to X : (g, x) \mapsto gx \tag{2.1}$$

is called a *left action* of the group G on the set X if the following property holds:

$$(g_1 g_2)x = g_1(g_2 x), \quad ex = x, \tag{2.2}$$

where $g_1, g_2 \in G, x \in X$ and e denotes the identity of the group G. The left action induces the map

$$\pi : G \to \operatorname{Aut} X; \quad \pi(g) : x \mapsto gx$$

of the group G into the group of all automorphisms of the set X. Property (2.2) means that the map π is a group homomorphism, i.e.

$$\pi(g_1 g_2) = \pi(g_1)\pi(g_2), \quad \pi(e) = \operatorname{id},$$

where the symbol id denotes the identity map. Together with the left action, the *right action* is also considered. It is the map

$$X \times G \to X : (x, g) \mapsto xg \tag{2.1'}$$

with the properties

$$x(g_1 g_2) = (xg_1)g_2, \quad xe = x. \tag{2.2'}$$

In this case, the map

$$\pi' : G \to \text{Aut } X; \quad \pi'(g) : x \mapsto xg$$

is an *antihomomorphism* of the group G into the group $\text{Aut } X$; this means that

$$\pi'(g_1 g_2) = \pi'(g_2)\,\pi'(g_1), \quad \pi'(e) = \text{id}.$$

A set X with a given left (resp. right) action of G on X is called a *left* (resp. *right*) *G-space*.

Example 2.1. Let $G = \text{GL}(n)$ be the group of invertible matrices of order n. The set of all vectors with n entries, considered as columns, is a left G-space and the set of all vectors with n entries, considered as rows, is a right G-space if the group action is given by the matrix multiplication.

1.2. The Category of G-spaces. The family of all left (right) G-spaces for a given group G forms a category with morphisms given by *G-maps*, i.e. by maps commuting with the group action.

The commutativity of two maps g and h is often conveniently expressed in the form of a *commutative diagram:*

$$
\begin{array}{ccc}
X_1 & \overset{g}{\longrightarrow} & X_1 \\
\downarrow{\scriptstyle h} & & \downarrow{\scriptstyle h} \\
X_2 & \overset{g}{\longrightarrow} & X_2
\end{array}
$$

The categories of left and right G-spaces are naturally isomorphic. Namely, for every left G-space X, there is the corresponding right G-space, which is the same set X with the action of G given by the formula $xg := g^{-1}x$ (here the sign := means that the right hand side defines the left hand side; similarly for =:).

Let X and Y be two G-spaces. The *product* of the spaces X and Y over G is defined as the factor space of the Cartesian product $X \times Y$ by the relation given by the action of the group G on $X \times Y$. For example, if X is a right G-space and Y is a left G-space, pairs (x, y) and (x', y') are equivalent if there is an element $g \in G$ such that $x' = xg$, $y' = g^{-1}y$. The product over G will be denoted by $X \times_G Y$.

If H is a subgroup of the group G, then every G-space is also an H-space. It defines the *restriction functor* Res_H^G mapping the category of G-spaces into the category of H-spaces. Less obvious, but very useful, is the dual functor Ind_H^G from the category of H-spaces into the category of G-spaces. The functor Ind_H^G is defined by the formula

$$\text{Ind}_H^G X = G \times_H X, \tag{2.3}$$

where G is considered as a left G-space and a right H-space. The notion of the *induction functor* is illustrated in Ex. 2.4 below.

We shall say that a G-space X is homogeneous or that G *acts transitively* on X if any point $x \in X$ can be mapped to any other point $y \in X$ by a transformation belonging to the group G. Any G-space is a union of homogeneous G-spaces. Namely, given a point $x \in X$, the set of all elements of the form $gx, g \in G$, will be called the *G-orbit* of the point x. It is clear that any G-orbit is a homogeneous G-space and that the set X is the union of the orbits of its points.

Example 2.2. Let G be the proper Lorentz group $SO_0(3,1)$ and let X be the Minkowski space $\mathbb{R}^{3,1}$. Then there are 4 types of orbits (see Fig. 1).

1) one-sheeted hyperboloid, formed by spacelike vectors with fixed length;
2) one half of the two-sheeted hyperboloid, formed by all timelike vectors with the same length and the same time direction;
3) one half of the cone, formed by all null vectors with the given time direction;
4) the zero vector.

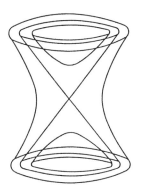

Fig. 1

· Given a point x of a G-space X, the set $G(x)$ of all elements $g \in G$ leaving x fixed will be called *the stabilizer* or *the stationary subgroup of the point x.* The set $G(x)$ is a subgroup of G. If the points x and y belong to the same G-orbit, then their stabilizers $G(x)$ and $G(y)$ are conjugate in G. Namely, if $y = gx$, then $G(y) = gG(x)g^{-1}$.

The set of all elements of the group G mapping a given point x to the point $y = gx$ has the form $gG(x) = \{gh | h \in G(x)\}$, i.e. it is a (left) coset in the quotient space $G/G(x)$. Any homogeneous left G-space is isomorphic to the space of all left cosets $G/G(x)$, where x is a point in X. Conversely, if H is any subgroup of G, then it is the stabilizer of the point $x = H$ in the space $X := G/H$ of left cosets of G with respect to H.

The same is true for right homogeneous G-spaces and for the corresponding right cosets with respect to a subgroup H; the set of all right cosets will be denoted by $H \backslash G$.

Homogeneous G-spaces G/H_1 and G/H_2 are isomorphic (in the category of G-spaces) if and only if the subgroups H_1 and H_2 are conjugate. The group $\operatorname{Aut} X$ of all automorphisms of the G-space $X = G/H$ is isomorphic to the factor group $N_G(H)/H$, where $N_G(H)$ is the normalizer of H in G.

Example 2.3. The family of subspaces

$$F : \{0\} = V_0 \subset V_1 \subset \ldots \subset V_n = \mathbb{C}^n, \; \dim V_k = k,$$

will be called a *flag* in $\mathbb{C}^n$. The set $\mathcal{F}$ of all flags F is a homogeneous $\mathrm{U}(n)$-space, where $\mathrm{U}(n)$ is the group of unitary transformations in $\mathbb{C}^n$ preserving the standard scalar product. The stabilizer of the "coordinate" flag F_0 (for which V_k is defined to be the span of the basis vectors $e_1, e_2, \ldots, e_k$) is the subgroup D of diagonal matrices in $\mathrm{U}(n)$; the normalizer of D in $\mathrm{U}(n)$ is the group M of monomial matrices (i.e. matrices with only one nonzero entry in every column and every row). The group $\operatorname{Aut} \mathcal{F} = M/D$ can be identified with the group S_n of all permutations of a set with n elements.

Let X and Y be G-spaces and let $p : X \mapsto Y$ be a G-map. Then the set $Z = p^{-1}(y)$ is a $G(y)$-space. If the space X is homogeneous, then the relation

$$X = \operatorname{Ind}_{G(y)}^{G} Z$$

holds in the category of G-spaces. In the case that the set Y is not just a point, the G-space X is called *imprimitive* and the decomposition of the set X into sets $p^{-1}(y), y \in Y$, is called the system of imprimitivity. A homogeneous space $X = G/H$ which has no system of imprimitivity will be called *primitive*. The primitivity of G/H is equivalent to the maximality of the subgroup H in G.

Example 2.4. Let Y be a smooth manifold, and let X be a fibre bundle over Y with fiber Z. Suppose that the group G acts on X preserving the structure of the fiber bundle (i.e. it maps fibers into fibers). If the action of G on the set of fibers is transitive, then $X = \operatorname{Ind}_H^G Z$, where H is the stabilizer of the fiber Z.

1.3. Actions of Topological Groups. The groups and spaces considered carry a topology in the majority of interesting examples.

A set G which is at the same time a group and a topological space is called a *topological group* if the group structure and the topological structure are connected by the condition:

the map $G \times G \to G : (g_1, g_2) \mapsto g_1 g_2^{-1}$ is continuous.

The condition is equivalent to the set of the following three conditions, which are much easier to verify:

1)   The map $(g_1, g_2) \mapsto g_1 g_2$ is continuous separately in g_1 and g_2.
2)   The map $g \mapsto g^{-1}$ is continuous at the point $g = e$.
3)   The map $(g_1, g_2) \mapsto g_1 g_2$ is continuous (as a map depending on two variables) at the point (e, e).

An *action of a topological group* G on a topological space X is defined in the same way as above with the additional requirement that the maps in (2.1) and (2.1') be continuous. If the topological G-space X is a Hausdorff space, then the stationary subgroup $G(x)$ is a closed subgroup in G for any point $x \in X$. Conversely, let H be a closed subgroup of a topological group G, then the set of all left cosets G/H equipped with the quotient topology will be a homogeneous Hausdorff topological G-space.

If X is another homogeneous topological space with the same stationary subgroup $H = G(x)$, then the natural map $\phi : G/H \mapsto X$ will be one-to-one and continuous. If X is locally compact, then ϕ is a homeomorphism.

In many important examples, the topological group G is a Lie group, i.e. it is a smooth manifold such that the group law is given by a smooth map. In such a case, any closed subgroup $H \subset G$ is also a Lie group, the space G/H is a smooth manifold and the action of G on it is smooth.

Let $p : G \mapsto X := G/H$, $p(g) = gH$, be the natural projection of the group G onto the homogeneous space X. Any map s from a subset of X into G with the property $p \circ s = \mathrm{id}$ (i.e. $s(x)H = x$) will be called a *section* of the projection p. There are usually no global continuous sections (i.e. continuous sections defined on all of X) but for locally compact groups G, there exist always local continuous sections defined in a neighborhood of a given point $x \in X$. In the case that the group G is a Lie group and X is a smooth manifold, there exist always smooth local sections.

Example 2.5. Consider the natural action of the group $G = \mathrm{SO}(3)$ on the two-dimensional sphere $S^2 = X$. Let us fix a unit tangent vector ξ at a point of the sphere. An element $g \in G$ is completely determined by the image $g\xi$ of the vector ξ and these images fill in the space $T_1 X$ of all unit tangent vectors to X. The projection $p : G \mapsto X$ has a simple geometrical meaning in this case: it maps the vector $\xi \in T_1 X$ to the point in which the vector is attached to the base. A section s is now a vector field on X or on part of X such that the vectors at individual points have unit length. The well-known "hedgehog theorem" then tells us that there are no global continuous sections in this case.

§2. Linear Representations

2.1. Basic Definitions. Let G be a group and let V be a vector space. A homomorphism T of the group G into the group $\operatorname{Aut} V = \operatorname{GL}(V)$ of invertible linear transformations of V will be called a *linear representation* of the group G in the space V. To say it another way, a linear representation is an operator-valued function $g \mapsto T(g)$ with the property

$$T(g_1 g_2) = T(g_1)T(g_2), \quad T(e) = \mathbb{1}_V, \tag{2.4}$$

where e is the identity of the group G and $\mathbb{1}_V$ is the identity map in the space V. In what follows, we shall consider, as a rule, only linear representations and we shall call them simply representations.

If a G-space X is given, it is possible to construct a linear representation T of the group G in the following way. Let $L(X)$ be the vector space of functions on X. (We shall not specify here where the values of the functions lie or what class of functions is considered.) Then the formula

$$[T(g)f](x) = \begin{cases} f(xg) & \text{for a right } G\text{-space} \\ f(g^{-1}x) & \text{for a left } G\text{-space} \end{cases} \tag{2.5}$$

gives a linear representation of the group G on the space $L(X)$, which is called the *geometrical* or the *permutational* representation. The described construction allows us to use linear representations for the study of general representations.

Example 2.6. Let μ be a measure on a space X and let $\{A_t\}$ be a one-parameter group of transformations of X preserving the measure μ. If we take the space $L_2(X, \mu)$ of square-integrable complex-valued functions on X for the space $L(X)$, then formula (2.5) defines a linear representation of the group $\mathbb{R}$ by unitary operators on $L(X)$. The study of such representations is the subject of the spectral theory of dynamical systems. The most interesting case for applications is the case in which X is the phase space of a mechanical system, μ is the Liouville measure and $\{A_t\}$ is the dynamical group describing the time evolution of the system.

If G is a topological group and V is a linear topological space, then, as a rule, we shall suppose (without mentioning it explicitly) that the considered representation T is *continuous* in the following sense: the map $(g, v) \mapsto T(g)v$ is jointly continuous in both variables $g \in G, v \in V$.

In the case when the group G is locally compact and the space V is a Fréchet (or even Banach or Hilbert) space, the continuity of T is equivalent to the formally weaker condition of separate continuity of $T(g)v$ in g and v, which, in turn, is equivalent to the continuity of T as a map from G into the group $\operatorname{Aut} V$ equipped with the strong operator topology.

The representation (T, V) is called a *Banach representation* if V is a Banach space, an *isometric representation* if the operators $T(g), g \in G$, preserve the norm in V, and a *unitary representation* if the space V is a Hilbert space and the operators $T(g), g \in G$, are unitary.

2.2. The Category of Linear Representations. The family of representations of a group G in linear spaces over a given field (or skew-field) K forms a category $\Pi(G, K)$. The morphisms of the category are the *intertwining operators*, i.e. the operators commuting with the action of the group. The set of intertwining operators for representations T_1 and T_2 acting in linear spaces V_1 and V_2 is denoted by $\mathcal{C}(T_1, T_2)$ or $\mathrm{Hom}_G(V_1, V_2)$. Instead of $\mathcal{C}(T, T)$ or $\mathrm{Hom}_G(V, V)$, we shall write simply $\mathcal{C}(T)$ or $\mathrm{End}_G(V)$. The dimension of the space $\mathcal{C}(T_1, T_2)$ is denoted by $c(T_1, T_2) = \dim \mathrm{Hom}_G(V_1, V_2)$ and is called the *intertwining number* of the representations T_1 and T_2. The relation $c(T_1, T_2) = c(T_2, T_1)$ holds for unitary representations.

Representations T_1 and T_2 are called *disjoint* if

$$c(T_1, T_2) = c(T_2, T_1) = 0.$$

Representations T_1 and T_2 are called *equivalent* if there is an invertible intertwining operator for them (other equivalent definitions are: T_1 and T_2 are equivalent objects of the category $\Pi(G, K)$ or $T_1(g)$ and $T_2(g)$ are represented by the same matrices for a suitable choice of bases in V_1 and V_2).

If T is a representation of G in a space V, then the formula

$$T^*(g) := T(g^{-1})^* \tag{2.6}$$

defines a representation T^* of the group G in the dual space V^*. The representation T^* is called the *dual* or *contragredient* to the representation T.

Let T be a representation of a group G in a space V. If there is a subspace V_1 invariant with respect to all operators $T(g), g \in G$, of the representation, we shall say that the representation T is *reducible*. The restrictions of the operators $T(g)$ to V_1 define a representation T_1 of the group G in the space V_1, called the *subrepresentation* of T. The representation T_2 induced in the quotient space $V_2 = V/V_1$ is called the *factor representation* of T. In the topological case, only closed subspaces V_1 are considered. Representations without nontrivial subrepresentations are called *algebraically irreducible* (resp. *topologically irreducible*). If an invariant subspace V_1 admits an invariant complement V_2, we shall call the representation T *decomposable* and we shall write $T = T_1 + T_2$.

Example 2.7. Let $G = \mathbb{R}$. For all $k = 1, 2, \ldots, \infty$, we shall define a representation T_k in the space V_k formed by all polynomials in x of degree less then k by the formula

$$[T_k(t)P](x) = P(x + t).$$

It is not difficult to show that all intertwining operators in $\mathcal{C}(T_k, T_l)$ are differential operators with constant coefficients. As a consequence, we have

$c(T_k, T_l) = \min(k, l)$ for $k < \infty$, $c(T_\infty, T_k) = 0$. The representations T_k, $k > 0$, are reducible but indecomposable: the space V_{k-1} has no G-invariant complement in V_k.

The operation of *tensor product* is defined for any pair of representations T_1 and T_2 acting on spaces V_1 and V_2. The representation $T = T_1 \otimes T_2$ acts on the tensor product $V = V_1 \otimes V_2$ by the formula

$$T(g)(v_1 \otimes v_2) = T_1(g)v_1 \otimes T_2(g)v_2. \tag{2.7}$$

The symbol $\otimes$ denotes here the ordinary (algebraic) tensor product of vectors, spaces and operators. In the infinite-dimensional situation, the construction is usually substituted by one or another variant of a topological tensor product given by the completion of the algebraic tensor product in a suitable topology.

For Banach representations, there are many ways to introduce a norm on the space $V = V_1 \otimes V_2$ even if we require (as is often done) the conditions

$$||v_1 \otimes v_2|| = ||v_1||.||v_2|| \text{ for } v_i \in V_i$$

and

$$||f_1 \otimes f_2|| = ||f_1||.||f_2|| \text{ for } f_i \in V_i^*$$

to be satisfied. Among the norms with these properties (they are called uniform crossnorms), there is the largest one $p_{\max}$ and the smallest one $p_{\min}$. The completions of V in these norms are denoted by $V_1 \hat{\otimes} V_2$, resp. $V_1 \check{\otimes} V_2$, and called *projective*, resp. *weak,* tensor products.

The explicit formulas for their norms are:

$$p_{\max}(v) = \inf \sum_k ||v_{1k}|| \cdot ||v_{2k}||,$$

where the infimum is taken over all presentations of $v \in V$ in the form $v = \sum_k v_{1k} \otimes v_{2k}$;

$$p_{\min}(v) = \sup(f_1 \otimes f_2)(v)$$

where the supremum is taken over all functionals $f_i \in V_i^*$ of unit norm.

The notions of the projective and weak tensor product can be generalized to locally convex spaces. The important class of *nuclear spaces* is characterized by the following distinguishing property: their projective and weak tensor products with any given locally convex space coincide.

Example 2.8. Let V be a Banach space and V^* its dual. Then the algebraic tensor product $V \otimes V^*$ can be identified with the space of linear operators of finite rank on V; the operator $w \mapsto f(w)v$ of rank 1 corresponds to the element $v \otimes f \in V \otimes V^*$. It is possible to check that in this case, the norm $p_{\min}$ coincides with the standard operator norm and the norm $p_{\max}$ coincides with the so-called *nuclear* (or *trace) norm.*

Let V_1 and V_2 be Hilbert spaces. Among the cross norms on $V = V_1 \otimes V_2$, there is the unique norm defined by the scalar product. The completion of V in this norm is called the *Hilbert tensor product* of V_1 and V_2.

In all cases mentioned above, the operators $T(g)$ described by the formula (2.7) can be extended as operators on the topological tensor product. In such a way the operations of the *projective* and *weak tensor product of Banach representations* and the *Hilbert tensor product of unitary representations* are defined.

Besides algebraic and topological irreducibility, there are several other notions of irreducibility for infinite-dimensional representations, which are useful in different circumstances.

A representation T of the group G in a space V is called *tensorially irreducible* if for every trivial representation of the group G in a nuclear space W, every closed G-invariant subspace in $V \hat{\otimes} W$ has the form $V \hat{\otimes} W_1$, where $W_1 \subset W$. In the special case when the space W is a finite-dimensional space of dimension k, the property of the representation T formulated above is called *k-irreducibility*.

It is clear that for $k = 1$ we get the standard topological irreducibility. For $k = 2$, it leads to the notion of *operator irreducibility* : the representation T is called operator irreducible if every closed operator A in V commuting with T is a scalar. (A linear operator A is said to be closed if the graph of A is a closed subspace in $V \oplus V = V \otimes W$.) For unitary representations T the properties of k-irreducibility for different k are equivalent.

2.3. Projective Representations. We shall consider one more class of representations of groups, close to the linear ones – the so-called *projective representations*. Let V be a linear space over a field K. The set $P(V)$ of all one-dimensional subspaces of V is called the projective space corresponding to V or the *projectivization* of V. It is clear that $P(V)$ coincides with the set of all orbits of the multiplicative group $K^* = K \setminus \{0\}$ in $V \setminus \{0\}$.

A map $\tilde{A} : P(V) \mapsto P(V)$ is called a *projective transformation* of $P(V)$ if there is an invertible linear operator A on V such that we have the commutative diagram

$$
\begin{array}{ccc}
V \setminus \{0\} & \xrightarrow{A} & V \setminus \{0\} \\
\downarrow & & \downarrow \\
P(V) & \xrightarrow{\tilde{A}} & P(V)
\end{array}
$$

The group of projective transformations of $P(V)$ is denoted by $\mathrm{PGL}(V)$. In the case when V is a topological linear space, the space $P(V)$ is endowed with the factor topology and then $\mathrm{PGL}(V)$ denotes the group of continuous projective transformations. If the space H is a complex or real Hilbert space, then the space $P(H)$ inherits the natural metric

$$d(l_1, l_2) = ||P_1 - P_2||,$$

where the P_i are orthogonal projections of H onto one-dimensional subspaces $l_i \in P(H)$. It can be shown that every isometry of $P(H)$ is induced from a unitary or antiunitary (i.e. antilinear isometric) operator on H. The group of projective unitary transformations is denoted be $\mathrm{PU}(H)$.

A *projective representation* of a group G is a homomorphism T of the group G into $\mathrm{PGL}(V)$. The notions of *reducible, decomposable, equivalent and unitary projective representations* are defined in the same way as for linear representations.

An element of the group $\mathrm{PGL}(V)$ can be considered as an invertible linear operator in V, defined up to a scalar multiple. A projective representation T of the group G is hence given by an operator–valued map $\tilde{T} : G \mapsto \mathrm{GL}(V)$ having the property

$$\tilde{T}(g_1)\tilde{T}(g_2) = c(g_1, g_2)\tilde{T}(g_1 g_2),$$

where $c(g_1, g_2)$ is a scalar function on $G \times G$ with values in K^*; it is sufficient to choose a representive $\tilde{T}(g)$ of the class $T(g)$. A different choice of representatives leads to the multiplication of $c(g_1, g_2)$ by a function of the form

$$c_0(g_1, g_2) = b(g_1)b(g_2)b(g_1 g_2)^{-1}. \tag{2.8}$$

Aside from that, the function $c(g_1, g_2)$ is not arbitrary. It satisfies the identity

$$c(g_1, g_2)c(g_1 g_2, g_3) = c(g_1, g_2 g_3)c(g_2, g_3), \tag{2.9}$$

which follows from the associativity of the multiplication in G. In the language of group cohomology, the relation (2.9) says that $c(g_1, g_2)$ is a 2-cocycle on G with values in K^* and (2.8) means that $c_0(g_1, g_2)$ is the coboundary of the 1-cochain $b(g)$. Thus the cohomology class $h_T \in H^2(G, K^*)$ containing the cocycle c is well–defined by the projective representation T. Any cocycle $c(g_1, g_2)$ belonging to the class h_T is called a *multiplier* of the projective representation T. If the class h_T is trivial, then we say that the projective representation T is equivalent to a linear one or that the representation is *linearized*. In such a case it is possible to choose representatives $\tilde{T}(g)$ such that the multiplier $c(g_1, g_2)$ is identically equal to 1.

To study projective representations of a group G, it is convenient to introduce a bigger group $\tilde{G}$ that is an extension of G by an Abelian group A:

$$1 \mapsto A \mapsto \tilde{G} \mapsto G \mapsto 1.$$

Every projective representation T of the group G can be considered as a representation of $\tilde{G}$, trivial on A. The image of the class h_T in $H^2(\tilde{G}, K^*)$ can be trivial (for a suitable choice of the group A and the extension $\tilde{G}$). Then we say that the representation T is linearized by the extension $\tilde{G}$. For a given projective representation T, there always exists an extension $\tilde{G}$ linearizing T. Namely, it is possible to define $\tilde{G}$ as the set of all pairs $(g, \tilde{T}(g))$, where $\tilde{T}(g) \in T(g)$. The map $(g, \tilde{T}(g)) \mapsto g$ defines a projection of $\tilde{G}$ onto G and the map $(g, \tilde{T}(g)) \mapsto \tilde{T}(g)$ is a linear representation of $\tilde{G}$. The role of A is played here by the group K^*.

The multiplier $c(g_1, g_2)$ has unit norm for unitary projective representations.

It is remarkable that for a given group G, there exists the *universal extension* $\tilde{G}$ linearizing all unitary projective representations of G. The group A in this case is the group dual to $Z^2(G, T)$.

§3. Noncommutative Harmonic Analysis

The etymology of the term "harmonic analysis" is connected with the physical interpretation of the decomposition of (periodic) functions using Fourier integral (Fourier series) as the representation of an arbitrary process in the form of a superposition of harmonic oscilations.

In noncommutative harmonic analysis, all mathematical objects with noncommutative symmetry group are studied. We shall list here basic problems of noncommutative harmonic analysis.

3.1. The Classification of Representations. The question is to describe, for a given group G, all its linear or projective representations of a certain type (finite-dimensional, unitary, irreducible, nondecomposable, real, integral, etc.). There are general methods to solve the problem for many types of groups and representations. They will be described in the following chapters of the presented review. It is necessary to remark that many specific problems are not covered by the available methods. Besides this, the procedures for the classification of representations suggested by the theory often lead to difficult algebraic, combinatorial and analytic problems. Computer techniques have recently been successfully applied to the classification problems connected with finite groups.

The classification of representations is usually reduced to the description of the "most simple" (indecomposable or irreducible) representations and to the study of their direct sums (integrals) and extensions.

3.2. The Computation of the Spectrum of a Representation. Representations interesting for applications are realized, as a rule, on spaces of functions, vector or tensor fields, differential operators and another geometric objects on G-manifolds. In the case of homogenenous spaces G/H, they are induced from the subgroup H (see Chap. 3 and Chap. 5 below).

The basic role in the solution of the problem is played by the study of intertwining operators for the representation T with itself or with other representations of the group. In particular, the "method of horospheres", introduced by I.M.Gel'fand and M.I.Graev, has proved to be very effective for infinite-dimensional representations of semisimple Lie groups. The heart of the method is the study of a special intertwining operator between the given representation and a representation on the space of functions defined on the so-called "basic affine space" of the semisimple group G having the form G/N, where N is a maximal unipotent subgroup. In its simplest variant, the method leads to the operator given by the integration over horospheres in Lobachevskij space, which gave the name to the method.

3.3. The Functors Res **and** Ind. The restriction functor Res_H^G from $\Pi(G, K)$ to $\Pi(H, K)$ is defined for any subgroup H of G (see 1.2 of Chap. 2). For some classes of groups and subgroups and for some subcategories of representations, it is also possible to define the functor Ind_H^G from $\Pi(H, K)$ to $\Pi(G, K)$ – see 4.2 of Chap. 3, 1.3 of Chap. 4 and 2.3 of Chap. 6 below. (Note that the induction functor, defined in 1.2 of Chap. 2, maps objects of $\Pi(H, K)$ into nonlinear G-spaces.)

For a given group G and a subgroup H, we need to compute the action of the functors Res_H^G and Ind_H^G on irreducible representations explicitly. It includes, as a special case, the problem of the decomposition of the induced representation and the problem of the decomposition of the tensor product of two irreducible representations. (The point is that the tensor product $T_1 \otimes T_2$ can be interpreted as $\mathrm{Res}_G^{G \times G}(T_1 \times T_2)$; see Chap. 3.)

The basic result here is the Frobenius reciprocity theorem and its various generalizations. In its simplest form (for compact groups), the theorem establishes a duality between the functors Res_H^G and Ind_H^G expressed by the coincidence of the intertwining numbers

$$c(\mathrm{Res}_H^G T, S) = c(T, \mathrm{Ind}_H^G S),$$

where T is a representation of the group G and S is a representation of its subgroup H. The theorem is no longer valid for infinite-dimensional representations of noncompact groups, but in some situations (e.g. if G is a Lie group and H is a discrete subgroup), a similar assertion, with a more complicated formulation, holds.

A simple geometrical interpretation of the functors Res_H^G and Ind_H^G is given by the method of orbits (see Chap. 7).

3.4. The Fourier Transform on a Group. For commutative groups, the Fourier transform maps functions on a group G into functions on the dual group $\tilde{G}$. The characteristic property of the transformation is the fact that it mutually exchanges two maximal commutative subalgebras of operators: one is generated by the multiplication by functions, the other by operators of translation on the group.

The convolution operators are included in the second subalgebra. If G is a Lie group, differential operators with constant coefficients are limits of convolution operators. This explains the role of the Fourier transform in the theory of differential equations: every differential equation with constant coefficients is reduced to an algebraic equation.

The Fourier transform on a noncommutative group is designed to play a similar role. However, the translations on the group now generate a noncommutative algebra. Even so, the Fourier transform carries over this algebra into the algebra of multiplication by functions, this time, however, by operator–valued functions instead of scalar valued functions. More precisely, the left translation by g is carried over to the left multiplication and the right translation by g is carried over to the right multiplication by the same operator-valued function on the set $\hat{G}$ (the *dual object* to the noncommutative group G).

For a broad class of groups, the dual object $\hat{G}$ is represented by a set of equivalence classes of irreducible unitary representations of the group G. An important and difficult problem is to describe the image of a given class of functions on the group G under the Fourier transform. In particular, an analogue of the Plancherel formula claims that the Fourier transform on a unimodular group G is a unitary operator from $L_2(G, dg)$ into $L_2(\hat{G}, \mu)$, where μ is the so-called Plancherel measure on $\hat{G}$.

For complex semisimple groups G, there is an analogue of the Paley–Wiener theorem, describing the image of the space $\mathcal{D}(G)$ of smooth functions on G with compact support under the Fourier transform.

Harmonic analysis on nilpotent Lie groups has recently found more and more applications in the theory of differential operators.

3.5. Special Functions and Representation Theory. Almost all special functions known nowadays were realized as matrix elements of operators of a representation and intertwining operators. Some of them were discovered in conection with a symmetry, for example, spherical functions, Bessel functions, Legendre polynomials, etc. For others, the connection with the theory of representations was found relatively recently. As a rule, it is natural to formulate various "addition theorems" and differential equations for special functions in the language of representation theory. The influence in the other direction is also substantial: the study of analytic properties of special functions brings important information to the theory of representation. A special part of one of the next issues of the series will be devoted to this subject.

3.6. The Computation of Generalized and Infinitesimal Characters of Representations of Lie Groups. The two problems mentioned are closely connected with each other because infinitesimal characters induce differential equations for generalized characters (see Chap. 6). Moreover, the knowledge of characters plays a substantial role in the solution of classification problems, the study of the functors Res and Ind and the study of the group algebra.

Explicit formulas for the characters of irreducible representations of compact and complex semisimple Lie groups are known. A universal integral representation of generalized characters and a simple expression for infinitesimal characters were obtained by using the orbit method (see Chap. 7).

Chapter 3
Representations of Finite Groups

§1. The General Theory of Complex Finite-Dimensional Representations

1.1. The Formulation of Basic Results.

Theorem 3.1.

a) *Every representation is equivalent to a unitary one (i.e. elements of the group are represented by unitary matrices in a suitable basis).*
b) *Every representation is a direct sum of irreducible representations.*
c) *For every finite group G, there are only a finite number of equivalence classes of irreducible representations. Their number is equal to the number of conjugacy classes of elements of G.*
d) *Let us denote the dimensions of the irreducible representations of the group G by $d_1, \ldots, d_k$ and the order of the group G by $|G|$. Then the number $|G|$ is divisible by all d_i and we have Burnside's identity*

$$d_1^2 + \ldots + d_k^2 = |G|.$$

(Stronger divisibility properties of numbers $d_1, \ldots, d_k$ can be found in Sect. 3 and Sect. 5 of Chap. 3.) In simple cases, the theorem gives enough information to compute the numbers $k, d_1, \ldots d_k$ for a given group G.

Example 3.1. Let $G = \mathbb{Z}_n$ be the cyclic group of order n. In this case, the number k is equal to n (in any commutative group, the conjugacy class of an element consists of one point). Hence all the numbers d_i are equal to 1 and all irreducible representations are one-dimensional, i.e. they are characters of the commutative group G. Let us denote them by $\chi_0, \chi_1, \ldots, \chi_{n-1}$. It is not difficult to find them explicitly. Because of $\chi_k(1)^n = \chi_k(n) = \chi_k(0) = 1$, the numbers $\chi_k(1)$ are n-th roots of unity. Set $\chi_k(1) = e^{2\pi i k/n}$. Then

$$\chi_k(m) = e^{2\pi i k m/n}, \ \ k = 0, 1, \ldots n - 1.$$

Example 3.2. Let $G = S_3$ be the group of all permutations of a set with 3 elements. It can be suitably realized as the group of symmetries of an equilateral triangle. The group has 6 elements and is formed by three conjugacy classes: the trivial map, 3 reflections and 2 rotations. The equation $d_1^2 + d_2^2 + d_3^2 = 6$ has a unique solution (up to permutations) in the set of all integers, namely $d_1 = d_2 = 1, d_3 = 2$. So the group has two one-dimensional representations (i.e. characters) χ_0 and χ_1 and one two-dimensional representation T. The explicit form of the characters χ_0 and χ_1 is easy to guess.

One of them is trivial: $\chi_0 \equiv 1$, and the other coincides with the sign of the permutation (or with the determinant of the transformation of the plane): $\chi_1(g) = \operatorname{sgn} g \ (= \det \ g)$. The explicit form of the two–dimensional representation will be shown below.

The proof of part a) of Theorem 3.1 is based on the important process of *averaging over the group*.

Lemma 3.1. *If a group G acts on a linear space W by a representation S, then the operator*

$$P = \frac{1}{|G|} \sum_{g \in G} S(g)$$

is the projector onto the subspace W^G of all vectors fixed or invariant with respect to the group G.

Applying the procedure to the space $W = H(V)$ of hermitian forms on V we shall get the following result. Let Q be a hermitian form on the space V of a representation T of the group G; then the form

$$\bar{Q} = \frac{1}{|G|} \sum_{g \in G} T(g)^* Q T(g)$$

is a G-invariant hermitian form on V. If $Q > 0$, then $\bar{Q} > 0$ and thus is regular.

Part a) can be deduced in a standard way: in a suitable basis, the matrix of the form $\bar{Q}$ will be the identity matrix; all operators of the representation will be represented by unitary matrices in such a basis.

Part b) follows from a) because all unitary representations are completely reducible (see Sect. 2 of Chap. 2).

To prove parts c) and d), the following simple, but very important, result is usually used. We shall describe it in the next subsection.

1.2. Schur's Lemma and Its Consequences.

Lemma 3.2. (Schur's Lemma) *Suppose that the representations T_1 and T_2 are irreducible; then any intertwining map A between them is either invertible or zero map.*

The proof follows immediately from the fact that the kernel and the range of the operator A are invariant subspaces.

Two important corollaries can be deduced from Schur's lemma: a formula connecting the intertwining numbers of two representations S and R with their decomposition into irreducible parts and orthogonality relations for matrix elements of irreducible representations.

Let $\hat{G} = \{T_1, \ldots, T_k\}$ be the set (of equivalence classes) of irreducible representations of G and let the representations S and R have the following decomposition into irreducible parts:

$$S = \sum_{i=1}^{k} s_i T_i, \ R = \sum_{i=1}^{k} r_i T_i.$$

Then

$$c(S, R) = \sum_{i=1}^{k} s_i r_i. \tag{3.1}$$

The proof is based on the possibility of finding an explicit form for an intertwining operator $A \in \mathcal{C}(S, R)$. (This fact is often used in applications, where the operators considered are intertwining operators of the corresponding representations of the group). Namely, in suitable bases, the representations S and R can be written in block-diagonal form with irreducible blocks along the main diagonal:

$$S = \begin{pmatrix} T_{i_1} & 0 & \cdots & 0 \\ 0 & T_{i_2} & \cdots & 0 \\ 0 & 0 & \cdots & T_{i_m} \end{pmatrix}, \quad R = \begin{pmatrix} T_{j_1} & 0 & \cdots & 0 \\ 0 & T_{j_2} & \cdots & 0 \\ 0 & 0 & \cdots & T_{j_n} \end{pmatrix}.$$

With the same choice of the basis, the operator A is also represented by block matrices such that the individual blocks A_{pq} belong to $\mathcal{C}(T_{i_p}, T_{j_q})$. Hence $c(S, R) = \sum_{i=1}^{k} s_i r_j c(T_i, T_j)$. It follows from Schur's lemma that we have $c(T_i, T_j) = 0$ for $i \neq j$. As for $\mathcal{C}(T_i, T_i)$, Schur's lemma claims that it is a field (i.e. an algebra such that all nonzero elements are invertible). Over the field of complex numbers $\mathbb{C}$, any finite-dimensional field coincides with $\mathbb{C}$. This means that $c(T_i, T_i) = 1$. It proves Eq. (3.1) and, what is even more important, it gives an explicit form for the operator $A \in \mathcal{C}(S, R)$ as a block matrix with scalar blocks.

Let us consider two irreducible representations T and S of the group G acting on spaces V and W, respectively. Let A be any operator from V to W. Using the averaging process, we can construct an intertwining operator $\bar{A} = |G|^{-1} \sum_{g \in G} S(g) A T(g)^{-1}$ from A. If T and S are not equivalent, then $\mathcal{C}(T, S) = 0$; hence the operator $\bar{A}$ should be the zero operator. In the language of matrix elements this means that

$$\sum_{g \in G} s_{ij}(g) t_{kl}(g^{-1}) = 0 \quad \text{for all} \quad i, j, k, l. \tag{3.2}$$

Choosing a basis in V in such a way that the matrices $T(g)$ are unitary, we can rewrite the relation (3.2) in the form

$$\sum_{g \in G} s_{ij}(g) \overline{t_{kl}(g)} = 0 \quad \text{for all} \quad i, j, l, k. \tag{3.2'}$$

If the representations T and S are equivalent, then they are represented in suitable bases by the same unitary matrices. In such a case, the operator $\bar{A}$ is necessarily a scalar operator. Its trace is clearly equal to the trace of A, so we have

$$|G|^{-1} \sum_{g \in G} T(g) A T(g)^{-1} = \frac{\text{tr } A}{\dim V} \mathbb{1}_V.$$

In terms of matrix elements this means that

$$|G|^{-1} \sum_{g \in G} t_{ij}(g) \overline{t_{lk}(g)} = (\dim T)^{-1} \delta_{il} \delta_{jk}. \tag{3.3}$$

Equations (3.2) and (3.3) prove

Lemma 3.3. *The matrix elements of unitary irreducible representations form an orthogonal system in the space $L_2(G)$ of functions on G with scalar product*

$$(f_1, f_2) = |G|^{-1} \sum_{g \in G} f_1(g) \overline{f_2(g)}.$$

Part b) of Theorem 3.1 follows from Lemma 3.3 because the space $L_2(G)$ is finite-dimensional.

The right and left hand sides of the Burnside identity can be interpreted as the dimension of the space $L_2(G)$ and the cardinality of the orthogonal system of matrix elements in it, respectively. At the same time, the identity is equivalent to the completeness of the system of matrix elements. There are several ways of proving the completeness. All of them, in essence, use one special representation R of the group G. It acts in the space $L_2(G)$ by the formula

$$[R(g)f](g_1) = f(g_1 g)$$

and is called the *right regular representation*. An equivalent version, called the *left regular representation,* is given by the formula

$$[L(g)f](g_1) = f(g^{-1}g_1).$$

The characteristic property of the regular representation R is described by

Lemma 3.4. *We have $c(R, T) = \dim T$ for any representation T.*

The proof of a more general statement will be given in Sect. 5 of Chap. 3 (Ex. 3.10).

Applying Lemma 3.4 to all irreducible representation $T_i \in \hat{G}$ and then to the regular representation R we get (because of Eq.(3.1)):

$$R = \sum_{i1}^{k} d_i T_i, \quad |G| = \dim R = \sum_{i=1}^{k} d_i{}^2,$$

where $d_i = \dim T_i$.

§2. The Theory of Characters and Group Algebras

2.1. Basic Properties of Characters. The notion of a *character* (trace) of a representation T :

$$\chi_T(g) = \operatorname{tr} T(g)$$

is a very important technical tool in the theory of representations.

Theorem 3.2.

a) *Characters are constant on conjugacy classes: $\chi(ghg^{-1}) = \chi(h)$. (In other words, characters are central functions on the group.)*

b) *Characters of irreducible representations form an orthonormal basis in the space of all central functions on G (considered as a subspace in $L_2(G)$).*

c) *The characters of two representations coincide if and only if the representations are equivalent.*

d) *We have the following relations:*

$$\chi_{T_1+T_2} = \chi_{T_1} + \chi_{T_2}, \ \chi_{T_1 \otimes T_2} = \chi_{T_1} \chi_{T_2}.$$

Part a) is a consequence of the well-known property of the trace of a matrix: $\operatorname{tr}(AB) = \operatorname{tr}(BA)$. If we take into account that, up to a multiple, there is a unique central function among all linear combinations of matrix elements of an irreducible representation T – the character χ_T (such a function necessarily has the form $f(g) = \operatorname{tr}(AT(g))$, where A is an intertwining operator), part b) follows from the orthogonality relations and from the completeness of the system of matrix elements. Note that the second part of assertion c) of Theorem 3.1 follows from part b) of Theorem 3.2 just proved.

Part c) follows from part d) and from Theorem 3.1.

To prove d), we note that, in a suitable basis, all the operators considered are diagonal and the assertion becomes trivial.

Example 3.3. Let X be a G-space. We shall compute the character of the "geometric" representation T acting on the space $L_2(X)$ by the formula

$$[T(g)f](x) = f(g^{-1}x).$$

If we consider the standard basis formed by δ-functions on X (i.e. by functions that are different from zero only at one point), then the operators $T(g)$ act by permutations of vectors of the basis. The trace of a permutation matrix is equal to the number of units on the diagonal, i.e. to the number of fixed points:

$$\chi_T(g) = |X^g|,$$

where X^g denotes the set of all points $x \in X$ such that $gx = x$.

The formula above gives in the special case of the regular representation

$$\chi_R(g) = |G| \cdot \delta(g) = \begin{cases} |G|, & \text{if } g = e, \\ 0, & \text{if } g \neq e. \end{cases}$$

Example 3.4. Let us find the character θ of the two-dimensional represen-tation T from Ex. 3.2. The regular representation of the group S_3 has the form $R = \chi_0 + \chi_1 + 2T$. It implies $\theta = \frac{1}{2}(\chi_R - \chi_0 - \chi_1)$. It follows from $\chi_R = 6\delta$ that the values of θ on the identity, reflection and rotation are equal to 2, 0 and -1, respectively.

We add a useful reformulation of Eq.(3.1) in terms of characters:

$$c(T, S) = (\chi_T, \chi_S)_{L_2(G)} \tag{3.1'}$$

and a criterion following from it.

Criterion of irreducibility: a representation T is irreducible if and only if $||\chi_T||_{L_2(G)} = 1$.

As the first application of the theory of characters, we shall consider the problem of the description of representations of the direct product of two groups. Suppose that $G = G_1 \times G_2$. If T_i are representations of the groups G_i on spaces V_i, then we can define a representation $T = T_1 \times T_2$ of the group G on the space $V_1 \otimes V_2$ by the formula

$$T(g_1, g_2) = T_1(g_1) \otimes T_2(g_2).$$

Lemma 3.5. *The map $(T_1, T_2) \mapsto T_1 \times T_2$ gives a bijection between $\hat{G}_1 \times \hat{G}_2$ and $\hat{G}$.*

In fact, it is easy to show (see Theorem 3.2 c)) that

$$\chi_{T_1 \times T_2}(g_1, g_2) = \chi_{T_1}(g_1)\chi_{T_2}(g_2).$$

Applying the criterion of irreducibility, we get that if T_1 and T_2 are irre-ducible, the same is true for $T_1 \times T_2$. The orthogonality relations for characters show that representations $T = T_1 \times T_2$ constructed from different pairs (T_1, T_2) are not equivalent and the Burnside identity implies the completeness of the constructed system of representations.

2.2. The Group Algebra. The group algebra of a finite group G over the field $\mathbb{C}$ is defined as the set $\mathbb{C}[G]$ of all formal linear combinations $\sum_{g \in G} c_g \cdot g$ of elements of the group with complex coefficients. The operations of addition and multiplication by a number are defined componentwise and multiplication is extended from the group law in G by linearity.

It is useful to interpret the set of coefficients $\{c_g\}$ as a function on G. Under such an interpretation, addition and multiplication by a number are the usual operations on functions and multiplication is given by the formula

$$(f_1 * f_2)(g) = \sum_{g_1 g_2 = g} f_1(g_1)f_2(g_2) = \sum_{h \in G} f_1(h)f_2(h^{-1}g).$$

The function $f_1 * f_2$ is called the *convolution of the functions f_1 and f_2*. It is possible to define an *involution* on the algebra $\mathbb{C}[G]$ by

$$f^*(g) = \overline{f(g^{-1})}.$$

The involution $f \mapsto f^*$ is an antilinear antiautomorphism of $\mathbb{C}[G]$, i.e.

$$(\lambda_1 f_1 + \lambda_2 f_2)^* = \bar{\lambda}_1 f_1^* + \bar{\lambda}_2 f_2^*, \quad (f_1 * f_2)^* = f_2^* * f_1^*.$$

For every complex representation T of the group G, there is a representation of the algebra $\mathbb{C}[G]$ denoted by the same letter and given by the formula

$$T(f) = \sum_{g \in G} f(g) T(g).$$

So the space V of the representation T is a $\mathbb{C}[G]$-module. It is easy to see that the category $\Pi(G, \mathbb{C})$ of complex representations of the group G can be identified with the category of nondegenerate modules over $\mathbb{C}[G]$ (a module V over an algebra A is called nondegenerate if $AV = V$). The subcategory of unitary representations is identified with the category of nondegenerate modules over $\mathbb{C}[G]$ considered as algebra with involution, i.e. with the category of nondegenerate *-representations of $\mathbb{C}[G]$. (A linear representation T of an algebra A with involution in a Hilbert space is called a *-representation* if $T(a^*) = T(a)^*$ for all $a \in A$.) Part d) of Theorem 3.1 shows that the algebra $\mathbb{C}[G]$ is semisimple and is isomorphic, as follows from the general theory of complex semisimple algebras, to a direct sum of matrix algebras. This can also be proved directly with the help of the Fourier transform discussed in the next subsection.

2.3. The Fourier Transform. The Fourier transform on a finite group G maps any function f on G to a function $\hat{f}$ on the set $\hat{G}$ by the formula:

$$\hat{f}(\lambda) = \sum_{g \in G} f(g) T_\lambda(g)^*, \tag{3.4}$$

where T_λ is an irreducible representation in the class $\lambda \in \hat{G}$. Let us denote $\dim T_\lambda$ by d_λ and let us fix an orthonormal basis in the space V_λ of the representation T_λ. Then $\hat{f}(\lambda)$ belongs to the matrix algebra $\mathrm{Mat}_{d_\lambda}(\mathbb{C})$.

The basic properties of the Fourier transform are described in the following theorem.

Theorem 3.3. *The Fourier transform defined above gives an isomorphism of the * - algebras $\mathbb{C}[G]$ and $\prod_{\lambda \in \hat{G}} \mathrm{Mat}_{d_\lambda}(\mathbb{C})$. The inversion formula has the form*

$$f(g) = \frac{1}{|G|} \sum_{\lambda \in \hat{G}} d_\lambda \, \mathrm{tr}\, [\hat{f}(\lambda) T_\lambda(g)]. \tag{3.5}$$

We have also (an analogue of the Plancherel formula):

$$(f_1, f_2)_{L_2(G)} = (\hat{f}_1, \hat{f}_2)_{L_2(\hat{G})} := \frac{1}{|G|} \sum_{\lambda \in \hat{G}} d_\lambda \, \mathrm{tr}\, [\hat{f}_1(\lambda)\hat{f}_2(\lambda)^*]. \qquad (3.6)$$

All parts of Theorem 3.3 are reformulations of already proved facts.

It is useful to compare Theorem 3.3 with well-known facts of commutative harmonic analysis. Let G be a commutative group. Then all its irreducible representations are one-dimensional and coincide with their characters. The set $\hat{G}$ itself is a group with respect to the tensor product of representations, which coincides here with the usual multiplication of characters as functions on the group. The group $\hat{G}$ is called the *dual* group of G. If we modify the definition of the Fourier transform a little bit by introducing a multiplier $|G|^{-\frac{1}{2}}$, then (3.4) – (3.6) obtain a more symmetric form:

$$\hat{f}(\chi) = |G|^{-\frac{1}{2}} \sum_{g \in G} f(g)\overline{\chi(g)}, \qquad (3.4')$$

$$f(g) = |G|^{-\frac{1}{2}} \sum_{\chi \in \hat{G}} \hat{f}(\chi)\chi(g), \qquad (3.5')$$

$$(f_1, f_2)_{L_2(G)} = (\hat{f}_1, \hat{f}_2)_{L_2(\hat{G})}. \qquad (3.6')$$

This is a particular case of the well-known *Pontryagin duality principle*, claiming that G and $\hat{\hat{G}}$ are canonically isomorphic for any locally compact Abelian groups. (Note that for finite groups, G and $\hat{G}$ are already isomorphic, see e.g. Ex. 3.1, but the isomorphism is not canonical.)

Let us return to the general case. The orthogonality relations implies that the Fourier transform maps the character χ_λ to

$$\hat{\chi}_\lambda(\mu) = \begin{cases} 0 & , \quad \text{if } \lambda \neq \mu, \\ \frac{|G|}{d_\lambda}.\mathbb{1}, & \text{if } \lambda = \mu. \end{cases}$$

The operator of convolution with χ_λ is mapped to the operator of multiplication by $\hat{\chi}_\lambda$ under the Fourier transform. It has an eigenvalue $\frac{|G|}{d_\lambda}$ (with multiplicity d_λ^2). In the natural basis consisting of δ-functions in $L_2(G)$, this operator is given by a matrix with algebraic integer entries (the values of a character are sums of roots of unity). Hence its eigenvalues are also algebraic integers. We have proved that the dimension of the representation d_λ divides the order of the group G (Theorem 3.1 d)). Next, if Z is the center of G, then irreducible representations $T^N = T \times \ldots \times T$ (N factors) of the group $G^N = G \times \ldots \times G$ are trivial on the subgroup $Z_0 \subset Z^N$ composed of N-tuples $(z_1, \ldots, z_N)$ with the property $z_1 \ldots z_N = e$ (the operators $T(z), z \in Z$ are scalars). Hence for any N, the number $\dim T^N = (\dim T)^N$ divides the order of the quotient group G^N/Z_0, which is equal to $|G|^N.|Z|^{1-N}$. As a consequence, $\dim T$ divides $|G|/|Z|$.

§3. The Decomposition of Representations

The decomposition into irreducible components (Theorem 3.1 d)) has one defect: it is not, generally speaking, unique. A coarser decomposition into *isotypic components* is already unique. The most popular example of such a decomposition is the decomposition of a function space into the direct sum of subspaces of even and odd functions. (The role of the group G is played by $\mathbb{Z}_2$ and the element different from the identity acts as the transformation $f(x) \to f(-x)$.)

In the general case of a representation S of a group G on a space W, isotypic components are indexed by points $\lambda \in \hat{G}$; we shall denote them by W^λ. The subrepresentation of S acting on W^λ is a multiple of the irreducible representation T_λ. Let n_λ denote the multiplicity of the component W^λ, i.e. the number of summands in a decomposition of W^λ into a sum of irreducible subspaces:

$$W^\lambda = \oplus_{i=1}^{n_\lambda} W_i^\lambda.$$

The search for an explicit form of the subspaces W^λ and W_i^λ is one of the most important problems in the theory of representations and its applications. A general solution of the problem is formulated in terms of intertwining operators. Let us first recall some facts from linear algebra.

For every decomposition of a Hilbert space W into a sum of orthogonal subspaces W_i, $1 \le i \le k$, there is a commutative *-subalgebra $A \subset \operatorname{End} W$ consisting of all linear combinations of orthogonal projections P_i onto the subspaces W_i. The converse is also true: every commutative *-algebra $A \subset \operatorname{End} W$ can be obtained in such a way from a decomposition $W = \oplus_{i=1}^k W_i$ (it is possible to set $W_i = P_i W$, where the P_i are minimal symmetric idempotents in A). The commutative subalgebra A and the decomposition $W = \oplus_{i=1}^k W_i$ are called *compatible* with each other.

Theorem 3.4. *Let S be a unitary representation of the group G in a space W.*

a) *An orthogonal decomposition $W = \oplus W_i$ is invariant with respect to operators of the representation S (i.e. it gives a decomposition of the representation of S) if and only if the *-algebra compatible with the decomposition belongs to $\mathcal{C}(S)$.*

b) *The decomposition in a) is a decomposition into irreducible components if and only if A is a maximal commutative *-subalgebra of $\mathcal{C}(S)$.*

c) *The decomposition in a) is a decomposition into isotypic components if and only if A coincides with the center of the algebra $\mathcal{C}(S)$. In such a case, the minimal idempotents P_λ have the form*

$$P_\lambda = S\left(\frac{d_\lambda}{|G|}\overline{\chi_\lambda}\right), \quad \lambda \in \hat{G}. \tag{3.7}$$

d) the $\mathbb{C}[G]$-module W is canonically isomorphic to $\oplus_{\lambda \in \hat{G}} H_\lambda \otimes V_\lambda$, where
    $V_\lambda, \lambda \in \hat{G}$, are irreducible $\mathbb{C}[G]$-modules and H_λ is the space $\mathrm{Hom}_G(V_\lambda, W)$
    considered as a trivial $\mathbb{C}[G]$-module.
e) The following properties of a representation S are equivalent:
    1) The spectrum of S is simple   (i.e. all multiplicities n_λ are less then
        or equal to 1);
    2) The algebra $\mathcal{C}(S)$ is commutative;
    3) The decomposition of the space W into irreducible spaces is unique.

The proof of part a) reduces to the verification of the property that a subspace $W' \subset W$ is invariant with respect to the representation S if and only if
the orthogonal projection from W onto W' is an intertwining operator. Part b)
follows from the "monotonicity" of the correspondence between decompositions and commutative subalgebras: if one subalgebra is contained in another,
then the second decomposition is obtained from the first one by a supplementary decomposition of some summands. Hence maximal subalgebras are
compatible with decompositions into minimal (i.e. irreducible) parts.

To prove part c), let us denote the algebra generated by the operators
$S(g), g \in G$, by B. Then B and $C := \mathcal{C}(S)$ are mutually commuting subalgebras in $\mathrm{End}\, W$:

$$B = C^!, C = B^!.$$

The center of the algebra C hence coincides with $B \cap C$ and, consequently, with
the center of the algebra B. It follows from Lemma 3.1 that the center of B
is the image of the center of $\mathbb{C}[G]$ under the representation S. Theorem 3.3
implies that the image is generated by idempotents $p_\lambda = d_\lambda/|G| \cdot \overline{\chi_\lambda}, \lambda \in \hat{G}$.
The element p_λ is mapped, under the representation T_μ, to the zero operator
for $\lambda \neq \mu$ and to the identity operator for $\lambda = \mu$. So $S(p_\lambda)$ coincides with the
projector P_λ onto the isotypic component W^λ.

The isomorphism in d) is defined in the following way: the element $h_\lambda \otimes w_\lambda$
is mapped to the vector $w = h_\lambda(w_\lambda) \in W$. It is easy to see that it is a
morphism of $\mathbb{C}[G]$-modules, because the h_λ are intertwining operators. Next,
the correspondence $W \mapsto \oplus_{\lambda \in \hat{G}} H_\lambda \otimes V_\lambda$ is clearly an additive functor in the
category of $\mathbb{C}[G]$-modules. To prove the bijectivity of the described morphism,
it is sufficient to check it for irreducible modules, but this is trivial.

Finally, the isomorphism $\mathcal{C}(S) = \prod_{\lambda \in \hat{G}} \mathrm{Mat}_{n_\lambda}(\mathbb{C})$ follows from d) and it
implies, together with b) and c), part e).

Example 3.5. Let the group $G = S_3$ act by permutations of the variables
x_1, x_2, x_3 and let W be any function space invariant with respect to the action
of G.

Knowing the characters of the group G (Exs. 3.2 and 3.4), it is possible
to exhibit explicitly the projectors onto the three isotopic components of the
space W :

$$P_0 f(x_1, x_2, x_3) = \frac{1}{6} \sum f(x_{\sigma(1)}, x_{\sigma(2)}, x_{\sigma(3)}),$$

$$P_1 f(x_1, x_2, x_3) = \frac{1}{6} \sum \operatorname{sgn} \sigma f(x_{\sigma(1)}, x_{\sigma(2)}, x_{\sigma(3)}),$$

$$P_2 f(x, y, z) = f(x, y, z) - \frac{1}{3}[f(x, y, z) + f(y, z, x) + f(z, x, y)].$$

In such a way, the component W^0 is formed by symmetric functions, W^1 is formed by antisymmetric functions and W^2 is formed by functions satisfying the Jacobi identity:

$$f(x, y, z) + f(y, z, x) + f(z, x, y) = 0.$$

Example 3.6. Let a homogeneous space $X = G/H$ have the following property:

(*) For any pair of points x_1, x_2 in X, there is an element g of the group G such that $gx_1 = x_2, gx_2 = x_1$.

Then the geometrical representation T of the group G on the space $L(X)$ of functions on X has simple spectrum.

Indeed, any linear operator in $L(X)$ is given by a matrix $K(x_1, x_2), x_i \in X$ in the basis consisting of δ-functions. It is easy to show that the kernel of an intertwining operator has the property $K(gx_1, gx_2) = K(x_1, x_2)$ for all $g \in G$. Hence (*) implies that matrices of intertwining operators are symmetric. But any algebra consisting of symmetric matrices is commutative, because

$$AB = C = C' = B'A' = BA.$$

Hence $C(T)$ is commutative and it follows from Theorem 3.4 d) that the representation T has simple spectrum.

§4. The Connection Between Representations of a Group and Its Subgroups

4.1. The Functors Res and Ind. Let T be a representation of a group G in a space V and let H be a subgroup of G. The family of operators $T(h), h \in H$, clearly forms a representation of the group H on the space V. It is called the *restriction* of T (not to be confused with a subrepresentation!) and is denoted by $\operatorname{Res}_H^G T$ or simply $\operatorname{Res} T$ if G and H are known from the context. Even if T is an irreducible representation, its restriction is, generally speaking, reducible and one of the basic problems in the theory of representations is to find a decomposition of $\operatorname{Res} T$. Let us note that a special case of this problem is the question of the decomposition of the tensor product of two irreducible representations T_1 and T_2. Namely, $T_1 \otimes T_2$ can be written as $\operatorname{Res}_G^{G \times G}(T_1 \times T_2)$.

The mapping $T \to \operatorname{Res}_H^G T$ is an additive and multiplicative functor from the category $\Pi(G, \mathbb{C})$ into $\Pi(H, \mathbb{C})$.

Theorem 3.5. (the Frobenius reciprocity theorem) *There is an additive functor* Ind_H^G *from* $\Pi(H,\mathbb{C})$ *into* $\Pi(G,\mathbb{C})$ *dual to* Res_H^G *in the following sense:*

$$c(\mathrm{Res}_H^G T, S) = c(T, \mathrm{Ind}_H^G S) \tag{3.8}$$

for any representation T *of the group* G *and any representation* S *of the subgroup* H.

The functor Ind_H^G is called the *induction functor* and the representation $\mathrm{Ind}_H^G S$ is called the *induced representation* (more precisely: the representation of the group G induced from the representation S of the subgroup H).

Let us note that (3.8) characterizes the representation $\mathrm{Ind}_H^G S$ up to an equivalence. If we substitute all irreducible representations $T_\lambda, \lambda \in \hat{G}$, one after another instead of T, we get

$$\mathrm{Ind}_H^G S = \sum_{\lambda \in \hat{G}} c(\mathrm{Res}_H^G T_\lambda, S).T_\lambda. \tag{3.9}$$

The formula (3.9) could equally well be taken as the definition of the induced representation (check that (3.8) follows from (3.9)). But there is still another more useful and more geometrical construction of the induced representation. Let W be the space of the representation S of the subgroup $H \subset G$. Let us consider the space $L(G, H, W)$ of all functions on G with values in W having the property

$$f(hg) = S(h)f(g), \quad h \in H, \quad g \in G. \tag{3.10}$$

The space $L(G, H, W)$ is clearly invariant with respect to right translations. Hence there is a representation R of G on it acting by the formula

$$[R(g)f](g_1) = f(g_1 g). \tag{3.11}$$

The representation R is (equivalent to) the induced representation Ind_H^G. In the language of $\mathbb{C}[G]$-modules, the described construction reduces to the tensor product:

$$L(G, H, W) = \mathbb{C}[G] \otimes_{\mathbb{C}[H]} W,$$

where $\mathbb{C}[G]$ is considered as a right $\mathbb{C}[G]$–module and a left $\mathbb{C}[H]$-module (with the operation of convolution). To show that the definitions (3.8) and (3.11) are equivalent, it is sufficient to verify that there is an isomorphism

$$\mathrm{Hom}_{\mathbb{C}[G]}(V, \mathbb{C}[G] \otimes_{\mathbb{C}[H]} W) = \mathrm{Hom}_{\mathbb{C}[H]}(V, W),$$

i.e. that

$$V^* \otimes_{\mathbb{C}[G]} (\mathbb{C}[G] \otimes_{\mathbb{C}[H]} W) = V^* \otimes_{\mathbb{C}[H]} W.$$

The last isomorphism follows from the associativity of the tensor product and from the isomorphism

$$M \otimes A = M$$

valid for any A-module M.

Hence it is also possible to establish the isomorphism of the spaces $C(T, R)$ and $C(\operatorname{Res} T, S)$ directly, showing in such a way that R and $\operatorname{Ind} S$ are equivalent. Namely, if $A \in C(T, R)$, then Av is a vector-valued function on G belonging to $L(G, H, W)$ for all $v \in V$. Let us define B by $Bv = Av(e)$. It is easy to verify that $B \in C(\operatorname{Res} T, S)$ and that the correspondence $A \mapsto B$ is the desired isomorphism.

Example 3.7. Let S_0 be the trivial one-dimensional representation of the subgroup H. Then the representation $T = \operatorname{Ind}_H^G S_0$ is equivalent to the geometrical representation of the group G in $L(X)$, where $X = G/H$.

In fact, the space $L(X)$ can be identified with the space of functions on G constant on H-cosets, which clearly coincides with $L(G, H, S_0)$ (see (3.10)).

The relation (3.8) gives in the case considered the following:

Lemma 3.6. *An irreducible representation T_λ appears in the decomposition of the geometrical representation on $L(G/H)$ with multiplicity equal to the number of independent H-invariant vectors in the space V_λ of the representation T_λ.*

It is immediate that the functor Res_H^G has the following property: if K is a subgroup of H, then $\operatorname{Res}_K^H \circ \operatorname{Res}_H^G = \operatorname{Res}_K^G$. Using (3.8) we have

Lemma 3.7. (The principle of induction in stages.)

$$\operatorname{Ind}_H^G \circ \operatorname{Ind}_K^H = \operatorname{Ind}_K^G .$$

4.2. Induced Representations. A representation T of a group G is called *monomial* if it is induced from a one-dimensional representation of a subgroup H. (*Monomial matrices* are matrices with one nonzero element in every row and every column. If we use the natural basis consisting of functions on G concentrated on one H-coset, then the operators of monomial representations are given by monomial matrices.)

A group is called *monomial* if all its irreducible representations are monomial. It is known that the class of monomial groups is properly contained between the classes of solvable and nilpotent groups.

The induced representation $T = \operatorname{Ind}_H^G S$ is a generalization of the geometrical representation $T_0 = \operatorname{Ind}_H^G S_0$ (where S_0 is the trivial one-dimensional representation of H.) It can be realized in the space of vector-valued functions on $X = H \backslash G$ in the following way. Every function $F \in L(G, H, W)$ is completely determined by its values on a set containing one point from every equivalence class Hg. Suppose that a section $s : X \to G$ of the natural projection of G on X is given. Let us set $f(x) = F(s(x))$. We identify in this way the space $L(G, H, W)$ with the space $L(X, W)$ of all W-valued functions on X. The representation T has, in this realization, the form

$$[T(g)f](x) = A(g,x)f(xg), \qquad (3.12)$$

where $A(g,x) = S(s(x)gs(xg)^{-1})$. (The elements $s(x)g$ and $s(xg)$ lie in the same H-coset, hence their quotient belongs to H.) As a consequence, we get the *Frobenius formula* for the character of the induced representation:

$$\chi_{\mathrm{Ind}_H^G S}(g) = \sum_{x \in H\backslash G} \chi_S\left(s(x)gs(x)^{-1}\right). \qquad (3.13)$$

The function χ_S on the right hand side of the formula is considered to be extended from H to G by zero. In particular, if an element g is not conjugate to any element of H, then the character of the induced representation at the point g is equal to zero.

Hence the operators of induced representations are given by block-monomial matrices (or, in other words, by the superposition of a translation and a multiplication by an operator-valued function).

In fact, the latter property characterizes induced representations.

Theorem 3.6. (Criterion of inducibility) *Let $X = H\backslash G$. Let the space V of a representation T admit a decomposition into a direct sum*

$$V = \oplus_{x \in X} V_x,$$

such that the relations

$$T(g)V_x \subset V_{xg^{-1}}, \qquad x \in X, \; g \in G \qquad (3.14)$$

hold. Then the representation T is induced by a representation S of the subgroup H acting in $W := V_{x_0}$ by $S(h) = T(h)|_W$, $h \in H$.

Proof. It follows from (1.4 of Chap. 3) that all spaces V_x have the same dimension. Let us choose a section $s : X \to G$ and identify the spaces V_x with the space $W = V_{x_0}$ using the operator $T(s(x))$. Then the space V is identified with $L(X, W)$ and the operator T has the form (3.12).

As an example of an application of the criterion of inducibility, we have the following result, which plays an important role in the study of irreducible representations.

Lemma 3.8. *Let T be an irreducible representation of the group G in the space V and let H be a normal subgroup of G. Then either the representation $\mathrm{Res}_H^G T$ is isotypic or there is a proper subgroup K containing H such that the representation T is induced by an irreducible representation of K.*

For the proof, it is sufficient to note that G acts on $\hat{H}$; if S belongs to a class $\lambda \in \hat{H}$, then $S^g(h) = S(ghg^{-1})$ is a representation of the class λg. Let

$$V = \sum_{\lambda \in \hat{H}} V^\lambda$$

be the decomposition of the representation $\mathrm{Res}_H^G T$ into isotypic components. It satisfies the property (3.14). The irreducibility of T implies that the set of all $\lambda \in \hat{H}$ for which $V^\lambda \neq \{0\}$ form an G-orbit in $\hat{H}$. The two cases in Lemma 3.8 correspond to the triviality (i.e. having cardinality one) or nontriviality of the orbit.

Corollary 3.1. *If the group G has an abelian normal subgroup A, then all the dimensions of irreducible representations of G divide the number $|G|/|A|$.*

Example 3.8. Let G be the group of symmetries of a regular n-sided polygon in the plane (the so-called dihedral group D_n). It consists of n rotations and n reflections. The subgroup H of all rotations is abelian and normal. Hence the dimensions of the irreducible representations of G are either 1 or 2.

The action of G on $\hat{H}$ can be described in this case in the following way. Let us identify H and $\hat{H}$ with $\mathbb{Z}_n$ (see Ex. 3.1.). Then any element of G which does not belong to H acts as multiplication by -1. For n odd, there are one trivial and $\frac{n-1}{2}$ nontrivial G-orbits in $\hat{H}$. Correspondingly, $\hat{G}$ consists of two one-dimensional and $\frac{n-1}{2}$ two-dimensional representations induced by nontrivial characters of H (χ_k and χ_{-k} induce the same representation of G). For n even, there are two trivial and $\frac{n-2}{2}$ nontrivial orbits. Correspondingly, the set $\hat{G}$ contains four one-dimensional and $\frac{n-2}{2}$ two-dimensional representations.

4.3. Big and Spherical Subgroups. A subgroup H of the group G is called a *big subgroup* if the condition

$$c(\mathrm{Res}_H^G T, S) = c(T, \mathrm{Ind}_H^G S) \leq 1$$

holds for all $T \in \hat{G}, S \in \hat{H}$. In other words, a subgroup H is big if any irreducible representation has simple spectrum after the restriction to H (or after the induction from it).

A subgroup H of G is called *spherical* if the geometrical representation T on functions on $X = G/H$ has a simple spectrum. (In such a case, the pair (G, H) is also called a *Gel'fand pair*.)

Lemma 3.9.

a) *A subgroup H in G is big if and only if the algebra $\mathbb{C}[G]^H$ consisting of all functions f on G having the property*

$$f(hg) = f(gh) \quad \text{for all} \quad h \in H, g \in G$$

is commutative.

b) *A subgroup H of G is spherical if and only if the algebra $\mathbb{C}[G]^{H \times H}$ consisting of functions f on G having the property*

$$f(h_1 g h_2) = f(g) \quad \text{for all} \quad h \in H, g \in G$$

is commutative.

c) A subgroup H of G is big if and only if its image in $G \times H$ under the map $h \mapsto (h, h)$ is a spherical subgroup of $G \times H$.

d) (Gel'fand's spherical condition) If there exists an antiautomorphism σ of the group G such that $\sigma(g) \in HgH$ for all $g \in G$, then (G, H) is a Gel'fand pair.

The proof is based on Lemma 3.1. and Theorem 3.3 d).

Example 3.9. Let $G = S_n$ denote the group of permutations of the set $X = \{1, 2, \ldots, n\}$, let $H = S_{n-1}$ be the subgroup leaving the point n fixed. Setting $\sigma(g) = g^{-1}$, we shall verify that the Gel'fand condition holds for the pair $(G \times H, H)$. To do so, it is necessary to verify whether g and g^{-1} are conjugate by an element $h \in H$. Since the lengths of cycles for g and g^{-1} are the same, the elements g and g^{-1} are conjugate in G. The corresponding element of G should reverse the order of every cycle. It is clear that we can choose it to lie in H. Hence S_{n-1} is a big subgroup of S_n. A similar consideration shows that $S_{n-2} \times S_2$ is also a big subgroup of S_n.

§5. The Representation Ring.
Operations on Representations

5.1. Virtual Representations. The operations of addition (direct sum) and multiplication (tensor product) are defined on the set of equivalence classes of representations of G. It is also useful to add the operation of formal subtraction to these operations to obtain a ring. More precisely, the *representation ring* of the group G is a ring $R(G)$ the generators of which are all finite-dimensional representations with relations generated by the expressions

$$T - CTC^{-1}, \quad T_1 + T_2 - (T_1 \oplus T_2), \quad T_1.T_2 - T_1 \otimes T_2.$$

It is clear that any element of $R(G)$ can be written in the form

$$T = \sum_{\lambda \in \hat{G}} n_\lambda T_\lambda, \; n_\lambda \in \mathbb{Z}.$$

Elements of $R(G)$ are called *virtual representations*. Any virtual representation is a difference of two ordinary ones. The functor Res_H^G generates a homomorphism of the ring $R(G)$ to the ring $R(H)$ and the functor Ind_H^G generates a homomorphism of the abelian group $R(H)$ to $R(G)$. The formula

$$\text{Ind}_H^G(\varphi . \text{Res}_H^G \psi) = (\text{Ind}_H^G \varphi) . \psi$$

implies that the image of $R(H)$ is an ideal in $R(G)$.

Example 3.10. If $H = \{e\}$, then $R(H)$ can be identified with $\mathbb{Z}$ and the image of $R(H)$ in $R(G)$ under the map Ind_H^G can be identified with the subring generated by the regular representation R. This implies, in particular, that the formula $T \cdot R = (\dim T) \cdot R$ holds for any representation T (see Lemma 3.4).

The *character* χ_T *of a virtual representation* T is defined in the natural way. At the same time, the map $T \mapsto \chi_T$ is an isomorphism of $R(G)$ onto a subring of the ring $\mathcal{C}(G)$ of central functions on G. There we endow $\mathcal{C}(G)$ with the ordinary multiplication of functions (not convolution!). The ring $R(G)$ is often identified with its image in $\mathcal{C}(G)$. Elements $\varphi \in R(G)$ are characterized in $\mathcal{C}(G)$ by integrality conditions:

$$(\varphi, \chi_T)_{L_2(G)} \in \mathbb{Z} \quad \text{for all} \quad T \in \hat{G}.$$

A family $\mathcal{H}$ of subgroups $H \subset G$ is called *ample*, if the following (equivalent) conditions are satisfied:

a)   The ring $R(G)$ is generated by the set $\bigcup_{H \in \mathcal{H}} \operatorname{Ind}_H^G R(H)$.
b)   A central function on G belongs to $R(G)$ if and only if its restriction to any subgroup $H \in \mathcal{H}$ belongs to $R(H)$.

A subgroup H is called *elementary* if it can be represented in the form $\mathbb{Z}_m \times P$, where P is a group of order p^k and p is a prime which does not divide m.

Theorem 3.7. (Brauer-Green). *A family $\mathcal{H}$ of subgroups of G is ample if and only if every elementary subgroup in G is conjugate to a subgroup of a group $H \in \mathcal{H}$.*

5.2. Operations on Representations. In addition to the ring operations in $R(G)$, the following maps are defined:

a)   *the symmetric power:* $T \mapsto S^k(T)$;
b)   *the exterior power:* $T \mapsto \Lambda^k(T)$;
c)   *the so-called ψ-operation or Adams operation:* $T \mapsto \psi_k(T)$.

The first two operations are defined for ordinary (i.e. not virtual) representations in the natural way: if T acts on the space V, then $S^k(T)$ (resp. $\Lambda^k(T)$) acts on $S^k(V)$ (resp. $\Lambda^k(V)$). To define them in the general case, it is useful to introduce the ring $R(G)[[t]]$ of formal power series in t with coefficients in $R(G)$ and to define maps S and Λ from $R(G)$ into $R(G)[[t]]$ by:

$$S : T \mapsto \sum_{k=0}^{\infty} t^k S^k(T),$$

$$\Lambda : T \mapsto \sum_{k=0}^{\infty} t^k \Lambda^k(T).$$

(The symbols S^0 and Λ^0 here denote the trivial operations sending any representation into the trivial one.)

It easy to see that the maps S and Λ have the following properties:

$$S(T_1 + T_2) = S(T_1)S(T_2), \ \Lambda(T_1 + T_2) = \Lambda(T_1)\Lambda(T_2). \tag{3.15}$$

This makes it possible to extend them in a unique way to virtual representations preserving the relations (3.15).

Example 3.11. Let us compute explicit formulas for the symmetric and exterior squares of the virtual representation $T_1 - T_2$. Writing $S(T_1 - T_2)$ in the form $S(T_1)/S(T_2)$, decomposing it into a series in t and computing the coefficient at t^2, we get (using the formula $S^2(T) + \Lambda^2(T) = T^2$):

$$S^2(T_1 - T_2) = S^2(T_1) - T_1T_2 + \Lambda^2(T_2).$$

A similar computation gives

$$\Lambda^2(T_1 - T_2) = \Lambda^2(T_1) - T_1T_2 + S^2(T_2).$$

In particular, $S^2(-T) = \Lambda^2(T)$, $\Lambda^2(-T) = S^2(T)$.

The symmetric and antisymmetric powers are special cases of the more general notion of *polynomial functor* in the category $\operatorname{Lin} K$ of linear spaces over a field K, i.e. a functor $F : \operatorname{Lin} K \mapsto \operatorname{Lin} K$ with the property that the matrix coefficients of the operator $F(A) : F(V) \mapsto F(W)$ are polynomials in the matrix coefficients of A. A general polynomial functor is a sum of homogeneous ones and a homogeneous functor of degree n is given by a representation ρ of the symmetric group S_n by the formula $F_\rho(V) = \operatorname{Hom}_{S_n}(V_\rho, V^{\otimes n})$, where V_ρ is the space of the representation ρ and S_n acts on $V^{\otimes n}$ by permutations of factors. If ρ is the trivial representation or the sign representation, then we get $F = S^n$ or $F = \Lambda^n$. Any such functor F defines an operation on (ordinary) representations by the formula $F(T)(g) := F(T(g))$.

To define the *Adams operation*, we consider the map $\psi_k, k \geq 1$, sending a function $f(g)$ into $f(g^k)$. It gives an endomorphism (with respect to the usual multiplication of functions) of the ring $\mathcal{C}(G)$ of central functions on G. It can be proved (by induction from the formulas (3.17) bellow) that the map ψ_k preserves the subring $R(G) \subset \mathcal{C}(G)$. It is just the Adams operation. The maps ψ_k have the following remarkable properties:

$$\psi_k(T_1 + T_2) = \psi_k(T_1) + \psi_k(T_2),$$
$$\psi_k(T_1T_2) = \psi_k(T_1)\psi_k(T_2), \tag{3.16}$$
$$\psi_k(\psi_l(T)) = \psi_{kl}(T).$$

The operation ψ_k on $R(G)$ depends only on the residue class of $k \bmod |G|$.

Set $\Psi = \sum_{k=1}^{\infty} t^k \psi_k$. The map $\Psi : R(G) \to R(G)[[t]]$ is linked with the maps S and Λ defined above by the formulas

$$d/dt \circ S = S \cdot \Psi, \ d/dt \circ \Lambda = -\Lambda \cdot \sigma \circ \Psi,$$

where σ denotes the substitution $t \mapsto -t$.

This implies, in particular, the following formulas useful for practical calculations

$$nS^n = \sum_{k=1}^{n} \psi_k \cdot S^{n-k},$$

$$\hspace{6cm} (3.17)$$

$$n\Lambda^n = \sum_{k=1}^{n} (-1)^{k-1} \psi_k \cdot \Lambda^{n-k},$$

which makes it possible to pass from one operation to another. A more general operation F_ρ (defined above only for ordinary representations) can be written using the ψ-operation by the formula:

$$F_\rho = \sum_{|\lambda|=n} \frac{|C_\lambda|}{n!} \chi_\rho(\lambda) \psi_{\lambda_1} \ldots \psi_{\lambda_k}, \hspace{2cm} (3.18)$$

where $\lambda = (\lambda_1, \ldots, \lambda_k)$, $|\lambda| = \lambda_1 + \ldots + \lambda_k$, the symbol C_λ denotes the conjugacy class in S_n given by the lengths of the cycles $\lambda_1, \ldots, \lambda_k$ and $\chi_\rho(\lambda)$ is the value of the character of the representation ρ on the class C_λ. Using the formula (3.18), it is possible to extend the operation F_ρ to the whole ring $R(G)$ of virtual representations.

Finally, there is the operation $*$ mapping a representation to the contragredient one. One can consider $*$ as ψ_{-1}. In matrix language, the operation maps $T(g)$ to $T(g^{-1})'$ (where prime denotes the transposition). Using Theorem 3.1 a), we can show that the operation $*$ is given by complex conjugation in a suitable basis. Even if the explicit form of the contragredient representation depends on the choice of the basis, the corresponding equivalence class is uniquely defined. In particular, the involution $*$ is defined on the set $\hat{G}$.

§6. Representations over Other Fields and Rings

6.1. Basic Definitions and Facts. Besides complex representations, also real, rational, integral, modular and other ones are used (a modular representation means a representation over a finite field). We shall briefly describe here relevant results.

Let K be a field or a commutative ring. A homomorphism of G into the group of automorphisms of a free K-module V of a finite type will be called a K-representation of G.

(It is more natural and appropriate to consider not only free modules, but all projective K-modules V. If K is either a field or the ring of integers, then all projective K-modules are free. Hence the above definition is sufficient for our purposes.)

The group ring $K[G]$ is defined in the same way as it was done for the group algebra $\mathbb{C}[G]$ (see 3.3 of Chap. 3); it consists of all K-valued functions on G. The category of K-representations of G is naturally isomorphic to the category of $K[G]$-modules of finite type which are free as K-modules. If K is a field, then $K[G]$ is an algebra over K and $K[G]$-modules are vector spaces over K.

Let m denote the *exponent* of the group G, i.e. the smallest integer having the property $g^m = e$ for all $g \in G$. Let L_m be the cyclotomic field $\mathbb{Q}(e^{\frac{2\pi i}{m}}) \subset \mathbb{C}$, where $\mathbb{Q}$ is the field of rational numbers. We shall say that the field K is *sufficiently large* (for a given group G) if it contains a subfield isomorphic to L_m.

It turns out that the theory of K-representations of the group G for a sufficiently large field K is completely analogous to the theory of complex representations explained above.

Lemma 3.10. *Every equivalence class of K-representations contains exactly one equivalence class of L_m-representations.*

In other words, every K-representation (in particular every complex representation) can be written with matrices with elements in L_m.

Let K be any field of characteristic 0. Let us denote the field constructed from the field K by adjoining all m-th roots of unity by L and the *Galois group* of the field L over K (i.e. the group of all automorphisms of the field L leaving the elements of K fixed) by $\Gamma_K = \mathrm{Gal}(L : K)$. Elements of the Galois group are completely determined by their action on m-th roots of unity. Hence Γ_K can be identified with the subgroup of $\mathbb{Z}_m^*$ of all invertible residue classes modulo m. Let us consider the element σ_s of Γ_K defined by $\sigma_s(\omega) = \omega^s$, $\omega^m = 1$. Let us note also that Lemma 3.10 implies that the group Γ_K acts on the set $\hat{G}$ and that $\sigma_s(T)$ is equivalent to $\psi_s(T)$ (see Sect. 5 of Chap. 3).

The elements $g_1, g_2 \in G$ are said to be Γ_K-equivalent if g_1 is conjugate to g_2^s by an element $s \in \Gamma_K \subset \mathbb{Z}_m^*$.

Theorem 3.8.

a) *The characters of K-representations are constant on Γ_K-equivalence classes. The number of irreducible K-representations of the group G is equal to the number of Γ_K-equivalence classes in G and also to the number of Γ_K-orbits in $\hat{G}$.*

b) *Let T be an irreducible K-representation of G on a space V, let us denote $D_T = \mathrm{Hom}_G(V, V)$ and let L_T be the center of the K-algebra D_T. Then L_T is isomorphic to a subfield of L and the L-algebra $D_T \otimes_{L_T} L$ is isomorphic to the matrix algebra of order m_T over L. The spectrum of the L - representation T^L acting on $V^L = V \otimes_K L$ consists of all points of the Γ_K-orbit $\mathcal{O} \subset \hat{G}$ considered with multiplicity m_T. The stabilizer of any point of the orbit $\mathcal{O}$ in Γ_K coincides with the subgroup $\mathrm{Gal}\,(L : L_T)$.*

c) *The group algebra $K[G]$ is isomorphic to $\prod_{T \in \hat{G}_K} \mathrm{Mat}_{d(T)} D_T^0$, where $\hat{G}_K$ is the set of all irreducible K-representations, $d(T) = \dim T$ and D^0 denotes the reversed algebra of D.*

The number m_T which appears in the theorem is called the *Schur index* of the K-representation T (and also of any irreducible representation belonging to the spectrum of T^L). Note that the ring $R_K(G)$ generated by the characters of K-representations does not coincide with the subring of K-valued functions in $R(G)$; it has index $\prod_{T \in \hat{G}_K} m_T$ in it. It is known that the Schur index of any K-representation divides the order of the group Γ_K, hence also $\varphi(m) = |\mathbb{Z}_m^*|$.

6.2. Real Representations. We shall discuss the case $K = \mathbb{R}$, important for many applications, in more detail. In this case, the field $\mathbb{C}$ plays the role of L (we shall not consider the trivial case of a group with exponent 2). The group $\Gamma_{\mathbb{R}}$ is isomorphic to $\mathbb{Z}_2$ and its nontrivial element acts on $\hat{G}$ mapping T to T^*. Elements g_1 and g_2 of G are $\Gamma_{\mathbb{R}}$-equivalent if and only if g_1 is conjugate either to g_2 or to g_2^{-1}. The Schur index of an irreducible real representation is equal either to 1 or to 2.

The Frobenius theorem implies that there exist exactly three division algebras over $\mathbb{R}$: the field $\mathbb{R}$ itself, the field $\mathbb{C}$ and the skew-field of quaternions $\mathbb{H}$. Hence irreducible real representations of G are of three different types: *real, complex* and *quaternionic* depending on the structure of the algebra $\mathcal{C}(T)$.

Let T be a complex irreducible representation of the group G. Then the number

$$\text{ind } T = \frac{1}{|G|} \sum_{g \in G} \chi_T(g^2) \qquad (3.19),$$

is called the *index* of T.

Lemma 3.11. *There exists a bijection $\hat{G}/\Gamma_{\mathbb{R}} \to \hat{G}_{\mathbb{R}} : T \mapsto T_{\mathbb{R}}$, where $T_{\mathbb{R}}$ is defined in the following table:*

ind T	$T_{\mathbb{R}}$	Type
1	T	real
0	$T + T^*$	complex
-1	2T	quaternionic

The proof is based on the fact that the right hand side of (3.19) can be written in the form $c(\psi_2(T), T_0)$, where T_0 is the trivial representation. It follows from $\psi_2 = S^2 - \Lambda^2$ (see (3.17) for $n = 2$) that $\text{ind } T$ is equal to the difference between the number of symmetric and antisymmetric invariants of the representation.

Example 3.12. Let us find rational representations of the dihedral group D_p (see Ex. 3.8) for an odd prime number p. In this case $|D_p| = 2p = m$. The field L coincides with $L_m = \mathbb{Q}(e^{\pi i/p})$ and the group $\Gamma_{\mathbb{Q}}$ coincides with $\mathbb{Z}_{2p}^* \simeq \mathbb{Z}_{p-1}$. There are three $\Gamma_{\mathbb{Q}}$-equivalence classes in the group D_p: the identity map, nontrivial rotations and reflections. The set $\hat{D}_p$ splits into three orbits: two of them consist of one element only (the characters χ_0 or χ_1)

and the third one contains $\frac{p-1}{2}$ two-dimensional representations. The geometrical representation of D_p (based on the space X of vertices of a p-sided
regular polygon) decomposes into the sum of the trivial character χ_0 and the
$(p-1)$-dimensional rational representation T. The spectrum of T^L is simple and consists of all two-dimensional representations of D_p. It is easy to
check that the algebra $\mathcal{C}(T)$ is generated over $\mathbb{Q}$ by the difference operator
$\Delta f(k) = f(k+1) - 2f(k) + f(k-1), k \in X \simeq \mathbb{Z}_p$. This algebra is isomorphic
to the field $\mathbb{Q}(e^{\pi i/p}) \cap \mathbb{R}$ and the Galois group of the field $\mathbb{Q}(e^{\pi i/p})$ over it consists of two elements and coincides with the stabilizer of any two-dimensional
representation of D_p in $\Gamma_{\mathbb{Q}}$.

6.3. Integer and Modular Representations. The case of a field K of charac-
teristic $p > 0$ such that p divides $|G|$ is much more complicated. The algebra
$K[G]$ is not semisimple, hence the group G has reducible, but indecomposable
representations. Even more difficulties arise in the case where the ring K is
the ring of integers (or, more generally, the ring of integers in a number field).
Such types of problems in the theory of representations are discussed more
extensively in the volume "Algebra" of the Encyclopaedia; the reader can find
more details there.

§7. Projective Representations of Finite Groups

Basic facts on the theory of projective representations of finite groups are
summarized in the following theorem.

Theorem 3.9. (Schur) *Let G be a finite group and let K be an algebraically
closed field (of any characteristic p).*

a) *Every 2-cocycle on G with values in K^* is equivalent to a cocycle with
    values in the group $\mathbb{Z}_m$ of m-th roots of unity in K, where the number m
    divides the order of the group G and is not divisible by the characteristic
    of the field K.*
b) *There exists a universal extension $\tilde{G}$ of the group G by a finite group A
    linearizing any projective representation of G over the field K. The group
    dual to $H^2(G, K^*)$ can be taken for A.*

The proof of part a) is based on two facts. First, if the left hand sides
and the right hand sides of the cocycle equation (2.9) are multiplied over all
$g_3 \in G$ and if we denote $\prod_{g_2} c(g_1, g_2)$ by $b(g_1)$, then we get

$$c(g_1, g_2)^n = b(g_1)b(g_2)b(g_1 g_2)^{-1}.$$

Secondly, there exist roots of any order in the field K and the p-th root is
uniquely defined. The proof of part b) uses the existence of a section of the
projection $Z^2(G, K^*) \to H^2(G, K^*)$.

Note that the correspondence between projective representations and co-homology groups can be used in "both directions". For example, it is easy to verify that any complex projective representation of the cyclic group $\mathbb{Z}_n$ is equivalent to a linear one. Hence $H^2(\mathbb{Z}_n, \mathbb{C}^*) = 0$.

Example 3.13. Let us find the universal extension of the symmetric group S_n linearizing all complex representations of the group. We shall use for this the fact that the group S_n is generated by the transpositions

$$\sigma_1 = (1,2), \sigma_2 = (2,3), \ldots, \sigma_{n-1} = (n-1, n)$$

with relations:

$$\sigma_i^2 = (\sigma_j \sigma_{j+1})^3 = (\sigma_k \sigma_l)^2 = e \quad \text{for} \quad |k - l| > 1.$$

Let T be a projective complex representation of S_n and let A_i be a rep-resentative of the class $T(\sigma_i)$. The operators A_i^2 are scalars and we can suppose (going to another representative, if necessary) that $A_i^2 = 1$ for $i = 1, 2, \ldots, n - 1$. Further, it follows from the formula $(A_j A_{j+1})^3 = \lambda_j \cdot 1$ that $A_j A_{j+1} A_j = \lambda_j A_{j+1} A_j A_{j+1}$. Taking squares of both sides and taking into account that $A_j^2 = A_{j+1}^2 = 1$, we get $\lambda_j^2 = 1$ and $\lambda_j = \pm 1$. Substituting, if necessary, A_j instead of $-A_j$, we get $A_i^2 = (A_j A_{j+1})^3 = 1$. Similar reasoning shows that relations $A_k A_l = \varepsilon_{k,l} A_l A_k$ with $\varepsilon_{k,l} = \pm 1$ hold for $|k - l| > 1$. Finally, applying the operation $\mathrm{Ad}\, A_k$ to the formula $(A_j A_{j+1})^3 = 1$ we find that all $\varepsilon_{k,l}$ coincide. In the case $n \leq 3$, this means that all projective representations of S_n are equivalent to linear ones. In the case $n \geq 4$, the universal extension of S_n is generated by $\sigma_1, \ldots, \sigma_{n-1}, z$ with relations

$$z^2 = \sigma_i^2 = (z\sigma_i)^2 = (\sigma_j \sigma_{j+1})^3 = e, \quad (\sigma_k \sigma_l)^2 = z \quad \text{for} \quad |k - l| > 1.$$

§8. Representations of the Symmetric Group

8.1. Notation and Subsidiary Constructions. The group S_n is the symmetry group of a simple mathematical object - a finite set consisting of n elements. Also, it is a Coxeter group, the Weyl group of a simple Lie group of type A_n, the Galois group of a general irreducible polynomial of degree n, and the symmetry group of any physical system consisting of n indistinguishable particles. There are hundreds of papers and several monographs devoted to the study of symmetric groups. But up to now many problems are still waiting for an answer and many answers wait for a clear and understandable presentation. We shall describe here only introductory facts on the structure of the group S_n and its representations.

It is useful to represent elements $\sigma \in S_n$ as transformations of the set $X_n = \{1, 2, .., n\}$. The subgroup leaving the point n fixed will be iden-tified with S_{n-1}. This implies that $|S_n| = |X_n| \cdot |S_{n-1}|$, hence we have

$|S_n| = n(n-1)\ldots 1 = n!$. Conjugacy classes of elements of S_n can be described in an intuitive way as follows. For a given $\sigma \in S_n$, we shall consider the graph Γ_σ whose vertices are points of the set X_n and whose edges are pairs $(k, \sigma(k)), k \in X_n$. It is clear that the graph splits into cycles, which are orbits of the group $\langle \sigma \rangle$ generated by the element σ. Elements σ_1 and σ_2 are conjugate in S_n if and only if the corresponding graphs are isomorphic. Another necessary and sufficient condition is that the partition of the number n into an (unordered) sum of lengths of cycles is the same for both graphs. Let us denote by $\mathcal{P}_n$ the set of all partitions of the number n into a sum of positive integers. An element $\lambda \in \mathcal{P}_n$ is given either by a nonincreasing sequence $(\lambda_1, \lambda_2, \ldots, \lambda_r, \ldots)$ of nonnegative integers or by a symbol $1^{m_1} 2^{m_2} \ldots r^{m_r} \ldots$, showing how many times individual positive integers appear in the partition or by a *Young diagram* (see Fig. 2) consisting of rows of lengths $\lambda_1, \lambda_2, \ldots$ or by *Frobenius coordinates* $(\alpha_1, \ldots, \alpha_r | \beta_1, \ldots, \beta_r)$, where the numbers $\alpha_1 > \alpha_2 > \ldots > \alpha_r \geq 0$ and $\beta_1 > \beta_2 > \ldots \beta_r \geq 0$ show the number of boxes of the table right of, resp. below the main diagonal.

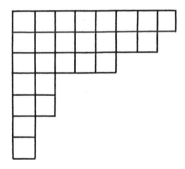

Fig. 2

For example, the partition shown in Fig. 2 is described either by the sequence $(8, 7, 5, 2, 2, 1, 1, 0, 0)$ or by the symbol $1^2 2^2 5^1 7^1 8^1$ (zeros in the partition are neglected) or by Frobenius coordinates $(7, 5, 2 | 6, 3, 0)$. There is a lexicographic order on the set $\mathcal{P}_n$: $\lambda > \mu$ if $\lambda_i = \mu_i$, $i = 1, 2, \ldots, k$, but $\lambda_{k+1} > \mu_{k+1}$. An important role is also played by the *partial order* $\lambda \succ \mu$ defined by the condition $\lambda_1 + \ldots + \lambda_k \geq \mu_1 + \ldots + \mu_k$ to be satisfied for any k (the expression "λ dominates μ" is also used for $\lambda \succ \mu$). Partitions λ and μ are said to be conjugate (denoted by $\lambda = \mu^*$ or $\mu = \lambda^*$) if Young diagrams are reflections of each other with respect to the main diagonal. In Frobenius coordinates, this means the exchange of α_i and β_i, $i = 1, 2, \ldots, r$. The condition $\lambda \succ \mu$ is equivalent to $\mu^* \succ \lambda^*$ and implies the relations $\lambda \geq \mu, \mu^* \geq \lambda^*$.

The cardinality of the class C_λ is equal to $n!(\prod_k k^{m_k}.m_k!)^{-1}$, because the centralizer of an element $\sigma \in C_\lambda$ has the form $\prod_k S_{m_k} \ltimes (\mathbb{Z}_k)^{m_k}$. We denote by $\operatorname{sgn} \lambda$ the sign of the permutation σ :

$$\operatorname{sgn} \sigma = \prod_{i<j} \frac{x_i - x_j}{x_{\sigma(i)} - x_{\sigma(j)}} = \operatorname{sgn} \prod_{i<j} (\sigma(j) - \sigma(i)).$$

It is clear that sgn is a character of the group S_n. Its kernel A_n is called the *alternating group* . The group A_n is simple for $n \geq 5$, hence the group S_n has only two characters in this case: sgn and the trivial character 1.

8.2. Irreducible Representations. Let $\lambda \in \mathcal{P}_n$ be given. The *Young subgroup* in S_n isomorphic to $S_{\lambda_1} \times \ldots \times S_{\lambda_r}$ will be denoted by S_λ; the factor-space $X_\lambda = S_n/S_\lambda$ can be identified with the set of all splittings of X_n into a disjoint union of sets with cardinalities $\lambda_1, \lambda_2, \ldots, \lambda_r$. The geometrical representation of the group S_n on the space of functions on X_λ will be denoted by Y_λ : $Y_\lambda = \operatorname{Ind}_{S_\lambda}^{S_n} 1$. The representation $\operatorname{Ind}_{S_\lambda}^{S_n}(\operatorname{sgn})$, which is isomorphic to $\operatorname{sgn} \otimes Y_\lambda$, will be denoted by Y_λ'.

Lemma 3.12. (von Neumann-Weyl) *The intertwining number* $c(Y_\lambda, Y_{\mu^*}')$ *is different from zero only for* $\lambda \prec \mu$ *and is equal to 1 for* $\lambda = \mu$.

Corollary 3.2. *The representations* Y_λ *and* Y_{λ^*}' *have exactly one common irreducible component* T_λ. *The family of representations* $T_\lambda, \lambda \in \mathcal{P}_n$, *exhausts the whole of the set* $\hat{S}_n$.

The proof of Lemma 3.12 follows from the explicit form of intertwining operators. To prove Corollary 3.2, it is necessary to note that T_λ and T_μ can be equivalent only if $\lambda \succ \mu$ and $\mu \succ \lambda$, i.e. for $\lambda = \mu$. On the other hand, the number of constructed representations is equal to $p_n := |\mathcal{P}_n|$, i.e. to the number of conjugacy classes in S_n.

Theorem 3.10. *The following conditions are equivalent for all* $\lambda, \mu \in \mathcal{P}_n$:

1) $\lambda \succ \mu$;
2) T_λ *belongs to the spectrum of* Y_μ;
2') T_{λ^*} *belongs to the spectrum of* Y_μ';
3) Y_λ *is equivalent to a subrepresentation of* Y_μ;
3') Y_λ' *is equivalent to a subrepresentation of* Y_μ';
4) $c(Y_{\lambda^*}, Y_\mu') > 0$.

The theorem shows, in particular, a constructive way of computing the index λ for a given irreducible representation. Namely, an irreducible representation $T \in \hat{S}_n$ acting on a space V is equivalent to T_λ if there exists an S_λ-invariant vector in the space V and there is no S_μ-invariant vector for $\mu > \lambda$. (It is possible to substitute λ^* and μ^* instead of λ and μ and a skew-invariant vector instead of an invariant one.)

8.3. Examples of Representations.

Example 3.14. Consider the representation T of the group S_n on the space $M_n = \mathbb{C}[x_1, \ldots, x_n]$ of polynomials in n variables given by permutations of the arguments:

$$[T(\sigma)P](x_1, \ldots, x_n) = P(x_{\sigma(1)}, \ldots, x_{\sigma(n)}).$$

The space M_n is graded by the degree of polynomials: $M_n = \oplus_{k=0}^{\infty} M_n^k$ and the action of S_n preserves the grading. An explicit decomposition of M_n^k into irreducible components is unknown up to now. To describe the spectrum of this representation we introduce the generating function

$$P_\lambda(t) = \sum_{k=0}^{\infty} c(T^k, T_\lambda) t^k,$$

where T^k denotes the subrepresentation of T on M_n^k. The above discussion implies that the series $P_\lambda(t)$ begins with the term $t^{n(\lambda)}$, where

$$n(\lambda) = \sum_i (i-1)\lambda_i = \sum_j \frac{\lambda_j^*(\lambda_j^* - 1)}{2}.$$

In fact, any polynomial skew-invariant with respect to S_{λ^*} is necessarily divisible by $\prod(x_i - x_j)$, where the product is taken over all pairs $i < j$ for which both i and j lie in the same set of the partition of the set X_n corresponding to λ^*.

It is possible to prove the formula

$$P_\lambda(t) = t^{n(\lambda)} \prod_{k=1}^{n} (1 - t^{h_k})^{-1}, \tag{3.20}$$

where h_k denotes the length of the hook belonging to the k-th box in the Young diagram of the partition λ (see Fig. 3). For example, for $n = 4$ and for the square table $\lambda = (2, 2)$, the lengths of hooks are equal to $3, 2, 2, 1$. Hence

$$P_{(2,2)}(t) = \frac{t^2}{(1 - t^3)(1 - t^2)^2(1 - t)}.$$

For the table $\lambda = (n)$ consisting from one row (corresponding to the trivial representation), we get $n(\lambda) = 0$; the lengths of the hooks are $n, n-1, \ldots, 1$ and

$$P_{(n)}(t) = \prod_{k=1}^{n} (1 - t^k)^{-1}.$$

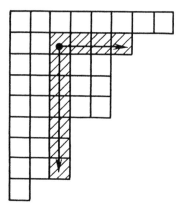

Fig. 3

This agrees with the fact that $M_n^{S_n}$ coincides with the algebra $\mathrm{Sym}[x_1, \ldots, x_n]$ of symmetric polynomials, which is isomorphic to $\mathbb{C}[\varphi_1, \ldots, \varphi_n]$, where φ_k is the elementary symmetric function of degree k :

$$\varphi_k = \sum_{i_1 < \ldots < i_k} x_{i_1} \ldots x_{i_k}.$$

A polynomial $P \in M_n$ is called *harmonic with respect to the group* S_n if it is annihilated by all differential operators of the form $Q(\frac{\partial}{\partial x_1}, \ldots, \frac{\partial}{\partial x_n})$, where $Q \in M_n^{S_n}$ and $Q(0) = 0$. The dimension of the space $\mathcal{H}_n$ of all harmonic polynomials is equal to $n!$ and the representation S_n on $\mathcal{H}_n$ is equivalent to the regular representation. There is an isomorphism of graded S_n-modules

$$M_n = \mathcal{H}_n \otimes \mathrm{Sym}[x_1, \ldots, x_n]. \tag{3.21}$$

Let us denote the polynomial $\sum_{k=0}^{\infty} c(\mathcal{H}_n^k, T_\lambda) t^k$ by $Q_\lambda(t)$. The relation

$$Q_\lambda(t) = t^{n(\lambda)} \prod_{k=1}^{n} \frac{1 - t^k}{1 - t^{h_k}}$$

follows from (3.7) and (3.8). For example,

$$Q_{(2,2)}(t) = t^2 \frac{1 - t^4}{1 - t^2} = t^2 + t^4.$$

Because of the fact that the multiplicity of a representation T_λ in the regular representation is equal to its dimension, we get a useful *"hook formula"*

$$\dim T_\lambda = Q_\lambda(1) = n! \prod_{k=1}^{n} h_k^{-1}.$$

Example 3.15. Let us study the so-called "superanalogue" of the representation T from Ex. 3.14. It is the representation S acting on the Grassmann algebra G_n with generators $\xi_1, \ldots, \xi_n$ and relations

$$\xi_i \xi_j + \xi_j \xi_i = 0.$$

The algebra G_n is finite-dimensional and graded by the degree of its elements considered as polynomials in the generators:

$$G_n = \oplus_{k=0}^n G_n^k, \quad \dim G_n^k = \binom{n}{k}.$$

The space of S_n-invariants in G_n is two-dimensional with a basis consisting of $1 \in G_n^0$ and $\xi_1 + \ldots + \xi_n \in G_n^1$. Let us denote by ∂_i the (super)-derivation in G_n defined by its action $\partial_i \xi_j = \delta_{ij}$ on generators. The algebra of operators in the space G_n is generated by the operators ξ_i and ∂_i with relations

$$\xi_i \xi_j + \xi_j \xi_i = 0, \quad \partial_i \partial_i + \partial_j \partial_i = 0,$$
$$\partial_i \xi_j + \xi_j \partial_i = \delta_{ij}.$$

The analogue of harmonic polynomials is defined by the equation

$$(\partial_1 + \ldots + \partial_n) f = 0.$$

The space $\mathcal{H}_n^k$ of harmonic elements in G_n^k is irreducible and it is a realization of the representation T_λ, where λ has $(n - k - 1|k)$ as Frobenius coordinates. The space G_n^k is the direct sum of the subspaces $\mathcal{H}_n^k$ and $(\xi_1 + \ldots + \xi_n) \mathcal{H}_n^{k-1}$. Hence the spectrum of the representation S has multiplicity two and contains exactly representations corresponding to Young diagrams having the form of a hook. It follows that, for this type of representations, there is a suitable model of the representation space making it possible to answer without difficulties all questions concerning the structure of the representations. There is no such model for general representations T_λ up to now.

8.4. Branching Rule. Let us now discuss a connection between representations of S_n and S_{n-1}. It has already been mentioned in 3.5 of Chap. 3 that S_{n-1} is a big subgroup of S_n. Hence the restriction of any representation $T_\lambda \in \hat{S}_n$ to the subgroup S_{n-1} has simple spectrum. There is the following *Young's branching rule*:

$$\operatorname{Res}_{S_{n-1}}^{S_n} T_\lambda = \sum_{\mu \subset \lambda} T_\mu,$$

where $\lambda \in \mathcal{P}_n, \mu \in \mathcal{P}_{n-1}$ and the symbol $\mu \subset \lambda$ denotes that the diagram μ is contained in the table λ (hence is obtained from λ by removing a box).

Let us define the *Young graph* in the following way. The vertices of the graph consist of all points of the set $\mathcal{P} = \cup_n \mathcal{P}_n$. Points $\lambda \in \mathcal{P}_n$ and $\mu \in \mathcal{P}_{n-1}$

are connected by an edge, if $\mu \subset \lambda$. A piece of the graph is shown in the Fig. 4 (the Young diagrams involved are, for convenience, rotated by 45°.)

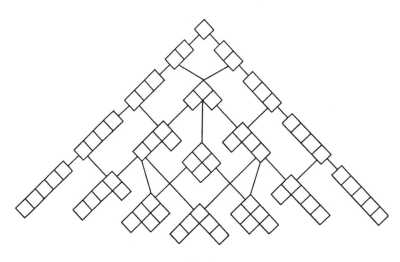

Fig. 4

Every path going from the top diagram (1) to a given diagram λ can be described by a *Young tableau*. This is, by definition, a Young diagram with numbers $1, 2, \ldots, n$ written in the boxes of the diagram in such a way that they form an increasing sequence in every row and every column. (The numbers in the boxes indicate the order of consecutive gluing of individual boxes on the path.) The Young's branching rule implies that the dimension of T_λ is equal to the number of Young tableaux of the form λ. Moreover, there is a correspondence assigning a one-dimensional subspace in the space V_λ of the representation T_λ to every indexed Young tableau (i.e. for every path $\mu_1 \subset \mu_2 \subset \ldots \subset \mu_n = \lambda$) in such a way that the space V_λ is a direct sum of these subspaces. Choosing a nonzero vector in every subspace we get the *Young basis* of the space V_λ.

8.5. The Ring R. Let $R_n = R(S_n)$ be the representation ring of the group S_n (see Sect. 5 of Chap. 3) and define $R = \oplus_{n=1}^{\infty} R_n$. It is possible to introduce a structure of a graded ring on the set R by defining an operation $\circ$ (the so-called "circle multiplication") in the following way. For $T \in R_m$ and $S \in R_n$ we define

$$T \circ S = \mathrm{Ind}_{S_m \times S_n}^{S_{m+n}} (T \times S) \in R_{m+n}.$$

The operation $\circ$ is commutative and associative. (Unfortunately, its connection with the multiplication in the rings R_n is very complicated. This is the reason why we shall consider the sets R_n as abelian groups in this section.)

Let us denote the identity of the ring R_n (i.e. the trivial one-dimensional representation of the group S_n) by $[n]$. It turns out that the ring R is freely generated as a commutative ring by the elements $[n], n = 1, 2, \ldots$. In particular, the representations Y_λ are very easily written using those generators:

$$Y_\lambda = [\lambda_1] \circ [\lambda_2] \circ \ldots \circ [\lambda_r].$$

It is useful to introduce a symbol $[0]$ freely generating an abelian group R_0. Then $R = \oplus_{n=0}^\infty R_n$ become a graded ring with unity $[0]$. Let us also agree that the symbol $[k]$ denotes zero for $k < 0$.

This leads to a remarkable Frobenius formula, expressing an irreducible representation $T_\lambda \in R_n$ using generators of the ring R, called the "*determinantal formula*" :

$$T_\lambda = \det ||[\lambda_i + j]||.$$

Example 3.16. If $\lambda = (p, q, 0, 0, \ldots)$ with $p \geq q > 0$, then

$$T_\lambda = \det \left|\left| \begin{matrix} [p] & [p+1] \\ [q-1] & [q] \end{matrix} \right|\right| = Y_{(p,q)} - Y_{(p+1,q-1)}.$$

It is possible to interpret the result as follows. The space $X_{(p,q)} = S_n/S_p \times S_q$ can be identified with the family of subsets of X_n having p elements. The representation $Y_{(p,q)}$ acts on the space $V_{(p,q)}$ of all functions on $X_{(p,q)}$. There exists a natural intertwining operator C from $V_{(p+1,q-1)}$ into $V_{(p,q)}$:

$$(Cf)(A) = \sum_{B \supset A} f(B), \ A \in X_{(p,q)}, \ B \in X_{(p+1,q-1)}.$$

If $p \geq q$, then the operator C is an imbedding of $V_{(p+1,q-1)}$ into $V_{(p,q)}$, the dual operator C^* is a projection of $V_{(p,q)}$ onto $V_{(p+1,q-1)}$ and the representation $T_{(p,q)}$ is realized in the space $\ker C^*$.

Chapter 4
Representations of Compact Groups

§1. Invariant Integration

1.1. The Haar Measure. A basic property of compact topological groups, making them similar to finite groups, is the existence of a left and right invariant finite measure on the group.

Lemma 4.1. *Let G be a compact topological group and let $\mathcal{A}$ denote the σ-algebra of Borel subsets of G. Then there exists a unique σ-additive nonnegative function μ on $\mathcal{A}$ with the properties*

1) $\mu(gE) = \mu(E)$ for all $g \in G$, $E \in \mathcal{A}$,
2) $\mu(G) = 1$.

We have moreover

3) $\mu(Eg) = \mu(E)$ for all $g \in G$, $E \in \mathcal{A}$,
4) $\mu(E^{-1}) = \mu(E)$.

The function μ *is called the* Haar measure *or* invariant measure *on the group G. The integral with respect to this measure is denoted by $\int_G f(g)d\mu(g)$ or simply by $\int f(g)dg$ and is called the* invariant mean *of a function f.*

The existence and uniqueness of the Haar measure is deduced from the following fact, which is of independent interest.

Lemma 4.2. (von Neumann) *Let G be a compact topological group and let f be a real continuous function on G. For every $\varepsilon > 0$, there exists a finite family $\{g_1, \ldots, g_N\}$ of elements of G such that the variation of the function*

$$\bar{f}(g) = \frac{1}{N} \sum_{i=1}^{N} f(g_i^{-1} g)$$

on G is smaller than ε (i.e. $\max_{g \in G} \bar{f}(g) - \min_{g \in G} \bar{f}(g) < \varepsilon$).

Corollary 4.1. *Let $X = G/H$ be a homogeneous space of a compact topological group G, let H be a closed subgroup of G. Then there exists a unique G-invariant Borel measure on X of total mass 1.*

In fact, to construct such a measure, it is possible to take the image of the Haar measure μ under the natural projection $g \mapsto gx_0$ of G onto X. The uniqueness follows from the Radon-Nikodým theorem: if ν_1 and ν_2 are two invariant measures, then

$$\frac{d\nu_1}{d(\nu_1 + \nu_2)}$$

is an invariant function.

Note that applying the corollary to the case of the group G itself, considered as a homogenous space with respect to left and right translations, we get Lemma 4.1.

1.2. Examples. Let us consider several examples showing how to compute invariant measures.

Example 4.1. The unit sphere $S^n \subset \mathbb{R}^{n+1}$ is the homogeneous space

$$O(n + 1)/O(n) \ (= SO(n + 1)/SO(n)),$$

where $O(n)$ (resp. $SO(n)$) is the group of (special) orthogonal transformations of $\mathbb{R}^n$. The invariant measure on S^n is given by the differential n-form

$$\omega = \sum_{k=0}^{n} (-1)^k x_k dx_0 \wedge \ldots \wedge \widehat{dx_k} \wedge \ldots \wedge dx_n.$$

The form ω is invariant because it is possible to write it in the form $\delta(x^2 - 1)$ or in the form $i_\xi v$, where

$$\xi = \sum_{k=0}^{n} x_k \partial_k, \; v = dx_0 \wedge \ldots \wedge dx_n$$

and i_ξ is the operator of inner multiplication of a differential form by a vector field. In the local coordinate system on S^n obtained by expressing a coordinate x_k through the other ones, the form ω looks like

$$\omega = (-1)^k \frac{dx_0 \wedge \ldots \wedge \widehat{dx_k} \wedge \ldots \wedge dx_n}{x_k}.$$

The volume of the sphere S^n is equal to

$$c_n = \int_{S^n} \omega = (n+1) \int_{B^{n+1}} v = \frac{2\pi^{(n+1)/2}}{\Gamma((n+1)/2)}.$$

Note that in the case $n = 1$ and $n = 3$, the sphere S^n is a group and the above formula gives the Haar measure multiplied by c_n.

Example 4.2. Let p be a prime number. The set $\mathcal{O}_p$ of *p-adic integers* is defined as the completion of $\mathbb{Z}$ with respect to the metric $d_p(m, n) = p^{-k}$, where k is the highest power of p dividing $m - n$. Elements of $\mathcal{O}_p$ are conveniently written in the form of series $\sum_{k=0}^{\infty} a_k p^k$, where $0 \le a_k < p$ or in the form of *integers with an infinite number of digits* $\ldots a_3 a_2 a_1 a_0$ in the number system based on p. (So for the number -1, for example, all a_k are equal to $p - 1$.) It is not difficult to verify that $\mathcal{O}_p$ is a compact topological ring. The Haar measure of its additive group is a direct product of a countable number of copies of the equidistributed probability measure on the set $\{0, 1, \ldots, p - 1\}$. The Haar measure of the "cylindrical set" given by fixing the digits $a_0, a_1, \ldots, a_k$ is equal to p^{-k-1}.

Let us consider the direct product $A = \prod_p \mathcal{O}_p$. It is a compact topological ring as well. It coincides with the completion of $\mathbb{Z}$ with respect to the uniform structure such that its open neighborhoods are all nontrivial subgroups of $\mathbb{Z}$ (they are all of the form $n\mathbb{Z}, n \ne 0$). Elements of A are called *adèles* (more precisely *integer adèles*). They can be written in the form of series $\sum_{k=1}^{\infty} a_k \cdot k!$, where $0 \le a_k \le k$, which are convergent in the sense described above. The Haar measure on the additive group A is the product of the Haar measures on the $\mathcal{O}_p$.

In many problemes connected with divisibility, the introduction of the group A and its Haar measure makes it possible to give a precise sense to statements similar to the next one (known already to D. Bernoulli): two arbitrarily chosen integers are relatively prime with probability $\frac{6}{\pi^2} \approx 0.61$.

Note that the group $\hat{A}$ dual to A is isomorphic to $\mathbb{Q}/\mathbb{Z}$, i.e. to the subgroup of all elements of finite order (roots of unity) in $\mathbb{C}^*$.

Example 4.3. Let us consider the group $G = U(n)$ of unitary matrices of rank n. The integral $\int_G f(g)dg$ is often computed for a function f constant on conjugacy classes (i.e. functions depending only on the set of eigenvalues of the matrix g). Let us recall that such functions are called central functions (compare Sect. 2 of Chap. 3). Let T be the subgroup of diagonal matrices with elements

$$
t = \begin{pmatrix}
e^{i\tau_1} & 0 & \cdots & 0 \\
0 & e^{i\tau_2} & \cdots & 0 \\
\vdots & \vdots & & \vdots \\
0 & 0 & \cdots & e^{i\tau_n}
\end{pmatrix}.
$$

Every central function is uniquely determined by its restriction to T and the following *Weyl integral formula* holds:

$$
\int_G f(g)dg = \int_T f(t)\rho(t)dt,
$$

where $dt = (2\pi)^{-n} \prod_{k=1}^{n} d\tau_k$ is the Haar measure on T and the density $\rho(t)$ is given by the formula:

$$
\rho(t) = \frac{1}{n!} \prod_{j<k} \sin^2\left(\frac{\tau_k - \tau_j}{2}\right).
$$

Similar formulas are also known for other compact Lie groups. In the general case, the density $\rho(t)$ has the form

$$
\rho(t) = \frac{1}{|W|} \prod_{\alpha>0} \left| \sinh \frac{\langle \alpha, \ln t \rangle}{2} \right|^2,
$$

where $|W|$ is the order of the Weyl group and the product is taken over all positive roots α (see Chap. 5).

1.3. Integration of Vector and Operator Valued Functions. It is possible to define integration with respect to the Haar measure not only for ordinary (real-valued) functions but also for functions with values in any complete locally convex linear topological space.

Lemma 4.3.

Let S be a representation of a compact group G in a complete locally convex linear topological space W. Then the operator P given by the formula

$$
Px = \int_G S(g)x\,dg
$$

is the projection onto the subspace W^G consisting of vectors invariant with respect to G.

The existence of an invariant measure μ on the quotient space $X = G/H$ makes it possible, for a compact group G, to define the *geometrical representation* T on the space $L_2(X, \mu)$ by

$$[T(g)f](x) = f(g^{-1}x).$$

The representation T is clearly a unitary representation. Moreover, it is possible to define the operation of *unitary induction* for unitary representations (in the same way as was done in Chap. 3). The role of the space $L(G, H, W)$ is now played by the space of all measurable vector-functions f on G with values in W having the following property:

a) $f(hg) = S(h)f(g)$, $h \in H$, $g \in G$;
b) $\int_X \|f(g)\|_W^2 \, d\mu(x) < \infty$.

The integral in the last condition is well-defined because the unitarity of the representation S and the condition a) imply that $\|f(g)\|_W^2$ depends only on the equivalence class $x = Hg$. Note that the induced representation is, as a rule, infinite-dimensional (with the exception of the case when H is a subgroup of G of finite index).

§2. General Properties of Representations

2.1. The Formulation of Results. Let us begin with the formulation of basic properties of complex representations of a compact topological group G.

Theorem 4.1.

a) *Any (not necessarily finite-dimensional) representation in a Hilbert space is equivalent to a unitary one.*
b) *The matrix elements of irreducible unitary representations form a complete orthogonal system in the space $L_2(G, dg)$.*

Part a) of the theorem is proved in the same way as was done in Sect. 1 of Chap. 3, i.e. we construct a G-invariant scalar product by integration over the group. Note that it is far from true that all infinite-dimensional linear topological spaces admit a Hilbert structure. Hence clearly not all infinite-dimensional representations of G are equivalent to unitary ones. Nevertheless, for compact groups, the study of general representations can to a high degree be reduced to the study of unitary ones.

The proof of part b) of the theorem needs a more refined use of integration with respect to the Haar measure. Namely, let us consider the operation of convolution with a central function φ from the space $C(G)$ of continuous functions on G : $S_\varphi f = \varphi * f$. The operator S_φ is compact (as an integral operator with continuous kernel) and it commutes with the operators of left and right translations. This implies that all eigenspaces corresponding to nonzero eigenvalues are finite-dimensional and invariant with respect to left and right translations. Furthermore, every irreducible representation T of the group G in a space V is contained in the right regular representation of G on the space $C(G)$. To see this, take a functional $f \in V^*$ and define a map sending a vector $v \in V$ to the function $g \to f(T(g)v)$. (The irreducibility of the representation V is needed for the proof of injectivity of the map from V to $C(G)$ defined above; it is clear that kernel is a closed invariant subspace of V.)

Applying the operator S_φ to the image $\tilde{V}$ of the space V in $C(G)$, we arrive at the following alternative possibilities: either the space V is finite-dimensional or the spectrum of the operator S_φ in $\tilde{V}$ consists only of the point 0. The second possibility cannot be true for all functions φ. (If $\varphi = \psi * \psi^*$, where ψ is concentrated in a sufficiently small neighbourhood of the identity, then the operator $S_\varphi = S_\psi S_\psi^*$ is selfadjoint in $L_2(G)$ and it does not annihilate an a priori given element in $\tilde{V}$.)

The orthogonality relations for matrix elements of irreducible unitary representations are proved in exactly same way as was done in the case of finite groups (Sect. 1 of Chap. 3). Let us note that it is possible to prove them without knowing a priori the finite-dimensionality of the representation, which would offer an independent proof of this fact (for this, however, it would be necessary to have a stronger version of the Schur lemma and to introduce nuclear operators or, operators of trace class, in infinite-dimensional spaces).

The completness of the system of matrix elements in $L_2(G)$ is proved as in Chap. 3 by using the regular representation. We can also prove a stronger version:

Theorem 4.2. (Peter-Weyl)   *The span of matrix elements of irreducible representations is dense in the space $C(G)$ of continuous functions on G.*

In the case $G = S^1$, the theorem reduces to the classical theorem of Weierstrass on the density of the set of trigonometric polynomials in $C(S^1)$.

There are two approaches to the proof of Theorem 4.2. The first one uses integral operators in $D(G)$, the second one uses the general Stone-Weierstrass theorem on the structure of closed subalgebras in $C(X)$, where X is a compact set.

The results of Sect. 4 and Sect. 5 of Chap. 3 and Lemma 3.11 in Sect. 6 of Chap. 3 can be transferred to the case of compact groups with only minor modifications. For example, the criterion of inducibility (Theorem 3.6) is formulated in the following way (compare also 2.3 of Chap. 4 below).

Theorem 4.3.   *A representation T of a group G is induced from a subgroup H if and only if it is possible to realize it in the space $L_2(X, W)$ of vector-functions on $X = H \backslash G$ with values in a Hilbert space W in such a way that operators of the representation have the form:*

$$[T(g)f](x) = A(g, x)f(xg),$$

where $A(g, x)$ is an operator-valued function on $G \times X$, the values of which are unitary operators in W.

To transfer the results of parts 3.2., 3.3. and 3.4. of Chap. 3 to the case of compact groups, it is necessary to introduce new notions. We introduce them below.

2.2. Characters. All the claims of Theorem 3.2 remain valid in the case of finite-dimensional representations of compact groups. The notion of a character as the trace of operators of the representation has no sense for infinite-dimensional representations but it is possible to define a generalized character as a distribution on the group G, i.e. as a continuous linear functional on a suitable space of test functions on the group. The formal definition of the character χ_T of the representation T has the form:

$$\langle \chi_T, \varphi \rangle = \mathrm{tr}\, T(\varphi), \tag{4.1}$$

where

$$T(\varphi) = \int_G \varphi(g) T(g) dg \tag{4.2}$$

and φ belongs to the space of test functions.

Example 4.4. We want to compute the character of the left regular representation. The operator $T(\varphi)$ coincides in this case with the operator of convolution $S_\varphi : f \mapsto \varphi * f$. It is an integral operator in $L_2(G, dg)$ with the kernel $K(g_1, g_2) = \varphi(g_1 g_2^{-1})$. Under some additional assumptions (e.g. if the group G is a Lie group and φ is sufficiently smooth), the operator is a nuclear operator and its trace can be computed (as for the trace of a matrix) as an integral of the kernel over the diagonal $g_1 = g_2$. In our case, it leads to the formula $\mathrm{tr}\, T(\varphi) = \varphi(e)$, i.e. the character χ_T is the δ-function concentrated at the identity of the group G. The result coincides with the one obtained in Ex. 3.3 taking into account that the Haar measure on a finite group G has the value $|G|^{-1}$ on a one-point set.

The formulas (3.1) and (3.1') in Sect. 1 and Sect. 2 of Chap. 3 also remain true. In the second case, it is possible to take one of the characters as a test function and the other as a distribution on G.

The criterion for irreducibility and Lemma 3.5 in Sect. 2 of Chap. 3 remain true without modification for the case of a compact group.

2.3. Group Algebras and the Fourier Transform. In the case of infinite groups, the notion of the group algebra can be introduced in several possible ways. Let $\mathbb{C}[G]$ denote, as before, the space of all formal linear combinations of elements of the group G with complex coefficients. The operation of convolution gives the structure of an algebra to $\mathbb{C}[G]$ but the category of (continuous) representations of the group G is no longer equivalent to the category of $\mathbb{C}[G]$-modules.

Other candidates for the role of the group algebra are the space $L_1(G, dg)$ of integrable functions on G, its subspace $C(G)$ of continuous functions as well as the space $M(G)$ of complex measures on G. The operation of convolution is defined for all of them. The most suitable possibility is, however, to define the group algebra as the completion of either of the algebras $L_1(G), C(G)$ in the norm

$$||\varphi|| = \sup_T ||T(\varphi)||, \qquad (4.3)$$

where T varies over all unitary representations of the group G. (It is sufficient to consider only irreducible representations.) The completion is denoted by $C^*(G)$ and is called the C^*-algebra of the group G.

One of the advantages of this variant of the group algebra is the natural isomorphism between the category of representations of the group G and the category of $C^*(G)$-modules.

The Fourier transform maps an integrable function f defined on the group G to a matrix-valued function $\hat{f}$ on $\hat{G}$:

$$\hat{f}(\lambda) = \int_G f(g) T_\lambda(g)^* dg. \qquad (4.4)$$

The Plancherel formula

$$\int_G f_1(g) \overline{f_2(g)} dg = \sum_{\lambda \in \hat{G}} d_\lambda \operatorname{tr} [\hat{f}_1(\lambda) \hat{f}_2^*(\lambda)] \qquad (4.5)$$

holds for all functions f_1, f_2 in $L_2(G, dg)$; it implies the inversion formula

$$f(g) = \sum_{\lambda \in \hat{G}} d_\lambda \operatorname{tr} [\hat{f}(\lambda) T_\lambda(g)] \qquad (4.6)$$

for all functions f that can be represented in the form $\sum_{i=1}^N \varphi_i * \psi_i$, where $\varphi_i, \psi_i \in L_2(G, dg)$.

The Fourier transform is also defined for elements of $M(G)$:

$$\hat{\mu}(\lambda) = \int_G T_\lambda(g)^* d\mu(g). \qquad (4.4')$$

All variants of the group algebra introduced above are transformed into various subalgebras of the direct product $\prod_{\lambda \in \hat{G}} \operatorname{Mat}_{d_\lambda}(\mathbb{C})$ under the Fourier transform but it is possible to describe its image only in the case of the algebra $C^*(G)$. It consists of all sets of matrices $\{A_\lambda\}_{\lambda \in \hat{G}}$ such that $||A_\lambda|| \to 0$ for $\lambda \to \infty$ (the last symbol denotes the limit with respect to the directed family of complements of finite sets in $\hat{G}$). The C^*-norm (4.3) corresponds to $\max_{\lambda \in \hat{G}} ||A_\lambda||$ under the Fourier transform.

Many results in the classical theory of Fourier series (e.g. Hausdorff-Young inequalities for the Fourier coefficients of a function in $L_p(G), 1 \le p \le 2$) can also be proved in the case of compact groups, although the theory is substantially less advanced in the noncommutative case.

Example 4.5. Let G be the group of all isometries of the circle S^1. We shall suppose that the circle S^1 is imbedded in the complex plane $\mathbb{C}$ as the set of points $z = e^{it}$, $t \in \mathbb{R}$ mod $2\pi\mathbb{Z}$. There are two types of elements of the group G: rotations $h_\alpha : z \mapsto e^{i\alpha}z$ and reflections $s_\beta : z \mapsto e^{i\beta}\bar{z}$, where $(\alpha, \beta \in \mathbb{R}$ mod $2\pi\mathbb{Z})$. The subgroup H of rotations is commutative and the corresponding dual group is isomorphic to $\mathbb{Z}$: there exists a character $\chi_n(h_\alpha) = e^{in\alpha}$ for every integer n. Orbits of the group G in $\hat{H}$ have the form $\{\chi_n, \chi_{-n}\}$ or $\{\chi_0\}$. It follows, as in Ex. 3.8, that the complete list of irreducible representations of G has the form:

1) one-dimensional representations χ_0 and χ_0', where

$$\chi_0(g) \equiv 1; \; \chi_0'(h_\alpha) = 1, \; \chi_0'(s_\beta) = -1.$$

2) two-dimensional representations $T_k, k = 1, 2, \ldots$, where

$$T_k(h_\alpha) = \begin{pmatrix} e^{ik\alpha} & 0 \\ 0 & e^{-ik\alpha} \end{pmatrix}, \; T_k(s_\beta) = \begin{pmatrix} 0 & e^{ik\beta} \\ e^{-ik\beta} & 0 \end{pmatrix}.$$

It is suitable to write a function f defined on the group G as a pair f_0, f_1 of periodic functions on the real line: $f(h_\alpha) = f_0(\alpha), f(s_\beta) = f_1(\beta)$. The Fourier transform maps a function $f = (f_0, f_1)$ to a sequence $\{a_0, a_0'; A_1, A_2, \ldots\}$, where

$$a_0 = \frac{b_0 + c_0}{2}, \; a_0' = \frac{b_0 - c_0}{2},$$

$$A_k = \begin{pmatrix} b_k & c_k \\ c_{-k} & b_{-k} \end{pmatrix}$$

and where b_k, c_k denote the Fourier coefficients of the functions f_0, f_1, respectively.

Central functions f are characterized by the conditions that f_0 is even and f_1 is a constant.

2.4. The Decomposition of Representations. The results of Sect. 4 of Chap. 3 remain valid for finite-dimensional representations of compact groups.

Let T be a representation of a compact group G in an infinite-dimensional linear topological space V. Suppose that the space V is locally convex and complete. Then we can define the operators

$$P_\lambda = d_\lambda \int_G T(g)\overline{\chi_\lambda(g)}dg. \tag{4.7}$$

The operator P_λ is the projection operator from V onto the isotypic component V^λ, which can then be represented in the form

$$V^\lambda = V_\lambda \otimes W_\lambda, \tag{4.8}$$

where V_λ is the representation space of the representation T_λ and

$$W_\lambda = \mathrm{Hom}_G(V_\lambda, V) = \mathrm{Hom}_G(V, V_\lambda)$$

is a space with the trivial action of G.

If T is a unitary representation, then the P_λ are orthogonal projections and

$$V = \oplus_{\lambda \in \hat{G}} V^\lambda \qquad (4.9)$$

is a Hilbert direct sum of Hilbert subspaces.

The only information available in the general case is that the algebraic direct sum $\oplus_{\lambda \in \hat{G}} V^\lambda$ is dense in V. There is a useful corollary of this fact. Let a representation T of the group G act on a space V. A vector $v \in V$ is called G-finite if the span of the vectors $T(g)v, g \in G$, is finite-dimensional.

Theorem 4.4. (Generalized Peter-Weyl theorem) *Let $X = G/H$ be a homogeneous space of a compact group G.*

a) *The set of G-finite functions is dense in $C(X)$.*
b) *If G is a Lie group (and, consequently, X is a smooth manifold), then the set of G-finite functions is dense in $C^k(X), k = 1, 2, \dots, \infty$.*

Example 4.6. Let $G = S^1, V = C(\mathbb{R}^2)$ and let T be the geometrical representation given by rotations of the plane. Here $\hat{G} = \mathbb{Z}$ and the isotypic component V^n consists of functions on the plane having the form

$$F(r, \varphi) = e^{in\varphi} f(r)$$

in polar coordinates.

§3. Representations of the Groups SU(2) and SO(3)

There are many reasons why the groups SU(2) and SO(3) are interesting. Firstly, they are sufficiently simple and all computations can be done for them in an explicit form up to the end. Secondly, they are involved in a substantial part of all applications. Finally, they are substantially used in the study of more complicated groups.

3.1. The Group SU(2). As a topological space, the group SU(2) is homeomorphic to the three-dimensional sphere $S^3 \subset \mathbb{R}^4$. This is immediately clear if we write a general element of it in the form

$$g = \begin{pmatrix} a & b \\ -\bar{b} & \bar{a} \end{pmatrix} = \begin{pmatrix} x + iy & z + it \\ -z + it & x - iy \end{pmatrix}, \quad x^2 + y^2 + z^2 + t^2 = 1.$$

Eigenvalues of the matrix g have the form $e^{\pm i\theta}, 0 \le \theta \le \pi$. Conjugacy classes are characterized by the number

$$x = \frac{1}{2} \operatorname{tr} g = \cos \theta \in [-1, 1].$$

This means that the central functions can be identified with functions on the interval $[-1, 1]$. The Haar measure on G was computed in Ex. 4.1. The projection of the Haar measure onto the set of conjugacy classes has the form $\frac{2}{\pi}\sqrt{1 - x^2}\,dx = \frac{2}{\pi}\sin^2\theta d\theta$.

The defining representation $T(g) = g$ of the group G in the two-dimensional space V is irreducible; its character is $\chi(g) = 2x$. Consider now the representation $T_n = S^n(T)$. Let us choose an orthonormal basis (u, v) in V. The n-th symmetric power of the space V has dimension equal to $n + 1$; it will be identified with the space $\mathcal{P}_n$ of homogeneous polynomials of degree n in the variables u and v. The explicit form of the representation T_n is:

$$[T_n(g)P](u, v) = P(au - \bar{b}v, bu + \bar{a}v).$$

Matrix elements of the representations T_n can be expressed using homogeneous polynomials of degree n in matrix elements of the representation $T_1 = T$, i.e. in $a, b, \bar{a}, \bar{b}$ (or, equivalently, in x, y, z, t). It is useful to define

$$(P_1, P_2) =$$
$$= -\frac{1}{4\pi^2} \int_{\mathbb{C}^2} P_1(u, v)\overline{P_2(u, v)}e^{-|u|^2 - |v|^2}\,du \wedge d\bar{u} \wedge dv \wedge d\bar{v}.$$

The subgroup H of diagonal matrices in G is commutative and the spectrum of the representation $\text{Res}_H^G T_n$ is simple. The corresponding orthonormal basis in $\mathcal{P}_n$ consists of the monomials

$$e_k^{(n)} = \frac{u^{n-k}v^k}{\sqrt{(n-k)!k!}}, 0 \leq k \leq n.$$

It is easy to deduce from this an explicit formula for the character of the representation T_n :

$$\chi_n(g) = \sum_{k=0}^{n} e^{i(n-2k)\theta} = \frac{\sin(n+1)\theta}{\sin\theta} = P_n(x),$$

where the polynomial $P_n(x)$ has degree n and can be computed using the recursive formulas

$$P_n(x) = 2xP_{n-1}(x) - P_{n-2}(x);$$
$$P_0(x) = 1, P_1(x) = 2x.$$

The functions χ_n, $n = 0, 1, \ldots$ form an orthonormal basis in the space of central functions on G. It follows that the set $\hat{G}$ is exhausted by the equivalence classes of representations T_n, $n = 0, 1, \ldots$.

The number $s = \frac{1}{2}n$ is called the *spin*. It is often used, especially in applications in quantum mechanics, as a parameter on $\hat{G}$.

It is easy to check that

$$\operatorname{ind} T_n = \int_G \chi_n(g^2)dg = (-1)^n.$$

Hence the representations T_n can be described by real matrices for n even (integer spin) and by quaternionic matrices for n odd (half-integer spin).

Let us discuss representations with small spins in more details.

1) $s = 0$. The representation T_0 is the trivial one-dimensional representation (which is, of course, real).

2) $s = \frac{1}{2}$. The representation $T_1 = T$ is the defining representation. It can be written in the form of a one-dimensional quaternionic representation. Namely, let i, j, k be standard quaternionic units with relations $i^2 = j^2 = k^2 = -1$. The set of all quaternions with unit norm forms a group (with respect to multiplication) isomorphic to $SU(2)$: the quaternion $a + bj = x + yi + zj + tk$ corresponds to the element $g \in SU(2)$. The two-dimensional complex vector space V with a basis u, v is identified with the one-dimensional quaternionic vector space with the basis $u + vj$.

3) $s = 1$. The character of the representation T has real values, so the representations T and T^* are equivalent. Hence the representations of G in the spaces $V \otimes V$ and $V^* \otimes V = \operatorname{End} V$ are equivalent as well. The decomposition of the space $V \otimes V$ into the sum of $S^2(V)$ and $\Lambda^2(V)$ corresponds to the decomposition of $\operatorname{End} V$ into the sum of the space of traceless operators and the space of scalar operators.

The action of G on $\operatorname{End} V$ has the form:

$$g : A \mapsto gAg^*.$$

It commutes with the conjugation $A \mapsto A^*$ and it preserves the quadratic form $\det A$. The determinant of a matrix induces a negative definite quadratic form on the space of all Hermitian traceless 2×2 matrices. This implies that the representation T_2 can be written using real orthogonal matrices in a suitable basis. The explicit formulas are:

$$T_2(g) = \begin{pmatrix} |a|^2 - |b|^2 & 2\operatorname{Re}(\bar{a}b) & 2\operatorname{Im}(\bar{a}b) \\ -2\operatorname{Re}(ab) & \operatorname{Re}(a^2 - b^2) & -\operatorname{Im}(a^2 + b^2) \\ -2\operatorname{Im}(ab) & \operatorname{Im}(a^2 - b^2) & \operatorname{Re}(a^2 + b^2) \end{pmatrix} =$$

$$= \begin{pmatrix} x^2 + y^2 - z^2 - t^2 & 2(xz + yt) & 2(xt - yz) \\ 2(xz - yt) & x^2 - y^2 - z^2 + t^2 & 2(xy + yz) \\ 2(xt + yz) & 2(xy - zt) & x^2 - y^2 + z^2 - t^2 \end{pmatrix}.$$

The homomorphism $T_2 : SU(2) \to O(3)$ is part of an exact sequence:

$$1 \to \mathbb{Z}_2 \xrightarrow{i} SU(2) \xrightarrow{T_2} O(3) \xrightarrow{p} \mathbb{Z}_2 \to 1,$$

where i is the imbedding of $\mathbb{Z}_2$ into $SU(2)$ in the form of scalar matrices and p is the projection of $O(3)$ onto $\mathbb{Z}_2$ given by the determinant. In other words, the range of T_2 coincides with $SO(3)$ and its kernel is the center of $SU(2)$.

4) $s = \frac{3}{2}$. The representation T_3 is equivalent to the two-dimensional quaternionic representation. It is possible to extend it to a multiplicative map of the whole space of quaternions $\mathbb{H}$ into the algebra $\mathrm{Mat}_2(\mathbb{H})$:

$$a + bj \mapsto \begin{pmatrix} a^3 + b^3 j & \sqrt{3}ab(a - bj) \\ \sqrt{3}(bj - a)a\bar{b} & (|a|^2 - 2|b|^2)a + (|b|^2 - 2|a|^2)bj \end{pmatrix}.$$

Let us consider the tensor product of the representations T_m and T_n. It is possible to check that the group $\mathrm{SU}(2)$, imbedded diagonally into the group $\mathrm{SU}(2) \times \mathrm{SU}(2)$, is a big subgroup in the sense of 4.3 of Chap. 3 (the Gel'fand criterion is satisfied if the map σ is defined by $\sigma(g) = g' = \bar{g}^{-1}$). Hence the representation $T_m \otimes T_n$ has simple spectrum. A simple computation with characters shows that

$$T_m \otimes T_n = \sum_{k=0}^{\min(m,n)} T_{m+n-2k}.$$

This formula is a mathematical expression of the law of composition of momenta in quantum mechanics. An explicit isomorphism between the representation spaces has the form:

$$e_r^{(m+n-2k)} \leftrightarrow \sum_{p,q} B_{p,q,r}^{m,n,k} e_p^{(m)} \otimes e_q^{(n)},$$

where $B_{p,q,r}^{m,n,k}$ are called the *Clebsch-Gordan coefficients*. They are real numbers and they are different from zero only for $p + q = k + r$. Explicit formulas are quite awkward; we give them explicitly only in the case $r = 0$. The other ones can be found from them easily using the lowering operators (see Sect. 5 of Chap. 5):

$$B_{p,q,0}^{m,n,k} = \delta(p + q - k)(-1)^q \sqrt{(m - q)!(n - p)!p!q!}.$$

This result becomes intuitively more understandable if the elements of the space of the representation T_{m+n-2k} are homogeneous polynomials of degree $m + n - 2k$ in u, v and the elements of the space of the representation $T_m \otimes T_n$ are bihomogenous polynomials of degree m in u_1, v_1 and of degree n in u_2, v_2. The image of the monomial u^{m+n-2k} is then the polynomial $c \cdot u_1^{m-k} u_2^{n-k} (u_1 v_2 - u_2 v_1)^k$.

We shall write down several formulas concerning operations on representations T_m :

$$S^k(T_m) = S^m(T_k) \quad \text{(Hermite's identity)}$$
$$S^2(T_m) = T_{2m} + T_{2m-4} + T_{2m-8} + \cdots$$
$$\Lambda^2(T_m) = T_{2m-2} + T_{2m-6} + T_{2m-10} + \cdots$$
$$S^n(T_m) = \sum_{l \geq 0} c_{m,n}^l T_{mn-2l},$$

where $c^l_{m,n} = p_{m,n}(l) - p_{m,n}(l-1)$ and $p_{m,n}(l)$ denotes the number of partitions of the number l into a sum of m summands, all of them less than or equal to n. (The sum $G_{m,n}(q) = \sum_l p_{m,n}(l)q^l$ is called the *q-analogue of the Gauss binomial coefficient* $\binom{m+n}{m}$ and is denoted by $\left[\begin{smallmatrix} m+n \\ m \end{smallmatrix}\right]_q$ or simply $\left[\begin{smallmatrix} m+n \\ m \end{smallmatrix}\right]$. If $q \to 1$, then $\left[\begin{smallmatrix} m+n \\ m \end{smallmatrix}\right]_q$ approaches $\binom{m+n}{m}$. The formula $G_{m,n}(q) = \frac{(q)_{m+n}}{(q)_m(q)_n}$ holds, where $(q)_n = \prod_{k=1}^n (1 - q^k)$ is the *q-analogue of the factorial*. More details on q-analogues of different classical notions and their applications in combinatorics can be found in Macdonald (1979).)

$$T_1^{\otimes m} = T_m + \sum_{k \geq 1} \left[\binom{m}{k} - \binom{m}{k-1} \right] T_{m-2k}.$$

In particular, the number $c_n := c(T_1^{\otimes 2n}, T_0)$, called the *n-th Catalan number*, is often useful in applications in combinatorics.

$$\psi_k(T_m) \cdot T_{k-1} = T_{km+k-1}.$$

3.2. The Group $SO(3)$. As was already mentioned above, the group $SO(3)$ is isomorphic to the quotient group of the group $SU(2)$ by its center $\{\pm 1\}$. Hence every representation of $SO(3)$ can be considered as a representation of $SU(2)$ which is trivial on its centre. This implies immediately that the set $\widehat{SO(3)}$ consists of the representations S_n which correspond to the representations T_{2n} of the group $SU(2)$ (i.e. the representations with integer spin n). The other representations of $SU(2)$ can be interpreted as projective representations of $SO(3)$ with multiplier $\{\pm 1\}$. The simplest one is the two-dimensional representation $S_{\frac{1}{2}}$ called the *spinor representation* of $SO(3)$. It is simply the (two-valued) inverse map to the map $T_2 : SU(2) \to SO(3)$.

Elements of the representation space V of the representation $S_{\frac{1}{2}}$ are called *spinors,* or more precisely, *spinors of rank 1.* Elements of the k-th tensor power $V^{\otimes k}$ are called *spinors of rank k.* Coordinates on V are usually denoted by $a^\alpha, \alpha = 0, 1$ in the physical literature; coordinates in $V^{\otimes k}$ then look like $a^{\alpha_1 \alpha_2 \cdots \alpha_k}, \alpha_i = 0, 1$. The fact that the representations V and V^* are equivalent is expressed by the existence of an invariant bilinear form ε on V : $\varepsilon(a,b) = \varepsilon_{\alpha\beta} a^\alpha b^\beta$ (we use here and below the well-known *Einstein summation convention:* there is a sum understood whenever the same indices appear both as upper and lower ones). If a_α are coordinates with respect to the basis $e_0 = u, e_1 = v$ introduced above, then the form $\varepsilon_{\alpha\beta}$ is described by the matrix $\begin{pmatrix} 0 & 1 \\ -1 & 0 \end{pmatrix}$.

The numbers $a_\alpha = \varepsilon_{\alpha\beta} a^\beta$ can be taken as coordinates for spinors instead of a^α. They are called *covariant coordinates,* while the ordinary ones a^α are called *contravariant coordinates.* Using them, the invariant form can be written as $\varepsilon(a,b) = a^\alpha b_\alpha = -a_\alpha b^\alpha$.

The space V_n of the irreducible representation T_n is identified with the subspace of *symmetric spinors* of rank n. Their coordinates $a^{\alpha_1 \cdots \alpha_n}$ are symmetric in the indices $\alpha_1, \ldots, \alpha_n$. In particular, the standard three-dimensional vectors can be identified with symmetric spinors of rank 2. (This is the reason why spinors are often characterized as "square roots" of vectors. But too straightforward and literal use of this expression sometimes leads to misunderstanding.)

3.3. Harmonic Analysis on the Two-Dimensional Sphere. Let us consider the space $V = L_2(S^2, \sigma)$, where S^2 is the standard two-dimensional sphere in $\mathbb{R}^3$ given by the equation $x^2 + y^2 + z^2 = 1$ and σ denotes the normalized invariant measure on S^2 (Ex. 4.1). The group $G = \mathrm{SO}(3)$ of rotations acts on this space. The stabilizer of the point $P = (0, 0, 1)$ is the subgroup $H = \mathrm{SO}(2)$ of rotations around the z axis. The representation S of the group G on the space V is induced by the trivial representation of the group H. The Frobenius reciprocity theorem implies that the spectrum of the representation S consists of irreducible representations of G the space of which contains an H-invariant vector. The results in 3.2 of Chap. 4 imply that all irreducible representations $S_n \in \hat{G}, n = 0, 1, \ldots$ have this property and that the H-invariant vector is unique up to a multiple. Hence every representation S_n is included in the decomposition of S with multiplicity 1. (The fact that the spectrum of S is simple also follows from the fact that H is a big subgroup of G, see 4.3 of Chap. 3) It is easy to describe the corresponding subspace in V. Let $\mathcal{P}_n$ be the set of all homogeneous polynomials of order n in the variables x, y, z and let Π_n denote their restrictions to S^2. It is clear that $\Pi_{n-2} \subset \Pi_n \subset V$. The orthogonal complement of Π_{n-2} in Π_n will be denoted by H_n. Then

$$\dim H_n = \dim \Pi_n - \dim \Pi_{n-2} = \frac{(n+1)(n+2)}{2} - \frac{n(n-1)}{2} = 2n + 1.$$

The space H_n is irrreducible and its preimage in $\mathcal{P}_n$ consists of *harmonic polynomials* satisfying the equation $\Delta f = 0$, where

$$\Delta = \partial^2/\partial x^2 + \partial^2/\partial y^2 + \partial^2/\partial z^2$$

is the Laplace operator in $\mathbb{R}^3$.

Let us describe an explicit formula for the orthonormal basis in H_n consisting of eigenvectors for the operators $S(h), h \in H$. It is useful to take the longitude $\theta \in [-\pi/2, \pi/2]$ and the latitude $\varphi \in [0, 2\pi)$ as coordinates on the sphere. They are related to the coordinates x, y, z by

$$z = \sin \theta, \qquad \frac{x + iy}{|x + iy|} = e^{i\varphi}.$$

In these coordinates, the form σ has the form:

$$\sigma = \frac{1}{4\pi} \cos \theta d\theta d\varphi$$

and the basis consists of spherical functions

$$Y_n^m(\theta, \varphi) = e^{im\phi} P_n^m(\sin \theta), |m| \leq n,$$

where

$$P_n^m(z) = \sqrt{(n + 1/2)\frac{(n + m)!}{(n - m)!}} \frac{(1 - z^2)^{-m/2}}{2^n n!} \left(\frac{d}{dz}\right)^{n-m} (1 - z^2)^n.$$

In particular, the H-invariant vector in H_n has the form:

$$L_n(z) = P_n^0(z) = \frac{\sqrt{n + 1/2}}{(2n)!!} \left(\frac{d}{dz}\right)^n (1 - z^2)^n.$$

Note that the numbering of elements of the basis in H_n is different from that used in the space V_{2n} in 3.1 of Chap. 4; the function $Y_n^m \in H_n$ corresponds to the vector $e_k^{(2n)} \in V_{2n}$, where $m = n - k$. We shall give two examples of how to use the described results (both of them can be generalized to the case of the n-dimensional sphere S^n).

Example 4.7. We shall compute the spectrum of the *Laplace-Beltrami operator* on S^2 given by the formula

$$\Delta f = \operatorname{div} \operatorname{grad} f.$$

It is clear that Δ is an intertwining operator for the representation S. Hence it reduces to multiplication by a scalar λ_n on every irreducible subspace H_n. To compute λ_n, it is sufficient to apply Δ to the simplest spherical function Y_n^0. In local coordinates (φ, θ) we have

$$\operatorname{grad} f = \frac{f'_\varphi}{\cos^2 \theta} \partial/\partial \varphi + f'_\theta \partial/\partial \theta, \quad ds^2 = \cos^2 \theta d\varphi^2 + d\theta^2,$$

$$\operatorname{div} \operatorname{grad} f = (L_{\operatorname{grad} f} \sigma)/\sigma = \frac{f''_{\varphi\varphi}}{\cos^2 \theta} + f''_{\theta\theta} - \tan \theta f'_\theta,$$

$$\Delta Y_n^0 = \left(\frac{d^2}{d\theta^2} - \tan \theta \frac{d}{d\theta}\right) L_n(\sin \theta) =$$

$$= \cos^2 \theta L_n''(\sin \theta) - 2 \sin \theta L_n'(\sin \theta) = -n(n + 1)Y_n^0.$$

So the Laplace-Beltrami operator Δ has eignevalues $-n(n + 1)$ with the multiplicity $2n + 1, n = 0, 1, 2, \ldots$.

Example 4.8. Let us prove the following assertion of integral geometry.

Lemma 4.4. *An even function on the sphere S^2 can be reconstructed from its integrals over all great circles.*

Proof. Consider the integral operator J in the space $C(S^2)$ given by

$$J f(x) = \int_{C_x} f(y) \, dy,$$

where C_x is the great circle on the sphere with epicentre in the point x. The lemma claims that the operator J has trivial kernel in the subspace $C_+(S^2)$ of even functions. It is clear that J commutes with the action of rotations of the sphere, so it reduces to multiplication by a scalar c_n in every subspace H_n. It is easy to compute the corresponding number by applying the operator J to the function Y_n^0. Namely

$$c_n = 2\pi \frac{L_n(0)}{L_n(1)} = \begin{cases} 0 & \text{for odd n,} \\ 2\pi \frac{(n-1)!!}{n!!} & \text{for even n.} \end{cases}$$

Note that $c_{2k} \to 0$ for $k \to \infty$. Hence the operator J on $C_+(S^2)$ does not have a bounded inverse, so the problem of reconstruction of an even function f from its integrals is not correctly posed.

Corollary 4.2. *Any convex centrally symmetric body in $\mathbb{R}^3$ is uniquely defined by the areas of its sections by all planes containing its center of symmetry.*

Proof of corollary. A convex centrally symmetric body $K \subset \mathbb{R}^3$ can be characterized by the function f on the unit sphere defined by $f(M) = 1/2r_M^2$, where r_M is the length of the "radius of the body" in the direction OM. The area of the section of the body K by the plane P can be computed using the function f quite simply:

$$\text{Area}(K \cap P) = \int_{C(P)} f(x)dx,$$

where $C(P)$ is the great circle on the sphere cut by the plane P.

Chapter 5
Finite-Dimensional Representations of a Lie Group

§1. Lie Groups and Lie Algebras

Let us recall that a *Lie group* is a smooth manifold G equipped with a group law such that the maps $(x, y) \mapsto xy$ and $x \mapsto x^{-1}$ are smooth maps from $G \times G$ to G and from G to G, respectively.

A *Lie algebra* over a field K is a vector space L over K equipped with a bilinear operation $x, y \mapsto [x, y]$, (called the *Lie bracket* or *Lie multiplication*), satisfying the conditions:

a) $[x, x] = 0$ (skew-symmetry);

b) $[x, [y, z]] + [y, [z, x]] + [z, [x, y]] = 0$ (Jacobi identity).

The last property can be written as

$$\operatorname{ad} x[y, z] = [\operatorname{ad} x\ y, z] + [y, \operatorname{ad} x\ z]$$

or as

$$\operatorname{ad} x \operatorname{ad} y - \operatorname{ad} y \operatorname{ad} x = \operatorname{ad}[x, y],$$

where $\operatorname{ad} x\ :\ y \mapsto [x, y]$ is an operator in L. It shows that the operator $\operatorname{ad} x$ is a derivation of the Lie algebra L and that the correspondence $x \mapsto \operatorname{ad} x$ is a homomorphism of the Lie algebra L into the Lie algebra of all its derivations (endowed with the operation $[D_1, D_2] := D_1 D_2 - D_2 D_1$).

The relation between Lie groups and Lie algebras is based on the existence of a canonical structure of a real Lie algebra on the tangent space $L(G)$ of the group G at the identity. The following property is often used for the description of this structure.

Lemma 5.1. *Suppose that a Lie group G is realized by linear operators on a space V (i.e. that an imbedding $G \subset \operatorname{GL}(V)$ is given). Then $L(G)$ is a linear subspace of* End V, *closed with respect to the operation*

$$[A, B] := AB - BA.$$

The proof follows from the relation

$$(1 + \varepsilon A)(1 + \varepsilon B)(1 + \varepsilon A)^{-1}(1 + \varepsilon B)^{-1} =$$
$$= 1 + \varepsilon^2[A, B] + o(\varepsilon^2).$$

Elements of the Lie algebra $L(G)$ are called *generators* of the Lie group G in the physics literature (recently also in the mathematical literature). Gothic letters $\mathfrak{g}, \mathfrak{h}, \mathfrak{k}$ are often used instead of $L(G), L(H), L(K)$ etc.

Let us introduce the structure of a Lie algebra in $L(G)$ which is not connected to any specific realization of the group G. For a given element $g \in G$, let us consider the map $G \to G : h \mapsto ghg^{-1}$ (an inner automorphism of G). Its tangent map at the point e is called the *adjoint representation* of the group G in the space $L(G)$ and is denoted by $\operatorname{Ad} g : \mathfrak{g} \to \mathfrak{g}$. The map $g \mapsto \operatorname{Ad} g$ is a smooth map of G into $\operatorname{GL}(\mathfrak{g})$. Its tangent map at the point e coincides with the map $\operatorname{ad} : \mathfrak{g} \to$ End $\mathfrak{g} : x \mapsto \operatorname{ad} x$.

The following coordinate description of the Lie algebra $L(G)$ is often useful for explicit calculations. Let $(x^1, \ldots, x^n)$ be a local coordinate system on G centered in the point e. The law of multiplication in G is then described by a smooth function $\varphi(x, y)$ (where $x = (x^1, \ldots, x^n)$, $y = (y^1, \ldots, y^n)$, $\varphi = (\varphi^1, \ldots, \varphi^n)$). The Taylor series of the map φ at the point $x = y = 0$ has the form $\varphi^k(x, y) = x^k + y^k + b_{ij}^k x^i y^j + \ldots$. The numbers $c_{ij}^k = b_{ij}^k - b_{ji}^k$ are components of a tensor in the space $L(G)$ (this is not true for b_{ij}^k). They are called the *structure constants* of the algebra $L(G)$. The commutation relations for generators have the form:

$$[e_i, e_j] = \sum_k c_{ij}^k e_k.$$

If φ is a homomorphism of a Lie group G_1 into a Lie group G_2, then its tangent map $\varphi_*(e)$ is a homomorphism of the Lie algebra $L(G_1)$ into the Lie algebra $L(G_2)$. It defines a functor from the category of Lie groups to the category of Lie algebras.

For every generator $x \in L(G)$, there exists a unique parametrized curve $t \mapsto g_x(t)$ in the group G having the properties:

$$g_x(t)g_x(s) = g_x(t + s),$$
$$g_x(0) = e, \; g_x'(0) = x.$$

The curve is called the *one-parameter subgroup* in G corresponding to the generator x. It is easy to see that the element $g_x(t)$ depends only on the product $tx \in L(G)$. Let us denote it by $\exp(tx)$. The correspondence $x \to \exp x$ is called the *exponential map* of $L(G)$ into G.

The property $\exp_*(0) = \mathrm{id}$ (id denoting the identity map) implies that the image of the exponential map covers a neighbourhood of the identity in G. The inverse map (generally defined only in a neighbourhood of e) is called the *logarithm* and denoted by $\ln g$. In the case of matrix Lie groups and algebras, the maps exp and ln coincide with the ordinary exponential and logarithm maps:

$$\exp x = \sum_{k=0}^{\infty} \frac{x^k}{k!}, \quad \ln(1 + x) = \sum_{k=1}^{\infty}(-1)^{k-1}\frac{x^k}{k}.$$

Example 5.1.   Consider transformations of the type

$$x \mapsto \varphi(x) = a_0 x + a_1 x^2 + \ldots + a_n x^{n+1} + o(x^{n+1}),$$

where $a_0 \neq 0$, i.e. smooth transformations of the real variable x defined in a neighbourhood of 0 and considered up to an infinitesimal of order higher than $n+1$. The set G_n of all such transformations together with the operation of composition

$$(\varphi \circ \psi)(x) := \varphi(\psi(x)).$$

forms a group. It is clear that G_n is an $(n + 1)$-dimensional Lie group with global coordinates $a_0, a_1, \ldots, a_n$. It acts on the space $\mathcal{F}_n$ of smooth functions F in the variable x defined in a neighbourhood of 0 and also considered up to terms of the form $o(x^{n+1})$. The Lie algebra $L(G_n)$ is realized on $\mathcal{F}_n$ by differential operators of the form

$$Lf = (\alpha_0 x + \ldots + \alpha_n x^{n+1})\frac{df}{dx} \bmod o(x^{n+1}).$$

The elements

$$L_k = x^{k+1}\frac{d}{dx}, \quad k = 0, 1, \ldots, n,$$

form a basis of the algebra; their commutation relations are

$$[L_k, L_j] = (j - k)L_{k+j} \quad (\text{resp. } [L_k, L_j] = 0 \text{ if } k + j > n.)$$

The corresponding one-parameter subgroups have the form

$$\exp(tL_0)f(x) = f(e^t x),$$
$$\exp(tL_k)f(x) = f\left(\frac{x}{\sqrt[k]{1 - ktx^k}}\right), \quad k \geq 1.$$

As an application of these simple formulas, let us solve the following problem.

Consider the limit of the sequence $\{x_n\}$ given by the recursion relation $x_{n+1} = \sin x_n$. It is easy to show that $x_n \to 0$; our problem is to estimate how quickly the sequence approaches zero.

The solution is based on the fact that $\exp(aL_2)x \leq \sin x \leq \exp(bL_2)x$ for small x, where $a < -\frac{1}{6} < b$. Hence the sequence

$$x_n = \sin(\sin(\ldots \sin x_0)\ldots)$$

lies between by $\exp(naL_2)x_0$ and $\exp(nbL_2)x_0$, implying that x_n is equivalent to $\sqrt{\frac{3}{n}}$.

Theorem 5.1.

a)   A homomorphism of a connected Lie group G_1 into a Lie group G_2 is fully determined by its tangent map from the Lie algebra $L(G_1)$ into $L(G_2)$.
b)   If the Lie group G_1 is connected and simply connected, then any homomorphism of the Lie algebra $L(G_1)$ into $L(G_2)$ is the tangent map of a homomorphism of G_1 into G_2.

The proof is based on two assertions that are of independent interest.

Lemma 5.2. *For every homomorphism* $\varphi : G_1 \to G_2$*, the diagram*

$$
\begin{array}{ccc}
G_1 & \xrightarrow{\varphi} & G_2 \\
\exp\uparrow & & \uparrow\exp \\
L(G_1) & \xrightarrow{\varphi_*(e)} & L(G_2)
\end{array}
$$

is commutative.

Lemma 5.3. *Let G_1 be a connected and simply connected Lie group. Then every local homomorphism of G_1 to G_2 (i.e. a map φ of a neighbourhood of $e \in G_1$ into G_2 having the property $\varphi(gh) = \varphi(g)\varphi(h)$ for all $g, h, gh \in V$) extends to a global homomorphism of G_1 into G_2.*

The results described show that the theory of finite-dimensional representations of connected and simply connected Lie groups reduces completely to the theory of representations of Lie algebras.

Theorem 5.2. *Let T be a finite-dimensional representation of a Lie group G on a real or complex vector space V. Then the operator-valued function $g \mapsto T(g)$ is differentiable and its tangent map $T_*(e)$ is a representation of the Lie algebra $L(G)$ on the space V. If the group G is connected and simply connected, then every representation of $L(G)$ is the tangent map of a representation of the group G.*

The representation $T_*(e)$ of the Lie algebra $L(G)$ is often denoted by dT or simply by T and is called the *representation of $L(G)$ corresponding to the representation T of the group G*. A representation of a Lie algebra L on the vector space V is completely determined by the images A_i of the elements e_i of a basis of L. The operators A_i should satisfy the commutation relations

$$[A_i, A_j] = \sum_k c_{ij}^k A_k,$$

where c_{ij}^k are the structure constants of the Lie algebra L with respect to the basis $\{e_i\}$.

Example 5.2. Let L be a two-dimensional commutative Lie algebra with basis x, y and the relation $[x, y] = 0$. A representation of this algebra is defined by a pair A, B of commuting operators. The problem of description of equivalence classes of representations of L reduces to the problem of the description of pairs of commuting operators up to similarity. It appears that this problem belongs to the class of so called "wild" classification problems, which do not have a reasonable solution. As was shown in Gel'fand and Ponomarev (1969), the problem of the classification of all sets of operators $C_1, \ldots, C_n$ (up to similarity) is contained in it as a subproblem. We shall now describe their beautiful construction. Denote $n(n+1)/2$ different copies of V by $V_{ij}, i \geq 1, j \geq 1, i + j \leq n + 1$. We shall construct operators A and B on the space $V \oplus \sum_{i,j} V_{ij}$ by the following rules. The operator A is an isomorphism of V_{ij} onto $V_{i+1,j}$ for $i + j < n$ and it coincides with C_i as a map from $V_{i,n-i+1}$ into V. The operator B is an isomorphism of V_{ij} onto $V_{i,j+1}$ for $i + j \leq n$ and it coincides with C_{i-1} as a map from $V_{i,n-i+1}$ into V (we define here $C_0 = \mathbb{1}_V$). It is easy to verify that the operators A, B constructed above commute with each other.

Lemma 5.4. (Gel'fand-Ponomarev) *Two pairs of operators (A, B) and $(\tilde{A}, \tilde{B})$ constructed in the way described above are similar if and only if the sets of operators $C_1, \ldots, C_n$ and $\tilde{C}_1, \ldots, \tilde{C}_n$ are similar to each other.*

§2. Representations of Solvable Lie groups

For a given G, the symbol $[G, G]$ or $G^{(1)}$ denotes the first derived group generated by elements $ghg^{-1}h^{-1}$, $g, h \in G$. By induction, the *n-th derived group* $G^{(n)}$ is defined by the formula $G^{(n)} = [G^{(n-1)}, G^{(n-1)}]$. A group G is called *solvable* if there exists a positive integer n such that $G^{(n)} = \{e\}$.

For a given Lie algebra L, the symbol $[L, L]$ or $L^{(1)}$ denotes the subspace in L generated by all commutators $[x, y], x, y \in L$. It is a Lie subalgebra called the *derived algebra*. The *n-th derived algebra* is defined by the formula $L^{(n)} = [L^{(n-1)}, L^{(n-1)}]$. A Lie algebra L is called *solvable* if there exists a positive integer n such that $L^{(n)} = \{0\}$.

It is known that a connected Lie group G is solvable if and only if its Lie algebra $L(G)$ is solvable.

There is another (equivalent) definition of the solvability of Lie groups and algebras. A Lie group G (resp. a Lie algebra L) is solvable if there exists a chain of normal subgroups of G :

$$G = G_0 \supset G_1 \supset \ldots \supset G_n = \{e\} \tag{5.1}$$

(resp. a chain of ideals in L :

$$L = L_0 \supset L_1 \supset \ldots \supset L_n = \{0\}) \tag{5.2}$$

with commutative factors G_k/G_{k+1} (resp. L_k/L_{k+1}).

The name "solvable groups" is due to the role played by finite solvable groups in the theory of algebraic equations. It is well-known that an algebraic equation is solvable in radicals if and only if its Galois group is solvable. Solvable Lie groups play a similar role in the study of the solvability of differential equations by quadratures.

The following theorem plays an important role in the theory of finite-dimensional representations of solvable Lie groups and in the study of the structure of the Lie groups themselves.

Theorem 5.3. (Sophus Lie) *All irreducible complex finite-dimensional representations of a solvable Lie group are one-dimensional.*

There are two equivalent versions of the theorem.

Theorem 5.3'. *In any complex finite-dimensional representation of a solvable Lie group, there exists a common eigenvector of all operators of the representation.*

Theorem 5.3". *For any complex finite-dimensional representation of a solvable Lie group, there exists a basis such that all operators of the representation correspond to upper triangular matrices.*

It follows from 5.1 that all three theorems (Theorems 5.3, 5.3' and 5.3") have equivalent versions for solvable Lie algebras instead of solvable Lie groups. It is easy to prove them for Lie algebras by induction with respect to the dimension of the algebra.

Example 5.3. The group $T(n, \mathbb{C})$ of invertible upper triangular matrices is a connected solvable subgroup in $GL(n, \mathbb{C})$. The Lie theorem implies that any maximal connected solvable subgroup of $GL(n, \mathbb{C})$ is conjugate to $T(n, \mathbb{C})$ and any connected solvable subgroup is conjugate to a subgroup of $T(n, \mathbb{C})$.

There is a subclass of *nilpotent Lie groups G* (resp. *nilpotent Lie algebras L*) in the class of solvable Lie groups (Lie algebras). They are defined by the property that there exists a chain of normal subgroups (5.1) (resp. a chain of ideals (5.2)) such that G_k/G_{k+1} belongs to the center of the group G/G_{k+1} (resp. $\mathfrak{g}_k/\mathfrak{g}_{k+1}$ belongs to the center of $\mathfrak{g}/\mathfrak{g}_{k+1}$). A connected Lie group G is nilpotent if and only if its Lie algebra $\mathfrak{g}$ is nilpotent. The name "nilpotent Lie algebra" is explained by the following Engel's theorem.

Theorem. (The Engel Criterion) *A Lie algebra $\mathfrak{g}$ is nilpotent if and only if the operator ad x is nilpotent (i.e. $(\text{ad } x)^n = 0$ for a suitable n) for every $x \in L$.*

Theorem 5.3 and the Engel criterion imply

Corollary 5.1. *The derived group (resp. the derived Lie algebra) of a solvable Lie group (resp. Lie algebra) is nilpotent.*

§3. The Enveloping Algebra

It is often useful to realize a given Lie algebra L by elements of an associative algebra A (for example, the algebra of linear operators) such that the Lie bracket is represented by the ordinary commutator $[a, b] = ab - ba$. There exists a universal object among them in the following sense. Let us consider a category the objects of which are all pairs (φ, A), where A is an associative algebra and φ is a linear map of the Lie algebra L into A having the property $\varphi([x, y]) = \varphi(x)\varphi(y) - \varphi(y)\varphi(x)$. The morphisms of the category are all commutative diagrams of the form

$$
\begin{array}{ccc}
L & = & L \\
\varphi_1 \downarrow & & \downarrow \varphi_2 \\
A_1 & \xrightarrow{\alpha} & A_2
\end{array} \quad ,
$$

where α is a homomorphism of associative algebras. In this category, there exists a *universal object* (i.e. an object with the property that there exists a unique morphism from it to any other object of the category; this implies, in particular, that any two universal objects are isomorphic). This universal object is an imbedding of the Lie algebra L into an associative algebra $U(L)$

called the *universal enveloping algebra* of the Lie algebra L. It follows immediately from the definition of $U(L)$ that for every representation of the Lie algebra L in a vector space V (i.e. for every linear map of L into $\text{End}\,(V)$ having
the property $T([x, y]) = T(x)T(y) - T(y)T(x))$ there exists a homomorphism
of $U(L)$ into $\text{End}\,(V)$, i.e. a representation of the associative algebra $U(L)$
on the space V, usually denoted by the same symbol. The correspondence
is one-to-one: every representation T of the algebra $U(L)$ is generated by a
representation of L, namely by the restriction of T to $L \subset U(L)$.

It is useful to complement the above definition of $U(L)$ by a constructive
description of $U(L)$.

Lemma 5.5.   *The algebra $U(L)$ is isomorphic to the quotient algebra of the
tensor algebra $T(L)$ by its ideal $I(L)$ generated by elements of the form*

$$x \otimes y - y \otimes x - [x, y], \ x, y \in L.$$

There is another useful description of $U(L)$.

Theorem 5.4. (L. Schwartz) *Let G be a Lie group and let $\mathfrak{g}$ be its Lie
algebra. Then the algebra $U(\mathfrak{g})$ is isomorphic to:*

a)   *the algebra of all differential operators on G commuting with right trans-
    lations;*
b)   *the algebra of all distributions on G with support in $\{e\}$ (with the operation
    of convolution).*

The claims a) and b) are equivalent because any differential operator on G
commuting with right translations has the form $Lf = \varphi * f$, where φ is a
distribution on G with support in the point $\{e\}$. The composition of operators
corresponds to the convolution of distributions. For every element $X \in \mathfrak{g}$,
there exists the corresponding right invariant vector field L_X on G, which can
be considered as a first order differential operator. In this way the linear map
of L into the algebra of differential operators on G is defined, commuting with
right translations. Its universality follows from the theorem on the structure
of distributions with support in a point and from the following important
theorem.

Theorem 5.5. (Poincaré-Birkhoff-Witt) *Let L be a Lie algebra over a field
K of characteristic zero. The canonical projection*

$$p : T(L) \to U(L) = T(L)/I(L)$$

*restricted to the symmetric algebra $S(L) \subset T(L)$ gives an isomorphism σ of
the vector spaces $S(L)$ and $U(L)$.*

*(In other words, $T(L)$ is the direct sum of the ideal $I(L)$ and the sub-
space $S(L)$.)*

Despite the easy and intuitive formulation of the theorem, all known proofs are very awkward.

Let us describe another variant of the Poincaré-Birkhoff-Witt theorem valid over a field of any characteristic.

Theorem 5.5'. *Consider any order on the index set I of a basis $x_i, i \in I$, of a Lie algebra L. Then the monomials $x_{i_1} x_{i_2} \ldots x_{i_k}$ with $i_1 \leq i_2 \leq \ldots \leq i_k$ form a basis of $U(L)$.*

Note that if L is a commutative Lie algebra, then $U(L)$ can be naturally identified with $S(L)$ not only as a vector space but also as a (commutative) algebra. In the general case, the algebra $U(L)$ is not commutative but the deviation from commutativity has the character of a "small perturbation" in the following sense.

Let $U_k(L)$ denote the set of all elements which can be written as polynomials of degree less than or equal to k in the generators of the Lie algebra L. It induces an *increasing filtration* (i.e. a chain of subspaces imbedded into each other)

$$K = U_0(L) \subset U_1(L) \subset \ldots \subset U_n(L) \subset \ldots$$

having the properties

$$U_m(L)U_n(L) \subset U_{m+n}(L), \quad [U_m(L), U_n(L)] \subset U_{m+n-1}(L). \tag{5.3}$$

We set

$$\mathrm{gr}^k U(L) = U_k(L)/U_{k-1}(L), \quad \mathrm{gr}\, U(L) = \oplus_{k=0}^{\infty} \mathrm{gr}^k U(L).$$

Elements of $\mathrm{gr}^k U(L)$ can be interpreted as the "highest order terms" of polynomials of order k in the generators of the Lie algebra. The space $\mathrm{gr}\, U(L)$ is endowed with two operations: the multiplication and the *Poisson bracket* defined in the following way. Let $X \in \mathrm{gr}^m U(L), Y \in \mathrm{gr}^n U(L)$ and let $\tilde{X}, \tilde{Y}$ be representatives of X and Y in $U_m(L)$ and $U_n(L)$, respectively. Then

$$X \cdot Y = \tilde{X}\tilde{Y} \mod U_{m+n-1}(L) \in \mathrm{gr}^{m+n} U(L), \tag{5.4}$$

$$\{X, Y\} = (\tilde{X}\tilde{Y} - \tilde{Y}\tilde{X}) \mod U_{m+n-2}(L) \in \mathrm{gr}^{m+n-1} U(L). \tag{5.5}$$

It follows from (5.3) that the operations (5.4) and (5.5) are well-defined and have the following properties:

a) the multiplication is commutative and the algebra $\mathrm{gr}\, U(L)$ is isomorphic (as a graded algebra) to the algebra $S(L)$;
b) the Poisson bracket is anticommutative and satisfies the Jacobi identity and the *Leibniz identity*

$$\{X, YZ\} = \{X, Y\}Z + Y\{X, Z\}. \tag{5.6}$$

In this way, the multiplication in $U(L)$ is commutative on the highest order level and is reduced to the Poisson bracket on the next level. It is useful to consider the algebra $S(L)$ as the algebra $P(L^*)$ of polynomial functions on the space L^* dual to the space L. Then the multiplication in $S(L)$ corresponds to the ordinary mutliplication of functions and the Poisson bracket has the following form in coordinates $x_1, \ldots, x_n$ on L^* :

$$\{f, g\}(x) = \sum_{i,j,k} c_{ij}^k x_k \frac{\partial f}{\partial x_i} \frac{\partial g}{\partial x_j}, \tag{5.7}$$

where c_{ij}^k are the structure constants of the Lie algebra with respect to the basis $\{x_i\}$. (We identify here a coordinate x_i, i.e. a linear function on L^*, with an element of L.) The isomorphism $\sigma : S(L) \to U(L)$ then corresponds to the map $\sigma : P(L^*) \to U(L)$, called *"symmetrization"*, which maps a monomial $x_1 x_2 \ldots x_n$, $x_i \in L$, to the element

$$\frac{1}{n!} \sum_{s \in S_n} x_{s(1)} x_{s(2)} \ldots x_{s(n)} \in U(L).$$

When the space $U(L)$ is identified with $P(L^*)$ using the symmetrization, the multiplication in $U(L)$ looks quite complicated. Let us describe here explicit formulas only in the cases of left and right multiplication by an element $x \in L$:

$$x \cdot f = xf + \frac{1}{2} \sum_p [x_p, x] \frac{\partial f}{\partial x_p} + \sum_{k=2}^{\infty} \sum_{p_1, \ldots, p_k} B_k \cdot$$
$$\cdot (\operatorname{ad} x_{p_1} \ldots \operatorname{ad} x_{p_k}) x \frac{\partial^k f}{\partial x_{p_1} \ldots \partial x_{p_k}};$$
$$f \cdot x = xf - \frac{1}{2} \sum_p [x_p, x] \frac{\partial f}{\partial x_p} + \sum_{k=2}^{\infty} \sum_{p_1, \ldots, p_k} B_k \cdot \tag{5.8}$$
$$\cdot (\operatorname{ad} x_{p_1} \ldots \operatorname{ad} x_{p_k}) x \frac{\partial^k f}{\partial x_{p_1} \ldots \partial x_{p_k}}.$$

Here B_k are the *Bernoulli numbers* defined by the identity $\sum_{k=0}^{\infty} B_k \frac{t^k}{k!} = \frac{t}{e^t-1}$, and the symbol $(A_1 A_2 \ldots A_k)$ denotes the *symmetric product* of the operators A_i, i.e. $\frac{1}{k!} \sum_{s \in S_k} A_{s(1)} A_{s(2)} \ldots A_{s(k)}$. For every polynomial $f \in P(L^*)$, the series on the right hand side of the equations (5.8) is a finite sum. In particular, if $\deg f = 1$, then we get formulas

$$x \cdot y = xy + \frac{1}{2}[x, y],$$
$$y \cdot x = xy - \frac{1}{2}[x, y].$$

There is still another algebraic structure on the space $U(L)$; it is the structure of a *commutative Hopf algebra*. This means that a homomorphism

$$\Delta : U(L) \to U(L) \otimes U(L)$$

is given defining the structure of a commutative associative algebra on the dual space $U(L)^*$ (with the multiplication in $U(L)^*$ given by the operator $\Delta^* : U(L)^* \otimes U(L)^* \to U(L)^*$). The universal property of the algebra $U(L)$ implies that to construct the operator Δ, it is sufficient to define it for elements $x \in L$ in such a way that the condition

$$\Delta([x, y]) = [\Delta(x), \Delta(y)]$$

is satisfied. Thus, it is sufficient to set

$$\Delta(x) = x \otimes 1 + 1 \otimes x.$$

If $L = L(G)$ is the Lie algebra of a Lie group G, then $U(L)^*$ can, using Schwartz's theorem, be identified with the space of infinite jets of functions on G in the point e. The multiplication on $U(L)^*$ introduced above with the help of Δ^* then corresponds to the ordinary multiplication of functions on G.

A new notion of *quantum group* was recently created in connection with the theory of the quantum Yang-Baxter equation. It is a noncommutative Hopf algebra which is a deformation of the commutative Hopf algebra $U(L)$.

Example 5.4. Let L be a simple three-dimensional Lie algebra with basis X, Y, Z and with the commutation relations

$$[X, Y] = Z, \; [Y, Z] = X, \; [Z, X] = Y.$$

It is the Lie algebra of the Lie groups SU(2) and SO(3), which was considered in Sect. 3 of Chap. 4. The enveloping algebra is generated by the generators X, Y, Z over $\mathbb{C}$ with the relations

$$XY - YX = Z, \; YZ - ZY = X, \; ZX - XZ = Y.$$

§4. Laplace (Casimir) Operators

Let us denote the center of the algebra $U(\mathfrak{g})$ by $Z(\mathfrak{g})$. If we realize elements of $U(\mathfrak{g})$ by differential operators on the Lie group G, then elements in $Z(\mathfrak{g})$ are represented by left and right invariant operators (i.e. operators commuting both with the left and right translations). They are called the *Laplace operators* in the mathematical literature and the *Casimir operators* in the physics literature (a particular case of the Laplace operators – the second order operator on a simple Lie group – can often be found under the name of Casimir operator). The role of the elements of $Z(\mathfrak{g})$ in the theory of representations is connected with the following simple fact following from the Schur lemma.

Theorem 5.6. *Let T be an irreducible finite-dimensional representation of a Lie group G in a complex vector space V. Then the corresponding representation of $U(\mathfrak{g})$ maps elements $z \in Z(\mathfrak{g})$ into scalar operators:*

$$T(z) = \lambda_T(z) \cdot \mathbb{1}_V.$$

The map $z \to \lambda_T(z)$ is clearly a homomorphism of the algebra $Z(\mathfrak{g})$ into $\mathbb{C}$. The homomorphism λ_T is called the *infinitesimal character* of the representation T.

Theorem 5.7. *Matrix elements f of a representation T (and, in particular, its character χ_T) satisfy the differential equations*

$$L_z \check{f} = \lambda_T(z)\check{f}, \quad \text{for all } z \in Z(\mathfrak{g}).$$

The symbol L_z here denotes the Laplace operator on the group G corresponding to an element $z \in Z(\mathfrak{g})$ and $\check{f}(g) = f(g^{-1})$.

The theorem follows from the formula $T(L_x \varphi) = T(x)T(\varphi)$ valid for any smooth function with compact support on G and for any $x \in U(\mathfrak{g})$.

It is possible to give a simple description of $Z(\mathfrak{g})$ in terms of functions on $\mathfrak{g}^*$.

Lemma 5.6. *Let G be a connected Lie group with Lie algebra $\mathfrak{g}$. If the space $U(\mathfrak{g})$ is identified with $P(\mathfrak{g}^*)$, then the space $Z(\mathfrak{g})$ corresponds to the space $P(\mathfrak{g}^*)^G$ of polynomials on $\mathfrak{g}^*$ invariant under the coadjoint representation of G on $\mathfrak{g}^*$.*

The idea behind the proof is that the symmetrization map $\sigma : S(\mathfrak{g}) \to U(\mathfrak{g})$ commutes with the adjoint representation of G on $\mathfrak{g}$ canonically extended to $S(\mathfrak{g})$ and $U(\mathfrak{g})$. This is a consequence of the fact that σ commutes with the action of the Lie algebra $\mathfrak{g}$ on $S(\mathfrak{g})$ and $U(\mathfrak{g})$ by differentiation. The proof of the last statement reduces to the verification of the identity

$$\sigma(k[X, Y]^{k-1}) = XY^k - Y^k X$$

for all $X, Y \in \mathfrak{g}$.

In this way, elements of $Z(\mathfrak{g})$ correspond to polynomials on $\mathfrak{g}^*$ which are constant on the orbits of G. In a favourable situation, this means that the number of linearly independent Laplace operators is equal to the codimension of a generic orbit. This is true for semisimple and nilpotent groups. In the case of algebraic groups, generic orbits are almost characterized by invariant rational functions (i.e. generic common level sets consist of a finite number of orbits).

This leads to the notion of the *generalized infinitesimal character*. Let R be a rational function on $\mathfrak{g}^*$ invariant with respect to the action of G. Let us represent it in the form PQ^{-1}, where P and Q are polynomials.

Then for any finite-dimensional irreducible representation T of the group G, the operators $T(\sigma(P))$ and $T(\sigma(Q))$ differ only by a constant multiplier λ. The multiplier λ depends only on the function R (and is independent of its representation in the form PQ^{-1}; we shall denote it by $\lambda_T(R)$. The correspondence $R \mapsto \lambda_T(R)$ is defined on a subring of the field of G-invariant rational functions on $\mathfrak{g}^*$ and it induces a homomorphism of this subring into the field $\mathbb{C}$.

Remark. We have identified the space $Z(\mathfrak{g})$ with the space $S(\mathfrak{g})^G$. Both these spaces are commutative algebras (they are subalgebras of $U(\mathfrak{g})$ and $S(\mathfrak{g})$, respectively). Homomorphisms of the first algebra into $\mathbb{C}$ are the infinitesimal characters corresponding to irreducible representations of G; homomorphisms of the second algebra into $\mathbb{C}$ correspond to G-orbits in $\mathfrak{g}^*$. Also, for many of specific examples the algebras are isomorphic to each other. This observation was one of the sources behind the method of orbits (see Chap. 7). In particular, using the method of orbits, it was possible to prove that $Z(\mathfrak{g})$ and $S(\mathfrak{g})^G$ are isomorphic in the general case.

Example 5.5. Let $\mathfrak{g} = \mathrm{gl}\,(n, \mathbb{R})$ be the Lie algebra of all real matrices of order n. The space $P(\mathfrak{g}^*)$ can be identified with the space $\mathbb{C}[x_{ij}, 1 \le i, j \le n]$ of polynomials in n^2 variables. The action of the group $G = \mathrm{GL}(n, \mathbb{R})$ reduces to the similarity $x = ||x_{ij}|| \mapsto gxg^{-1}$. It is well-known that the algebra of polynomials in x_{ij} invariant with respect to this action is generated by the coefficients $c_k(x)$ of the characteristic polynomial

$$det(x - \lambda \cdot \mathbb{1}) = (-\lambda)^n + \sum_{k=1}^{n} c_k(x)(-\lambda)^{n-k}.$$

It is then easy to show that $Z(\mathfrak{g})$ is generated by n independent generators $\Delta_k = \sigma(c_k)$ of degree k, $1 \le k \le n$. Note that the symmetrization map $\sigma : S(\mathfrak{g})^G \to Z(\mathfrak{g})$ is *not a homomorphism of algebras*. So for the element $c_2 = \sum_{i<j}(x_{ii}x_{jj} - x_{ij}x_{ji})$, we have

$$\sigma(c_2)^2 = \sigma(c_2^2 - \frac{1}{3}c_1^2 + \frac{n}{3}c_2).$$

§5. Representations of the Group SU(2)
(Infinitesimal Approach)

Let T be a finite-dimensional representation of SU(2) in a complex vector space V. Let the same symbol T denote the corresponding representations of the Lie algebra $\mathfrak{g} = L(\mathrm{SU}(2))$ and the enveloping algebra $U(\mathfrak{g})$. Instead of the generators

$$X = \begin{pmatrix} 0 & 1/2 \\ -1/2 & 0 \end{pmatrix}, Y = \begin{pmatrix} 0 & -i/2 \\ -i/2 & 0 \end{pmatrix}, Z = \begin{pmatrix} -i/2 & 0 \\ 0 & i/2 \end{pmatrix},$$

used in Ex. 5.4 and Ex. 5.5, it is useful to use the elements $X_+ = X + iY$, $X_- = X - iY$, $X_0 = 2iZ$ of $\mathfrak{g}^c \subset U(\mathfrak{g})$. Their commutation relations are

$$[X_0, X_\pm] = \pm 2X_\pm, \quad [X_+, X_-] = X_0.$$

Under the representation T, they correspond to operators A_+, A_- and A_0 with the same commutation relations. As an immediate consequence, we get

Lemma 5.7. *The operators $A_\pm$ map an eigenvector of the operator A_0 corresponding to an eigenvalue λ either to an eigenvector corresponding to the eigenvalue $\lambda \pm 2$ or to the zero vector.*

This is the reason why the operators A_+ and A_- are called *raising* and *lowering* operators. An eigenvector of the operator A_0 is called a *weight vector* and its eigenvalue is called its *weight*. The space V being finite-dimensional, there are only a finite number of weights. Let us choose the weight with the biggest real part in the set of all weights. It will be denoted by λ_0 and called the *highest weight of the representation* T. Let v_0 be a corresponding weight vector and let us set $v_k = A_-^k v_0$. Lemma 5.7 implies that the vector v_k is a weight vector corresponding to the weight $\lambda_0 - 2k$. Hence the vectors v_k form a linearly independent set if they are all nonzero. Let n be the highest index such that $v_k \neq 0$ for all $0 \leq k \leq n$.

Lemma 5.8. *The span of the vectors $v_0, v_1, \ldots, v_n$ is invariant under the action of the operators $A_\pm, A_0$. We have*

$$A_+ v_k = k(\lambda_0 - k + 1)v_{k-1}.$$

(The first part of Lemma 5.8 follows from the second, which can easily be proved by induction.)

Suppose now that the representation T is irreducible. Then the vectors $v_0, v_1, \ldots, v_n$ form basis of V and the representation T is characterized, up to isomorphism, by the number λ_0. It follows from

$$\sum_{k=0}^{n} (\lambda_0 - 2k) = \operatorname{tr} A_0 = \operatorname{tr}[A_+, A_-] = 0$$

that $\lambda_0 = n$. This proves at the same time that an irreducible representation T is characterized, up to isomorphism, by its dimension. In the basis given by $e_k = \sqrt{\frac{(n-k+1)!}{k!}} v_k$, the operators $A_\pm, A_0$ have the form

$$A_+ e_k = \sqrt{k(n+1-k)} e_{k-1},$$
$$A_- e_k = \sqrt{(k+1)(n-k)} e_{k+1},$$
$$A_0 e_k = (n-2k) e_k.$$

Introducing a scalar product on V by demanding that $\{e_k\}$ be an orthonormal basis, we get

$$A_\pm^* = A_\mp, A_0^* = A_0.$$

Hence the operators

$$T(X) = \frac{1}{2}(A_+ - A_-), \; T(Y) = \frac{1}{2i}(A_+ + A_-), \; T(Z) = \frac{1}{2i}A_0$$

are skew-hermitian and the operators $T(g), g \in \mathrm{SU}(2)$, are unitary.

Let us find the infinitesimal character of the representation T. The algebra $Z(\mathfrak{g})$ is generated by the element

$$C = X^2 + Y^2 + Z^2 = -\frac{1}{4}X_0^2 - \frac{1}{2}X_0 + X_-X_+.$$

The action of the operator $T(C)$ on the highest weight vector gives

$$\lambda_T(C) = -\frac{n}{2}\left(\frac{n}{2}+1\right).$$

We would like to direct the attention of the reader to the coincidence of the result just described with the formula described in 3.3 of Chap. 4 for the spectrum of the Laplace-Beltrami operator on the two-dimensional sphere. This is, of course, no accident, the Laplace-Beltrami operator is the image of the element C under the representation of $\mathrm{SU}(2)$ on the space of functions on the sphere.

The approach to the classification of finite-dimensional representations of the group $G = \mathrm{SU}(2)$ was based more on the complexification $\mathfrak{g}^c$ of $\mathfrak{g}$ than on the Lie algebra $\mathfrak{g}$ itself. Hence all the described results (except the unitarity of the operators of the representation) also hold for any Lie group with the Lie algebra $\mathfrak{g}_1$ such that $\mathfrak{g}_1^c$ is isomorphic to $\mathfrak{g}^c$. In such a case, the Lie algebras $\mathfrak{g}$ and $\mathfrak{g}_1$ are called *real forms of the complex Lie algebra* $\mathfrak{g}^c$. In fact, the Lie algebra $\mathfrak{g}^c$ considered above has, apart from $L(\mathrm{SU}(2))$, just one more real form: the Lie algebra $\mathfrak{g}_1$ of the group $\mathrm{SL}(2, \mathbb{R})$. The algebra $\mathfrak{g}_1$ is generated by the elements $X_\pm, X_0$, introduced above, as is easily seen from their explicit form

$$X_+ = \begin{pmatrix} 0 & 1 \\ 0 & 0 \end{pmatrix}, \; X_- = \begin{pmatrix} 0 & 0 \\ 1 & 0 \end{pmatrix}, \; X_0 = \begin{pmatrix} 1 & 0 \\ 0 & -1 \end{pmatrix}.$$

The group $G_1 = \mathrm{SL}(2, \mathbb{R})$ is connected but not simply connected. Hence it is not possible to say a priori that all representations of the Lie algebra $\mathfrak{g}_1$ described above arise from representations of the group G_1. Nevertheless, this is so a posteriori; it is possible to check that all representations T_n of dimension $n + 1$ are realized as n-th symmetric powers of the standard two-dimensional representation of $\mathrm{SL}(2, \mathbb{R})$. An unexpected result can be deduced from it, namely that the simply connected group $\tilde{G}_1$ corresponding to the Lie algebra $\mathfrak{g}_1$ has no faithful finite-dimensional linear representations at all.

§6. Representations of Semisimple Lie Groups

The infinitesimal approach applied to representations of the groups SU(2) and SL(2, ℝ) in the previous section generalizes to a broad class of semisimple Lie groups, important for applications. It will be treated in more detail in the part written by A.V. Zelevinskij and G.I.Ol'shanskij, which will appear in future volumes of the series "The theory of representations and non-commutative harmonic analysis" of the Encyclopaedia.

Our aim here is only to make the reader familiar with the terminology and the basic results of the theory of finite-dimensional representations of semisimple Lie groups.

6.1. Semisimple Lie Groups and Algebras. A Lie group G is called *semisimple* if it has no commutative normal subgroup except discrete ones. An equivalent description is that the Lie algebra $\mathfrak{g} = L(G)$ has no nontrivial commutative ideals; such a Lie algebra is called *semisimple* as well.

Theorem. (Cartan's criterion) *A Lie algebra* $\mathfrak{g}$ *is semisimple if and only if the Killing form* $B(x, y) = \operatorname{tr}(\operatorname{ad} x \cdot \operatorname{ad} y)$ *is nonsingular on* $\mathfrak{g}$.

It is known that a semisimple Lie algebra is a direct sum of *simple Lie algebras* having no nonzero ideals at all. Hence every simply connected semisimple Lie group is a direct product of *almost simple Lie groups* having no normal subgroups except discrete ones. There is a complete list of all simple Lie algebras. Over the field ℂ, there are four infinite series $A_n, n \geq 1$; $B_n, n \geq 2$; $C_n, n \geq 3$ and $D_n, n \geq 4$ as well as the five exceptional simple Lie algebras E_6, E_7, E_8, F_4 and G_2. Over the field ℝ, there are 12 infinite series and 23 exceptional algebras. All the algebras included in the infinite series (called *classical simple Lie algebras*) are realized as matrix algebras with elements in ℝ, ℂ or ℍ subjected to simple algebraic conditions.

Let $\mathbb{1}_n$ denote the unit matrix of order n and let us introduce the matrices

$$J_{2n} = \begin{pmatrix} 0 & \mathbb{1}_n \\ -\mathbb{1}_n & 0 \end{pmatrix}, \quad I_{p,q} = \begin{pmatrix} \mathbb{1}_p & 0 \\ 0 & -\mathbb{1}_q \end{pmatrix}.$$

The list of all classical complex simple Lie algebras looks like:

1) A_n – the algebra $\operatorname{sl}(n + 1, \mathbb{C})$ of all complex traceless matrices of order $n + 1$, $n \geq 1$;
2) B_n – the algebra $\operatorname{so}(2n + 1, \mathbb{C})$ of all complex skew-symmetric matrices of order $2n + 1, n \geq 2$ (B_1 being isomorphic to A_1);
3) C_n – the algebra $\operatorname{sp}(2n, \mathbb{C})$ of all complex matrices X of order $2n$ satisfying the condition

$$X'J_{2n} + J_{2n}X = 0,$$
$$n \geq 3 \ (C_1 \approx A_1 \approx B_1, C_2 \approx B_2).$$

4) D_n – the algebra so $(2n, \mathbb{C})$ of all complex skew-symmetric matrices of order $2n, n \geq 4$ (D_1 is commutative, $D_2 \approx A_1 + A_1, D_3 \approx A_3$).

All these algebras are simple Lie algebras over $\mathbb{R}$ as well. The remaining classical real simple Lie algebras are real forms of them. Their list follows. The real forms of A_n :

a)   the algebra sl $(n + 1, \mathbb{R})$ of all real traceless matrices of order $n + 1$;
b)   the algebra su (p, q) of all complex traceless matrices X of order $n + 1$ satisfying the condition $X^* I_{p,q} + I_{p,q} X = 0$, where $p \geq q, p + q = n + 1$;
c)   for n odd, the algebra sl $\left(\frac{n+1}{2}, \mathbb{H}\right)$ (denoted also by su $*(2n)$) of all quaternionic matrices X of order $\frac{n+1}{2}$ having the property tr $(X + X^*) = 0$.

The real forms of B_n are algebras so(p, q) of all real matrices X of order $2n+1$ having the property $X' I_{p,q} + I_{p,q} X = 0, p > q, p + q = 2n + 1$.
The real forms of C_n :

a)   the algebra sp $(2n, \mathbb{R})$ of all real matrices X of order $2n$ with the property $X' J_{2n} + J_{2n} X = 0$;
b)   the algebra sp (p, q) of all quaternionic matrices X of order n with the property $X^* I_{p,q} + I_{p,q} X = 0, p \geq q, p + q = n$.

The real forms of D_n :

a)   the algebras so (p, q) of all real matrices of order $2n$ having the property $X' I_{p,q} + I_{p,q} X = 0, p \geq q, p + q = 2n$;
b)   for n even, the algebra so $*(2n)$ of all quaternionic matrices X of order n having the property $X^* J_n + J_n X = 0$.

Among all real forms of a given semisimple complex Lie algebra $\mathfrak{g}$, there are two most remarkable ones. One of them is the *compact real form* $\mathfrak{g}_u$. The Killing form is negative definite on $\mathfrak{g}_u$ and the connected and simply connected Lie group G_u corresponding to the Lie algebra $\mathfrak{g}_u$ is compact. The other one, called *the normal real form* $\mathfrak{g}_n$, is (as a vector space) a direct sum of three sub-algebras: $\mathfrak{g}_n = \mathfrak{n}_- + \mathfrak{h} + \mathfrak{n}_+$, where $\mathfrak{h}$ is commutative and normalizes both $\mathfrak{n}_+$ and $\mathfrak{n}_-$, moreover, the subalgebras $\mathfrak{n}_\pm$ are nilpotent. The subalgebra $\mathfrak{h}$ is called *Cartan subalgebra* of $\mathfrak{g}$ and is defined as a maximal commutative subalgebra of $\mathfrak{g}$ containing only *semisimple elements* (i.e. elements $x \in \mathfrak{g}$ such that the matrix of the operator ad x is diagonalizable). The dimension of $\mathfrak{h}$ is called the *rank of the semisimple Lie algebra* $\mathfrak{g}$. In the cases described above, the rank of the Lie algebra $\mathfrak{g}$ is equal to the index n in the notation $A_n, B_n, \ldots$. The restriction of the Killing form to $\mathfrak{h}$ is positive definite.

6.2. Weights and Roots. It is possible to choose generators of $\mathfrak{n}_\pm$ among the common eigenvectors of all the operators ad h, $h \in \mathfrak{h}$. In fact, the joint spectrum of all these operators is simple; hence the generators can be indexed by vectors $\alpha \in \mathfrak{h}^*$ such that

$$[h, x_\alpha] = \langle \alpha, h \rangle x_\alpha, \ h \in \mathfrak{h}. \tag{5.9}$$

The generators x_α are called *root vectors* and the corresponding linear forms α on $\mathfrak{h}$ are called *roots*. The set of all roots is denoted by R. Then we have the following relations

$$[x_\alpha, x_\beta] = \begin{cases} c_{\alpha,\beta} x_{\alpha+\beta}, & \text{if } \alpha + \beta \in R, \\ h_\alpha \in \mathfrak{h}, & \text{if } \alpha + \beta = 0, \\ 0, & \text{if } \alpha + \beta \notin R \cup \{0\}. \end{cases}$$

Roots α with $x_\alpha \in \mathfrak{n}_+ (\mathfrak{n}_-)$ are called *positive (negative)*. The set R of all roots has remarkable geometrical properties. The axiomatization of these properties leads to the following definition.

A finite set R of vectors in Euclidean space $\mathbb{R}^n$ is called a *root system* if

1) $2\frac{(x,y)}{(x,x)} \in \mathbb{Z}$ for all $x, y \in R$;

2) $y - 2\frac{(x,y)}{(x,x)} x \in R$ for all $x, y \in R$.

This means geometrically that the angles among vectors in R can only have the form $\frac{\pi k}{12}$, where $k = 0, 2, 3, 4, 6, 8, 9, 10, 12$, and that the set R is left invariant by maps σ_x, $x \in R$ consisting of reflections with respect to hyperplanes orthogonal to x.

The finite group W generated by the maps σ_x, $x \in R$ is called the *Weyl group* of the system R. The determinant of the linear map in $\mathbb{R}^n$ corresponding to an element $w \in W$ will be denoted by sgn w. Clearly sgn $\sigma_x = -1$.

For any root system R, there is the *dual root system* $\check{R}$ consisting of vectors $\check{\alpha} = \frac{2\alpha}{(\alpha,\alpha)}$, where $\alpha \in R$. The Weyl groups of R and $\check{R}$ coincide.

The root systems arising in connection with semisimple Lie groups have two additional properties:

3) vectors from R generate the space $\mathfrak{h}^* \cong \mathbb{R}^n$;

4) if $x \in R$, then $kx \notin R$ for $k \neq \pm 1$:

Root systems having these properties are called *nonsingular and reduced*. A choice of positive roots (see above) induces a partial order on R. Minimal positive roots with respect to this order are called *simple*. Every positive (negative) root is a linear combination of simple roots, the coefficients of which are positive (negative) integers. The family $\Delta = \{x_1, \ldots, x_n\}$ of simple roots determines the system R and is characterized by the properties:

1) the vectors $\{x_i\}$ form a basis of $\mathbb{R}^n$;

2) the numbers $a_{ij} = \frac{2(x_i, x_j)}{(x_i, x_i)}$ are nonpositive integers for all $i \neq j$.

The matrix $A = ||a_{ij}||$ is called the *Cartan matrix* of the system R.

In this way, a semisimple Lie algebra $\mathfrak{g}_n$ is then determined by the Cartan matrix A. It is possible to construct $\mathfrak{g}_n$ directly, starting from A. Namely, the algebra $\mathfrak{g}_n$ is generated by $3n$ generators e_k, f_k, h_k, $1 \leq k \leq n$, with relations:

$$\begin{array}{ll} 1) \ [h_i, h_j] = 0, & 2) \ [e_i, f_j] = \delta_{ij} h_j, \\ 3) \ [h_i, e_j] = a_{ij} e_j, & 4) \ [h_i, f_j] = -a_{ij} f_j, \\ 5) \ (\text{ad } e_i)^{1-a_{ij}} e_j = 0, & 6) \ (\text{ad } f_i)^{1-a_{ij}} f_j = 0. \end{array}$$

To avoid misunderstanding, note that there is no summation with respect to j on the right hand sides of 2), 3) and 4).

Example 5.6.   Consider the Lie algebra $sl(n+1, \mathbb{C})$. Its compact real form is the algebra $su(n+1)$ and its normal real form is the algebra $sl(n+1, \mathbb{R})$.

The Cartan subalgebra $\mathfrak{h}$ is formed by diagonal traceless matrices and the algebras $\mathfrak{n}_+(\mathfrak{n}_-)$ consist of upper (lower) triangular matrices with zero on the main diagonal. If E_{ij} denotes the matrix having 1 in the corresponding place as the only nontrivial entry, it is possible to take the matrices $E_{ij}, i \neq j$, as root vectors. If we identify $\mathfrak{h}$ with $\mathfrak{h}^*$ using the Killing form, then the corresponding roots α_{ij} are given by the diagonal matrices

$$\alpha_{ij} = \frac{1}{2n+2}(E_{ii} - E_{jj}).$$

The root α_{ij} is positive if $i < j$. The simple roots are $\alpha_i := \alpha_{i,i+1}, 1 \le i \le n$. The Cartan matrix has the form

$$A = \begin{pmatrix} 2 & -1 & 0 & \cdots & 0 \\ -1 & 2 & -1 & \cdots & 0 \\ 0 & -1 & 2 & \cdots & 0 \\ \cdot & \cdot & \cdot & \cdots & \cdot \\ 0 & 0 & 0 & \cdots & 2 \end{pmatrix}.$$

The canonical generators e_k, f_k, h_k are given by the formulas

$$e_k = E_{k,k+1}, \quad f_k = E_{k+1,k}, \quad h_k = E_{k,k} - E_{k+1,k+1}.$$

Example 5.7.   The Cartan matrix of the simple Lie algebra of type G_2 has the form $A = \begin{pmatrix} 2 & -1 \\ -3 & 2 \end{pmatrix}$. So the angle between the simple roots α, β is $150°$ and the ratio of the squares of their lengths is $1 : 3$. The corresponding root system is shown at the Fig. 5.

In particular, it is seen from this that the root vectors corresponding to the long roots generate a subalgebra of type A_2. It is known that the compact Lie group of type G_2 can be realized as the group of automorphisms of the algebra $\mathbb{O}$ of *octonions* or *Cayley numbers* (which is the only eight-dimensional division algebra over $\mathbb{R}$). The group Aut $\mathbb{O}$ preserves the identity of the algebra $\mathbb{O}$ and its orthogonal complement, i.e. the space of all purely imaginary octonions. It acts transitively on the sphere S^6 consisting of purely imaginary octonions of unit norm and the stabilizer of a point under this action is isomorphic to SU(3), which is the compact group of type A_2.

6.3. Representations of Semisimple Lie Groups and Algebras. Let T be a representation of a Lie group G on a finite-dimensional complex space V.

As was shown in Theorem 5.2, the representation T induces representations of the Lie algebra $\mathfrak{g} = L(G)$ and its universal enveloping algebra $U(\mathfrak{g})$, also

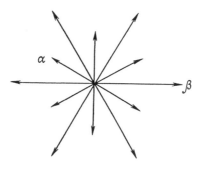

Fig. 5

denoted by T. The structure of the universal enveloping algebra does not
depend on a choice of real form of the algebra $\mathfrak{g}$, so we can suppose as well
that $\mathfrak{g}$ is the compact or the normal real form. This very useful idea is usually
called the *Weyl unitary trick*. So we get

Theorem 5.8. *Every finite-dimensional representation of a connected se-*
misimple Lie group is completely reducible.

In fact, we know this for compact groups (see Chap. 4) and the general
case is reduced to it by the Weyl unitary trick.

From now on we shall use the normal real form.

A *weight vector* in V is a joint eigenvector v for all operators $T(h)$,
$h \in \mathfrak{h}$. A linear functional $\lambda \in (\mathfrak{h}^*)^c$ such that $T(h)v = \lambda(h)v$ for all $h \in \mathfrak{h}$ is
called the *weight* of the vector v.

Operators $A_\alpha = T(x_\alpha)$, where $x_\alpha \in \mathfrak{n}_+(\mathfrak{n}_-)$ are called the *raising (lower-*
ing) operators.

The commutation relations (5.9) implies

Lemma 5.9. *The operator A_α maps a weight vector with weight λ either*
to a weight vector with weight $\lambda + \alpha$ or to the zero vector.

Let us introduce a partial order on the space $(\mathfrak{h}^*)^\mathbb{C}$ by defining $\lambda \geq \mu$ if $\lambda - \mu$
is a linear combination of simple roots the coefficients of which are nonnegative
integers. Then the raising (lowering) operators increase (decrease) the weight
of a weight vector.

The highest weight of a representation T is a maximal element of the set
of all weights. (This set is finite, because the space V is finite-dimensional
and because weight vectors corresponding to different weights are linearly
independent.)

Theorem 5.11. (É. Cartan) *Every finite-dimensional complex irreducible*
representation T of a connected semisimple group G has a unique highest
weight and is characterized by it up to equivalence. If G is a simply connected

group of rank l, then the set P_+ of all highest weights of irreducible represen-
tations is a free semigroup with l generators ω_i defined by the equations

$$(\omega_i, \check{\alpha}_j) = \delta_{ij}, \ 1 \le i, j \le l,$$

where α_j are the simple roots of the Lie algebra $\mathfrak{g}$ and $\check{\alpha}_j$ are the dual roots.

The elements ω_i, $1 \le i \le l$ are called the *fundamental weights* and the corresponding representations of its Lie group G and its Lie algebra $\mathfrak{g}$ are called the *fundamental representations*. The weights in P_+ are called *dominant weights*.

The scheme of the proof.   The irreducibility of the representation implies that every vector $v \in V$ can be written in the form $T(x)v_0$, where v_0 is a weight vector corresponding to the highest weight λ_0 and $x \in U(\mathfrak{g})$. Let us choose any ordering of the set of generators of the Lie algebra $\mathfrak{g}$ with the property that the group of weight vectors $x_\alpha, \alpha < 0$, is followed by $x_i \in \mathfrak{h}$ and by $x_\alpha, \alpha > 0$. Then Theorem 5.6' implies immediately that there is a basis of the space V consisting of vectors of the form $A_1 \dots A_k v_0$, where the A_k are annihilation operators. As a consequence, all weights λ of the representation T have the form $\lambda_0 - \alpha_1 - \dots - \alpha_k$, where α_i are positive roots. Moreover, if $Q(\mu)$ denotes the so called *Kostant function* (the value of which is equal to the number of partitions of the vector μ into a sum of positive roots), then the multiplicity of the weight λ is less than or equal to $Q(\lambda_0 - \lambda)$. This proves the first part of the theorem. The second part follows from the fact that the kernel of the map $U(\mathfrak{n}_-) \to V : x \mapsto T(x)v_0$ is completely determined by the highest weight λ_0. In fact, the vector $T(x)v_0$ is different from zero if and only if there exists an element $y \in U(\mathfrak{n}_+)$ such that $T(y) \cdot T(x)v_0 = v_0$; to check this formula means (due to Theorem 5.6) to write the element $yx \in U(\mathfrak{g})$ in canonical form, but it uses only commutation relations in $\mathfrak{g}$ and the weight of v_0. The proof of the last part of the theorem reduces to the verification of the fact that there exists an irreducible finite-dimensional representation of the Lie algebra $\mathfrak{g}$ with highest weight λ_0 if and only if $\lambda_0 \in P_+$.

If infinite-dimensional representations are allowed, then for every weight λ there exists such a representation. Namely, let I_λ be the left ideal in $U(\mathfrak{g})$ generated by elements $x \in \mathfrak{n}_+$ and $h - \lambda(h) \cdot 1$, $h \in \mathfrak{h}$. Then $U(\mathfrak{g})$ acts on the space $M_\lambda := U(\mathfrak{g})/I_\lambda$ and the element $1 \mod I_\lambda$ is clearly the vector corresponding to the highest weight λ. The $U(\mathfrak{g})$-module M_λ constructed above is called the *Verma module*. It is a universal object in the category of all $U(\mathfrak{g})$-modules with highest weight λ. Next, if W_λ denotes the sum of all proper submodules of M_λ, then the factor module $L_\lambda := M_\lambda/W_\lambda$ is irreducible and it is the unique (up to equivalence) irreducible $U(\mathfrak{g})$-module with highest weight λ. So we are left with the problem of finding when $\dim L_\lambda < \infty$.

Lemma 5.10.   *If T is a finite-dimensional complex representation of a semisimple Lie algebra $\mathfrak{g}$, then the set Λ of all weights of T is invariant with* respect to the Weyl group W of the algebra $\mathfrak{g}$, and moreover the action of W preserves the multiplicities of weights.

Proof. The group W is generated by reflections σ_α corresponding to roots α of the algebra $\mathfrak{g}$, so it is sufficient to prove the theorem for these elements. This is a consequence of the results proved in Sect. 5, applied to the three-dimensional subalgebra $\mathfrak{g}_\alpha$, generated by roots $x_{\pm\alpha}$.

Hence Lemma 5.10 and the results of Sect. 5 show that if the representation L_λ is finite-dimensional, then the numbers $\frac{2(\lambda,\alpha)}{(\alpha,\alpha)} = (\lambda, \check{\alpha})$ are non-negative integers for all positive α. Conversely, if this condition is satisfied, then L_λ is a sum of finite-dimensional $\mathfrak{g}_\alpha$-submodules (as follows from the fact that L_λ is irreducible and that the highest weight vector generates a finite-dimensional $\mathfrak{g}_\alpha$-module). Hence the set Λ_λ of all weights of the module L_λ is invariant under the action of σ_α, hence also under the full Weyl group W. There exists an element $w_0 \in W$ mapping all positive roots into negative ones. So the set Λ_λ has λ as an upper bound and $w_0(\lambda)$ as a lower bound, and hence is finite. The multiplicity of a weight μ is also finite (bounded by $Q(\lambda - \mu)$). So $\dim L_\lambda < \infty$.

Example 5.8. Let us consider the group $G = \mathrm{SU}(n+1)$. Then (see Ex. 5.6) $\check{\alpha}_i = E_{ii} - E_{i+1,i+1}$. For a diagonal matrix $h \in \mathfrak{h}$ with eigenvalues $h_1, \ldots, h_{n+1}$, we have $h = \sum_{k=1}^{n} s_k \check{\alpha}_k$, where $s_k = h_1 + \ldots + h_k$. Hence $\omega_k(h) = h_1 + \ldots + h_k$.

It is easy to show that the irreducible representation of G with the highest weight ω_k is the k-th exterior power of the standard representation.

Now let $G_1 = \mathrm{PSU}(n + 1)$ be the quotient group of $\mathrm{SU}(n + 1)$ by the subgroup C of scalar matrices. It can be realized either as the group of projective maps of n-dimensional complex projective space or as the image of G under the adjoint representation Ad. The irreducible representations of G_1 are the irreducible representations of G which are trivial on C. So the highest weights of these representations have the form $\sum_{k=1}^{n} m_k \omega_k$, where $m_k \geq 0$ and

$$\sum_{k=1}^{n} k m_k \equiv 0 \bmod (n + 1).$$

If $n > 1$, the set of all these weights is not a free semigroup.

6.4. Some Formulas. To finish, let us give a list of basic formulas.

The *Weyl formula for the character* of an irreducible representation T_λ with highest weight λ :

$$\chi_\lambda(\exp h) = \frac{E_{\lambda+\rho}(h)}{E_\rho(h)}, h \in \mathfrak{h},$$

where $E_\lambda(h) = \sum_{w \in W} \mathrm{sgn}\, w\, e^{\langle w(\lambda), h \rangle}$, ρ is the half-sum of the positive roots of the Lie algebra $\mathfrak{g}$ (and at the same time the sum of the fundamental weights ω_k, $1 \leq k \leq l$).

The multiplicative formula for $E_\rho(h)$:

$$E_\rho(h) = \prod_{\alpha>0} 2\sinh \frac{\langle \alpha, h \rangle}{2}.$$

The formula for the dimension of T_λ :

$$\dim T_\lambda = \prod_{\alpha>0} \frac{(\lambda + \rho, \alpha)}{(\rho, \alpha)}.$$

The *Kostant formula* for the multiplicity of the weight μ in the representation T_λ :

$$m_\lambda(\mu) = \sum_{w\in W} \operatorname{sgn} w\, Q(w(\lambda + \rho) - \mu - \rho).$$

The recursive *Freudenthal formula* for the multiplicity of a weight:

$$m_\lambda(\mu) = \frac{2}{(\lambda + \rho, \lambda + \rho) - (\mu + \rho, \mu + \rho)} \sum_{k>0, \alpha>0} m_\lambda(\mu + k\alpha) \cdot (\mu + k\alpha, \alpha).$$

The *Brauer formula* for the decomposition of the tensor product of two irreducible representations:

$$T_\lambda \otimes T_\nu = \sum_{\mu\in P_+} \sum_{w\in W} \operatorname{sgn} w\, m_\lambda(\mu) T_{w(\mu)+\nu},$$

where the symbol T_λ on the right hand side denotes:

a)   the irreducible representation with highest weight λ, if $\lambda \in P_+$;
b)   the virtual representation $\operatorname{sgn} w\, T_{w(\lambda+\rho)-\rho}$, if $w(\lambda + \rho) - \rho \in P_+$;
c)   zero in the other cases.

The *Harish-Chandra formula* for the infinitesimal character. If X is a primitive element (one of homogeneous generators) of the algebra $P(\mathfrak{g}^*)^G$ and T is an irreducible representation with highest weight μ, then

$$\lambda_T(\sigma(X)) = X(\mu + \rho) - X(\rho).$$

The *Demazure formula* for the character χ_λ :

$$\chi_\lambda(\exp h) = A_{i_1} \dots A_{i_r} e^{\langle \lambda, h \rangle},$$

where

$$A_i f(h) = \frac{f(h) - e^{-\langle \alpha_i, h \rangle} f(\sigma_i h)}{1 - e^{-\langle \alpha_i, h \rangle}}$$

and $w_0 = \sigma_{i_1} \dots \sigma_{i_r}$ is the reduced decomposition of an element $w_0 \in W$ of maximal length into a product of generators $\sigma_i := \sigma_{\alpha_i}$, corresponding to simple roots.

Chapter 6
General Theory of Infinite-Dimensional Unitary Representations

§1. Algebras of Operators in a Hilbert Space and the Decomposition of Unitary Representations

1.1. C^*-algebras. Algebras of operators used in representation theory usually have the following two properties:

a) symmetry: the algebra contains the adjoint operator A^* for any operator A of the algebra

b) the algebra is closed in the norm topology.

Such algebras have a simple axiomatic description which is motivation for the definition of the class of C^*-algebras.

A Banach algebra $\mathfrak{A}$ with an involution $a \mapsto a^*$ is called a C^*-algebra if for every $a \in \mathfrak{A}$, the condition $||a^*a|| = ||a^2||$ is satisfied.

Theorem 6.1. (Gel'fand-Najmark)

1) *Every C^*-algebra admits an isometric operator *-representation.*

2) *For every Banach algebra $\mathfrak{A}$ with involution, there exists a C^*-algebra $C^*(\mathfrak{A})$ and a homomorphism $\varphi : \mathfrak{A} \to C^*(\mathfrak{A})$ such that every operator *-representation T of the algebra $\mathfrak{A}$ has form $\tilde{T} \circ \varphi$, where $\tilde{T}$ is a representation of $C^*(\mathfrak{A})$.*

C^*-algebras have a number of remarkable properties. We list them now:

1) If $\varphi : \mathfrak{A}_1 \to \mathfrak{A}_2$ is a morphism of C^*-algebras (i.e. a homomorphism commuting with the involution), then $||\varphi(a)|| \le ||a||$ for all $a \in \mathfrak{A}_1$; if, moreover, $\ker \varphi = \{0\}$, then $||\varphi(a)|| = ||a||$. The main principle behind this fact is that $||a||^2$ coincides with the spectral radius of the element a^*a, and hence is determined by the algebraic structure of the algebra $\mathfrak{A}$.

2) A commutative C^*-algebra $\mathfrak{A}$ with unity is isomorphic to an algebra $C(X)$ of continuous complex functions on a compact set X with involution given by complex conjugation. The compact set X can be reconstructed from the algebra $\mathfrak{A}$ up to homeomorphism. Namely, it is possible to take the set $\mathfrak{M}(\mathfrak{A})$ of maximal ideals as the set X. The isomorphism $\mathfrak{A} \approx C(\mathfrak{M}(\mathfrak{A}))$ is called the *Gel'fand map*.

3) If a is a hermitian element of the C^*-algebra $\mathfrak{A}$ and if f is a continuous function on the spectrum $\mathrm{Sp}\, a$, then there is an element $f(a) \in \mathfrak{A}$ such that the correspondence $f \mapsto f(a)$ is a morphism of the algebra $C(\mathrm{Sp}\, a)$ of all continuous functions on the spectrum of a into the algebra $\mathfrak{A}$. (An element a is called *hermitian* if $a^* = a$. The *spectrum of the element a* is defined as the subset $\mathrm{Sp}\, a \subset \mathbb{C}$ consisting of all numbers λ such that $a - \lambda \cdot 1$

is not invertible in $\mathfrak{A}$. If a is hermitian, then the spectrum is a subset of the real line.)

4) The generalized Stone-Weierstrass theorem. Let $\mathfrak{A}$ be a C^*-algebra with unity. Then every C^*-subalgebra $\mathfrak{B}$ with identity can be defined by a condition of the form $\mathfrak{B} = \{a \in \mathfrak{A} \,|\, T(a) = S(a)\}$ where T, S are a pair of representations of the algebra $\mathfrak{A}$ in a common Hilbert space H. (If $\mathfrak{A} = C(X)$, then this implies the standard Stone-Weierstrass theorem on the structure of closed subalgebras of $C(X)$.)

5) Every topologically irreducible representation T of the C^*-algebra $\mathfrak{A}$ is algebraically irreducible. Even more is true.

Theorem. (Density Theorem) *Let $T_1, \ldots, T_k$ be irreducible, pairwise non-equivalent representations of the C^*-algebra $\mathfrak{A}$ in the spaces $H_1, \ldots, H_k$. For every given family of finite-dimensional subspaces V_i of H_i and linear operators $A_i : V_i \to H_i$, there exists an element $a \in \mathfrak{A}$ such that $T_i(a)|_{V_i} = A_i$.*

Example 6.1. The algebra $\mathcal{K}(H)$ of all compact operators in a given Hilbert space H is a C^*-algebra (without unity). A more general example is the algebra $C(X, \mathcal{K}(H))$ of all continuous (with respect to the norm) functions f on the compact set X with values in $\mathcal{K}(H)$. It is known that every non-zero irreducible representation of this algebra is equivalent to one of the *evaluation representations* $T_x : f \mapsto f(x)$, $x \in X$.

Example 6.2. Let $\nu = (n_1, n_2, \ldots, n_k, \ldots)$ be an increasing sequence of positive integers, every member n_k of which divides the next number n_{k+1}. The matrix algebra $\mathrm{Mat}_{n_k}(\mathbb{C})$ admits a unique imbedding (up to equivalence) into $\mathrm{Mat}_{n_{k+1}}(\mathbb{C})$ (a matrix a is mapped to a block-diagonal matrix with n_{k+1}/n_k blocks equal to a on the diagonal). Let us fix such an imbedding and define $\mathfrak{A}_\nu^0 = \cup_k \mathrm{Mat}_{n_k}(\mathbb{C})$. The completion of $\mathfrak{A}_\nu^0$ with respect to the matrix norm (coinciding with the algebra $C^*(\mathfrak{A}_\nu^0)$, see Theorem 6.1) will be denoted by $\mathfrak{A}_\nu$. It turns out that the C^*-algebra A_ν constructed in such a way depends on arithmetical properties of the sequence ν. Namely, for every such sequence ν we define a "supernatural number" $\prod p^{a_p}$ (the product is taken over all prime numbers p and the exponent a_p is a non-negative integer), where $a_p = \lim_{k \to \infty} a_p^{(k)}$ and $a_p^{(k)}$ is the corresponding exponent in the decomposition of n_k into prime numbers. The algebras $\mathfrak{A}_\nu$ and $\mathfrak{A}_\mu$ are isomorphic if and only if the corresponding supernatural numbers coincide.(Half of the result is easy to understand taking into account that countable tensor products are well-defined in the category of C^*-algebras and that the algebra $\mathfrak{A}_\nu$ is isomorphic to the product of the matrix algebras $\mathrm{Mat}_{m_k}(\mathbb{C})$, where $m_k = n_k/n_{k-1}$.) The algebra $\mathfrak{A}_\nu$ for $\nu = \{2^k\}$ is particularly well-known. It is called the *algebra of anticommutation relations* (also called the *Heisenberg algebra*) and it can be described by generators $a_k, a_k^*, 1 \leq k < \infty$, with relations

$$a_k a_l + a_l a_k = a_k^* a_l^* + a_l^* a_k^* = 0,$$
$$a_k a_l^* + a_l^* a_k = \delta_{kl}.$$

In fact, it is not difficult to verify that the algebra generated by a pair a, a^* of generators with the relations $a^2 = (a^*)^2 = 0$, $aa^* + a^*a = 1$ is isomorphic to the C^*-algebra $\text{Mat}_2(\mathbb{C})$. (Cf. Example 3.15).

1.2. States and Representations of C^*-algebras. A linear functional f on a C^*-algebra $\mathfrak{A}$ is called *positive* if $f(a^*a) \geq 0$ for every $a \in \mathfrak{A}$. A positive functional f is called a *state of the algebra* $\mathfrak{A}$ if its norm is equal to 1 (an equivalent condition for a C^* - algebra with unity is $f(1) = 1$). It is known that every continuous linear functional on a C^*-algebra is a linear combination of states. In particular, for any $a \in \mathfrak{A}$, there is a state different from zero on the element a.

Theorem 6.2. (Gel'fand-Najmark-Segal) *For any state f on a C^* - algebra $\mathfrak{A}$, there is a representation T of the algebra $\mathfrak{A}$ in a Hilbert space H and a vector $\xi \in H$ of unit norm such that $f(a) = (T(a)\xi, \xi)$ for all $a \in \mathfrak{A}$.*

Proof. The standard construction of adjoining of unity shows that we can suppose that $\mathfrak{A}$ is a C^*-algebra with unity. If we define a scalar product in $\mathfrak{A}$ by $(a, b) = f(b^*a)$, then $\mathfrak{A}$ is a pre-Hilbert space. Let us denote the corresponding Hilbert space by H and let ξ be the image of the unity of the algebra $\mathfrak{A}$ in H. The extension of the action of $\mathfrak{A}$ on itself by left translations then gives the representation T.

The construction of representations of C^*-algebras described above is called the *GNS-construction*.

The set of all states of a C^*-algebra $\mathfrak{A}$ forms a convex set $E(\mathfrak{A})$. The extremal points of this set are called *pure states* (a point of a convex set K is called an *extremal point* if it is not the center of any segment belonging to K). If $\mathfrak{A}$ has the unity, then the set $E(\mathfrak{A})$ is compact in the weak topology and the well-known Krejn-Mil'man theorem implies that $E(\mathfrak{A})$ is the closure of the convex hull of its extremal points. The role of pure states in the theory of representations is connected with the following simple, but important, fact giving additional information to Theorem 6.2.

Theorem 6.3. *A representation T of the algebra $\mathfrak{A}$ corresponding to a state f is irreducible if and only if f is a pure state.*

1.3. Von Neumann Algebras. The space $\mathcal{L}(H)$ of all bounded linear operators in a Hilbert space H can be endowed, apart from the norm topology, with many other useful topologies.

The *weak topology* is given by a family of seminorms

$$||A||_{\xi,\nu} = |(A\xi, \nu)|, \quad \xi, \nu \in H.$$

The *strong topology* is given by a family of seminorms

$$||A||_{\xi} = ||A\xi||, \quad \xi \in H.$$

The *stronger* (or *ultrastrong) topology* is given by a family of seminorms $||A||_\Xi^2 = \sum_k ||A\xi_k||^2$, where $\Xi = \{\xi_k\}$ is a countable set of vectors in H with the property $\sum_k ||\xi_k||^2 < \infty$.

Theorem 6.4. (von Neumann) *Let $\mathfrak{A}$ be a symmetric algebra of operators on H. Then the following conditions are equivalent:*

1) $\mathfrak{A}$ *is closed in the weak topology*
2) $\mathfrak{A}$ *is closed in the strong topology*
3) $\mathfrak{A}$ *is closed in the ultrastrong topology.*

Moreover, if $\mathfrak{A}$ is an algebra with unity, then the above topological conditions are equivalent to the algebraic condition

4) $\mathfrak{A} = (\mathfrak{A}^!)^!$.

(Recall that the symbol $\mathfrak{A}^!$ denotes the set of operators commuting with all operators in $\mathfrak{A}$.) Algebras having the properties described in the above theorem are called *von Neumann algebras*. While C^*-algebras are defined as abstract algebras, von Neumann algebras are algebras of operators on a Hilbert space. (There is a simple criterion to find when a C^*-algebra is isomorphic to a von Neumann algebra, but we shall not need it.)

Example 6.3. Let μ be a measure on a space X. Denoting $H = L_2(X, \mu)$, $\mathfrak{A} = L_\infty(X, \mu)$, the algebra $\mathfrak{A}$ can be realized as an algebra of operators on the Hilbert space H; for every function $f \in L_\infty(X, \mu)$ there is the corresponding operator in $L_2(X, \mu)$ given by multiplication by the function f. The algebra $\mathfrak{A}$ is commutative and coincides, as can be checked, with its commutant $\mathfrak{A}^!$. Hence $(\mathfrak{A}^!)^! = \mathfrak{A}$ and $\mathfrak{A}$ is a von Neumann algebra. The weak, strong and ultrastrong topologies on $\mathfrak{A}$ coincide with the weak-$*$ topology on the space $L_\infty(X, \mu)$, considered as the dual space to $L_1(X, \mu)$. (Note that the algebra $\mathfrak{A}$ is isomorphic, as a commutative C^*-algebra, to the algebra of continuous functions on the compact set $\mathfrak{M}(\mathfrak{A})$. Unfortunately, there is no simple description of this compact set.)

It is possible to define two versions of isomorphism between two von Neumann algebras:

1) *algebraic isomorphism* (the isomorphism in the category of C^*-algebras)
2) *spatial isomorphism* (which is induced by an isomorphism of Hilbert spaces, acted upon by the algebras considered).

Example 6.4. A finite-dimensional von Neumann algebra is algebraically isomorphic to a direct sum of matrix algebras: $\mathfrak{A} = \oplus_k \mathrm{Mat}_{m_k}(\mathbb{C})$. A class of spatially isomorphic algebras is given by a set of "multiplicities" $d_1, \ldots, d_k$. For a given element $a = \oplus_k a_k, a_k \in \mathrm{Mat}_{m_k}(\mathbb{C})$, the corresponding block-diagonal operator consists of d_1 blocks of a_1, d_2 blocks of $a_2, \ldots, d_k$ blocks of a_k. The symbols d_i here denotes any cardinality (finite or infinite).

A von Neumann algebra is called a *factor* if its center $\mathfrak{A} \cap \mathfrak{A}'$ contains only scalar operators. An example of a factor is the algebra $\mathcal{L}(H)$ of all bounded linear operators on a Hilbert space H.

A *factor of type I* is a factor which is algebraically isomorphic to $\mathcal{L}(H)$ for a Hilbert space H. Such a factor is spatially isomorphic to the algebra $\mathcal{L}(H) \otimes 1$ consisting of all operators of the form $A \otimes 1$ acting on the Hilbert tensor product $H \otimes H_1$. The Hilbert dimension of H_1 determines the class (up to a spatial isomorphism) of the factor. Its commutant coincides with the algebra $1 \otimes \mathcal{L}(H_1)$ consisting of all operators of the form $1 \otimes B$.

A remarkable discovery of von Neumann was the existence of factors which are not of type I. They are called *factors of type II and III*. For factors of type II, the equivalence class with respect to spatial isomorphisms is determined by a positive real number, while for factors of type III, the equivalence with respect to spatial isomorphisms coincides with the one with respect to algebraic isomorphisms.

For factors of type II, it is possible to define a specific analogue of the notion of a trace. It has the usual properties of additivity and unitary equivalence but the value of a projector on the factor can be either any real number (for a factor of type II_∞) or, after a suitable normalization, any real number between 0 and 1 (for a factor of type II_1). More details on traces on factors can be found in 6.4.

It is possible to subdivide factors of type III into the classes III_λ, $0 \le \lambda \le 1$, depending on the structure of their group of automorphisms.

Example 6.5. Let us consider the algebra $\mathfrak{A}$ of anticommutation relations (see Ex. 6.4) and its state ω_λ defined by the conditions

$$\omega_\lambda(a_k) = \omega_\lambda(a_k^*) = 0, \; \omega_\lambda(a_k a_k^*) = \lambda, \; 0 \le \lambda \le 1/2$$

and

$$\omega_\lambda(xy) = \omega_\lambda(x)\omega_\lambda(y),$$

where x and y are polynomials in the generators with different indices. Let T_λ be a representation of the algebra $\mathfrak{A}$ corresponding to the state ω_λ by Theorem 6.2. Then, as was shown by R.Powers, von Neumann algebras $\mathfrak{A}_\lambda$ generated by operators of the representation T_λ are pairwise non-isomorphic factors and the factor $\mathfrak{A}_{1/2}$ is of type II_1, while the factors $\mathfrak{A}_\lambda$, $0 < \lambda < 1/2$, are of type III_λ.

An almost complete classification of factors in a separable Hilbert space is known at present. (The so called factors of type III_0 are an exception. Their classification reduces basically to the classification of ergodic dynamical systems.)

1.4. Direct Integrals of Hilbert Spaces and von Neumann Algebras.

Consider a measure μ on a space X, and suppose that for every point $x \in X$, a Hilbert space H_x is given. We shall suppose that all spaces H_x are subspaces of one fixed Hilbert space H and that the orthogonal projector P_x

onto the subspace H_x is a weakly measurable operator-valued function on X (i.e. the functions $x \mapsto (P_x \nu, \xi)$ are measurable for every $\xi, \nu \in H$). Then it is possible to define the *direct integral* or the *continuous direct sum* of a family $\{H_x\}, x \in X$, denoted by

$$\int_X H_x d\mu(x).$$

This space consists of (equivalence classes of) weakly measurable H-valued vector-functions on X, having the properties

a) $f(x) \in H_x$ for almost all $x \in X$;
b) $\int_X \|f(x)\|^2_{H_x} d\mu(x) < \infty$. (Two functions are said to be equivalent if they coincide almost everywhere with respect to the measure μ.)

An operator A on the space $\mathcal{H} := \int_X H_x d\mu(x)$ is called *decomposable* if there is a family of operators $\{A_x\}, A_x \in \mathcal{L}(H_x)$ for $x \in X$, such that

$$(Af)(x) = A_x f(x)$$

for all $f \in \mathcal{H}$ almost everywhere on X.

A special case of decomposable operators are the *diagonal operators*, for which almost all operators A_x are scalars: $A_x = a(x) \cdot \mathbb{1}$.

It is clear that decomposable operators generate a subalgebra $\mathfrak{R}$ in $\mathcal{L}(\mathcal{H})$ and that diagonal operators generate a commutative subalgebra $\mathfrak{D}$ in $\mathfrak{R}$. If X is compact and μ is a regular measure on X (i.e. a positive linear functional on the algebra $C(X)$), then the subalgebra of $\mathfrak{D}$ consisting of *continuously diagonal operators* A (for which the function $a(x)$ is continuous on X) will be denoted by $\mathfrak{D}_0$.

Two decompositions $\mathcal{H} := \int_X H_x d\mu(x)$ and $\mathcal{H} := \int_Y V_y d\nu(y)$ are called *strongly (weakly) isomorphic* if there exists a continuous (measurable) mapping $\varphi : Y \to X$ and a family of linear maps $L_y : V_y \to H_{\varphi(y)}, y \in Y$ depending continuously (measurably) on y such that if an element h is given in the first decomposition by a vector-valued function $f(x)$, then h is given in the second decomposition by the vector-valued function $g(y) = L_y^{-1} f(\varphi(y))$.

Theorem 6.5. *Let H be a separable Hilbert space and let $\mathfrak{R} \supset \mathfrak{D} \supset \mathfrak{D}_0$ be a triple of algebras in $\mathcal{L}(H)$.*

a) *There is a realization of H as a direct integral $\int_X H_x d\mu(x)$ such that $\mathfrak{R}, \mathfrak{D}$ and $\mathfrak{D}_0$ are the algebras of decomposable, diagonal and continuous diagonal operators if and only if the following conditions are satisfied:*
    1) $\mathfrak{D}_0$ *is a commutative C^*-algebra;*
    2) $\mathfrak{D} = (\mathfrak{D}_0')' (=$ *the closure of $\mathfrak{D}_0$ in the weak, strong or stronger topology);*
    3) $\mathfrak{R} = \mathfrak{D}_0'(= \mathfrak{D}').$
b) *The decomposition $\mathcal{H} := \int_X H_x d\mu(x)$ is determined by the algebra $\mathfrak{D}_0$ ($\mathfrak{D}$) up to strong (weak) isomorphism.*

Note that the algebras $\mathfrak{D}_0$ play only a secondary role in the formulation of the theorem, but they are a very useful technical tool in its proof. The reason behind this is that it is always possible to identify $\mathfrak{D}_0$ with the completion of the algebra of polynomials in one variable in a suitable norm, and hence the problem of the construction of the direct integral is reduced to the spectral decomposition of a selfadjoint operator.

Consider now an arbitrary von Neumann algebra $\mathfrak{A}$ in a separable Hilbert space H. Choose any commutative von Neumann subalgebra $\mathfrak{D}$ in $\mathfrak{A}^!$ and let $H := \int_X H_x d\mu(x)$ be the corresponding decomposition of H into a direct integral such that $\mathfrak{D}$ is the algebra of diagonal operators. The algebra of decomposable operators $\mathfrak{R} = \mathfrak{D}^!$ contains $\mathfrak{A}$, so all elements $A \in \mathfrak{A}$ can be written using the operator-valued functions A_x on X. Let $\{A^{(k)}\}$ be a countable family of operators generating the von Neumann algebra $\mathfrak{A}$. (Such a family always exists for a separable Hilbert space H.) We denote the von Neumann algebra generated by the operators $\mathfrak{A}_x^{(k)}$ in H_x by $\mathfrak{A}_x$. In fact, the algebras $\mathfrak{A}_x$ are determined by the algebra $\mathfrak{A}$ (i.e. it does not depend on the choice of $\{A^{(k)}\}$) for almost all $x \in X$.

Let us now consider the family $\tilde{\mathfrak{A}}$ of all decomposable operators A such that $A_x \in \mathfrak{A}_x$ for almost all $x \in X$. It is possible to show that $\tilde{\mathfrak{A}}$ is a von Neumann subalgebra in $\mathfrak{R}$, called the *direct integral* or the *continuous direct sum of the algebras* $\mathfrak{A}_x$, and denoted by

$$\tilde{\mathfrak{A}} = \int_X \mathfrak{A}_x d\mu(x).$$

In particular, the algebra $\mathfrak{R}$ of all decomposable operators is a direct integral of the algebras $\mathcal{L}(H_x)$ and the algebra $\mathfrak{D}$ of diagonal operators is a direct integral of one-dimensional subalgebras $\mathbb{C} \cdot \mathbb{1}_{H_x}$. In the general case, the algebra $\mathfrak{A}$ forms only a part of the algebra $\tilde{\mathfrak{A}} = \int_X \mathfrak{A}_x d\mu(x)$. It is possible to show that $\tilde{\mathfrak{A}}$ is generated by the subalgebras $\mathfrak{A}$ and $\mathfrak{D}$. Hence $\mathfrak{A} = \tilde{\mathfrak{A}}$ is equivalent to the inclusion $\mathfrak{D} \subset \mathfrak{A}$, so $\mathfrak{D} \subset \mathfrak{A} \cap \mathfrak{A}^!$.

This means that every subalgebra $\mathfrak{D} \subset \mathfrak{A} \cap \mathfrak{A}^!$ induces the representation of the algebra $\mathfrak{A}$ in the form of the direct integral $\mathfrak{A} = \int_X \mathfrak{A}_x d\mu(x)$. The algebras $\mathfrak{A}$ and $\mathfrak{A}^!$ play a symmetric role in the construction, so we get at the same time the representation of the algebra $\mathfrak{A}^!$ in the form of the integral $\int_X \mathfrak{A}_x^! d\mu(x)$ as well as the representation of the algebra $\mathfrak{A} \cap \mathfrak{A}^!$ in the form of the integral $\mathfrak{A} \cap \mathfrak{A}^! = \int_X (\mathfrak{A}_x \cap \mathfrak{A}_x^!) d\mu(x)$. The most interesting case is the one with $\mathfrak{D} = \mathfrak{A} \cap \mathfrak{A}^!$. Then $\mathfrak{A}_x \cap \mathfrak{A}_x^!$ consists of scalar operators in H_x for almost all $x \in X$. This implies that the decomposition of H connected with the algebra $\mathfrak{A} \cap \mathfrak{A}^!$ gives a representation of the algebras $\mathfrak{A}$ and $\mathfrak{A}^!$ in the form of a direct integral of factors. The described representation is called the *canonical* one. An algebra $\mathfrak{A}$ is called an *algebra of type I* if almost all factors $\mathfrak{A}_x$ in its canonical representation $\mathfrak{A} = \int_X \mathfrak{A}_x d\mu(x)$ are of type I.

The definition implies that either both algebras $\mathfrak{A}$ and $\mathfrak{A}^!$ are of type I or both are not. An algebra $\mathfrak{A}$ of type I is called *simple* if $\mathfrak{A}^!$ is a commutative algebra and $\mathfrak{A}$ is called *homogeneous of degree n* if $\mathfrak{A}$ is isomorphic to a direct sum of n copies of a simple algebra $\mathfrak{B}, n = 1, 2, \ldots, \infty$ (the notation for it being $\mathfrak{A} = n\mathfrak{B}$).

Let $\mathfrak{B}$ be a simple algebra. Almost all factors $\mathfrak{B}_x$ in its canonical representation $\mathfrak{B} = \int_X \mathfrak{B}_x d\mu(x)$ coincide with $\mathcal{L}(H_x)$. Let us write the set X as a disjoint union of subsets $X_k = \{x \in X | \dim H_x = k\}, k = 1, 2, \ldots, \infty$. The algebra $\mathfrak{B}$ is isomorphic to the direct sum of the algebras $\mathfrak{B}_k = \int_{X_k} \mathfrak{B}_x d\mu(x)$ and the algebra $\mathfrak{B}_k$ is the tensor product of the commutative algebra $L_\infty(X_k, \mu)$ and the algebra $\mathcal{L}_k$ of all bounded operators on a k-dimensional Hilbert space. Now we are prepared to give the full description of algebras of type I up to algebraic and spatial isomorphism (see Ex. 6.4).

Theorem 6.6. *Let $\mathfrak{A}$ be a von Neumann algebra of type I in a separable Hilbert space H. Then there exists a unique decomposition of H into an orthogonal sum of subspaces H_k invariant with respect to $\mathfrak{A}$ and having the property that the restriction of $\mathfrak{A}$ to H_k generates a homogeneous algebra $\mathfrak{A}_k = k\mathfrak{B}_k$ of degree k. The algebra $\mathfrak{A}$ is algebraically isomorphic to a simple algebra $\mathfrak{B} = \sum_{k=1}^\infty \mathfrak{B}_k$.*

1.5. The Decomposition of Unitary Representations. Let T be a unitary representation of a group G in a separable Hilbert space H. If the representation T is reducible and if H_1 is an invariant subspace, then its orthogonal complement $H_2 = H_1^\perp$ is also invariant and the representation T splits into the sum of two subrepresentations T_1 and T_2 acting on H_1 and H_2. If one or both representations T_1, T_2 are reducible, then it is possible to repeat the process. Unlike the finite-dimensional case, it is possible that the process never stops.

Example 6.6. Let T be the representation of the additive group $\mathbb{R}$ in the space $L_2(\mathbb{R}, dx)$ given by the formula

$$[T(t)f](x) = f(x+t).$$

The Fourier transform translates it into an equivalent representation $\tilde{T}$ acting in $L_2(\mathbb{R}, d\lambda)$ by the formula

$$[\tilde{T}(t)\varphi](\lambda) = e^{it\lambda}\varphi(\lambda).$$

For any decomposition of the real line $\mathbb{R}$ into a finite or countable number of measurable subsets E_i, there is the corresponding decomposition of the representation into a sum of invariant subspaces $H_i = L_2(E_i, d\lambda)$. But all these spaces are reducible. It is not difficult to verify that this representation has no irreducible subrepresentations at all.

The above example shows that the decomposition of a representation into irreducible parts does not necessarily exist in the infinite-dimensional situation. At the same time, it is possible to interpret the defining formula of the representation $\tilde{T}$ as the decomposition of the representation into a direct integral of one-dimensional (hence irreducible) representations $T_\lambda(t) = e^{it\lambda}$ in the sense of the following definition.

Let $H = \int_X H_x d\mu(x)$ be a decomposition of the Hilbert space H into a direct integral. We say that a representation T of a group G decomposes into the *direct integral of representations* T_x in the spaces H_x if all operators $T(g), g \in G$, are decomposable and if the formula

$$T(g) = \int_X T_x(g)d\mu(x), \quad g \in G$$

holds.

The existence and uniqueness problem of the decomposition of a representation T into a direct integral of irreducible representations is mainly connected with the structure of the von Neumann algebra $\mathfrak{A}(T)$ generated by the operators of the representation T.

A representation T is called a *factor* representation (a representation of type I , homogeneous representation of order n, etc.) if the von Neumann algebra $\mathfrak{A}(T)$ is a factor (an algebra of type I, homogeneous of order n etc.). The decomposition of the representation into a direct integral determines the commutative von Neumann subalgebra $\mathfrak{D}$ in $\mathfrak{A}(T)$ corresponding to the algebra of diagonal operators under the decomposition. The decomposition itself is determined by the algebra $\mathfrak{D}$ up to weak isomorphism (Theorem 6.5 b)).

If a commutative von Neumann subalgebra $\mathfrak{D}$ in $\mathfrak{A}(T)^!$ is given, then there is a corresponding decomposition of H into a direct integral $H = \int_X H_x d\mu(x)$ such that all operators of the representation $T(g), g \in G$, are decomposable and are described by operator-valued functions $x \mapsto T_x(g)$ on X. The relation

$$T_x(g_1)T_x(g_2) = T_x(g_1 g_2)$$

holds for all pairs $g_1, g_2 \in G$ almost everywhere on X. If the group G is countable, then it implies that for almost all points $x \in X$, the correspondence $g \mapsto T_x(g)$ is a representation of G.

In the general case, some additional reasoning is necessary to claim that a commutative von Neumann subalgebra $\mathfrak{D} \subset \mathfrak{A}(T)^!$ determines a decomposition of T into a direct integral of representations. The case of a locally compact separable group G will be discussed in 6.2.

Theorem 6.7. *Let $T = \int_X T_x d\mu(x)$ be a decomposition of the representation T into a direct integral of representations $T_x, x \in X$, corresponding to a commutative subalgebra $\mathfrak{D} \subset \mathfrak{A}(T)^!$.*

a) *The representations T_x are irreducible (resp. are factor representations) for almost all $x \in X$ if and only if the subalgebra $\mathfrak{D}$ is a maximal commutative von Neumann subalgebra in $\mathfrak{A}(T)^!$ (resp. $\mathfrak{D} \supset \mathfrak{A}(T) \cap \mathfrak{A}(T)^!$).*

b) *If $\mathfrak{A}(T)$ is a von Neumann algebra of type I and if $\mathfrak{D} = \mathfrak{A}(T) \cap \mathfrak{A}(T)^!$, then the set $\hat{G}$ of equivalence classes of irreducible unitary representations of G can be taken for X so that the representations T_x on the* *spaces H_x are multiples of the irreducible representation of the equivalence class of $x \in \hat{G}$. The corresponding measure μ on $\hat{G}$ is determined by the representation T up to equivalence.*

c) *If T is a factor representation of type II or III, then there exist two decompositions of T into a direct integral of irreducible representations such that no component of the first decomposition is equivalent to any component of the second one.*

This means that an analogue of the unique decomposition into isotypic components exists only for representations of type I.

A group G (C^*-algebra $\mathfrak{A}$) is called a *group (algebra) of type I* if all its representations are of type I (groups of type I are also called *tame* and groups which are not of type I are called *wild* groups).

The class of groups of type I is sufficiently broad. It contains all compact, connected semisimple and nilpotent groups. It is also true that any linear algebraic group (i.e. a subgroup of a matrix group, given by algebraic equations) over a locally compact non-discrete field is of type I.

There are groups of type I as well as wild groups among the solvable groups. A criterion making it possible to distinguish these cases can be found in Chap. 7. Discrete groups are usually wild groups. The only exceptions are extensions of a finite group by a commutative one.

§2. Group Algebras of Locally Compact Groups

2.1. Integration on Groups and Homogeneous Spaces. The operation of invariant mean plays an important role in the theory of representations. Let L be a function space on a G-space X which is invariant under translations. An *invariant mean* is a positive linear functional on L which is invariant with respect to the group G. (A functional F is called positive if $F(f) \geq 0$ for all functions f such that $f(x) \geq 0$ for all $x \in X$.)

The most important case is the case when X is a topological group and G is the group of left, right, or two-sided translations on X. The corresponding means are also called left, right or two-sided. The most interesting cases among the function spaces L are: the space $B(G)$ of all bounded functions on G, the space $BC(G)$ of all bounded continuous functions and the space $C_0(G)$ of all continuous functions with compact support.

A topological group G is called *amenable* if there exists a left invariant mean on the space $BC(G)$. (Note that for discrete groups G, the space $BC(G)$ coincides with $B(G)$. Hence the existence of an invariant mean on $B(G)$ is equivalent to the amenability of the group G, considered as a discrete topological group.)

The class of amenable groups is sufficiently broad; it contains all solvable groups and is closed with respect to extensions.

A closed subgroup of an amenable group and its quotient group with respect to a closed normal subgroup are amenable. An example of a nonamenable group is the free group with two generators (with the discrete topology) as well as any noncompact semisimple Lie group. The notion of an amenable group has been successfully used in the theory of dynamical systems, operator algebras and in harmonic analysis.

The consideration of an invariant mean on $C_0(G)$ leads to the important concept of the left (right) *Haar measure*, i.e. a left (right) invariant regular Borel measure on G. An invariant mean on $C_0(G)$ can be written in the form of an integral with respect to the Haar measure μ :

$$I(f) = \int f(g)d\mu(g).$$

Theorem 6.8. (A. Weil) *A complete topological group G admits a nonzero left (right) Haar measure if and only if the group G is locally compact. The Haar measure is uniquely defined up to a numerical factor.*

The left (right) Haar measure is usually denoted by the symbol $d_l g$ ($d_r g$), omitting the indices l and r in those cases where a two-sided invariant measure exists. In the general case, the left Haar measure is not invariant under right translations, it is mapped onto another left-invariant measure. Hence we have

$$d_l(xy^{-1}) = \Delta_G(y)d_l(x), \tag{6.1}$$

where Δ_G is a homomorphism of the group G into a multiplicative group of positive real numbers. The following relations hold:

$$d_r(yx) = \Delta_G(y)d_r(x),$$
$$d_l(x^{-1}) = \Delta_G(x)d_l(x) = \text{const} \cdot d_r(x). \tag{6.2}$$

If G is a Lie group, the left (right) Haar measure is given by a differential form of highest degree invariant with respect to the left (right) translations. The homomorphism Δ_G then coincides with $\det \text{Ad}$.

Example 6.7. Let G be the group of affine transformations of the real line (also called the "$ax + b$"-group) which is isomorphic to the group of matrices of the form

$$g_{a,b} = \begin{pmatrix} a & b \\ 0 & 1 \end{pmatrix}, \ a > 0.$$

In this case, $d_l(g_{a,b}) = \frac{da \wedge db}{a^2}$, $d_r(g_{a,b}) = \frac{da \wedge db}{a}$ and $\Delta_G(g_{a,b}) = a$.

If G is a locally compact group and X is a homogeneous G-space, then there exists a *quasi-invariant measure* on X, which is unique up to equivalence (a measure is said to be quasi-invariant if it goes over into an equivalent measure under all translations).

Suppose for definiteness that $X = H \backslash G$ is a right G-space and that s is a Borel section of the projection $p : G \to H : g \mapsto Hg$. (For Lie groups, such a mapping s can be chosen to be smooth almost everywhere.) Then every element $g \in G$ can be uniquely written in the form

$$g = h \cdot s(x), \ h \in H, \ x \in X, \tag{6.3}$$

and thus G (as a set) can be identified with $H \times X$. Under this identification, the Haar measure on G goes over into a measure equivalent to the product of a quasi-invariant measure on X and the Haar measure on H. More precisely, if a quasi-invariant measure μ_s on X is appropriately chosen, then the following equalities are valid:

$$d_r(g) = \frac{\Delta_G(h)}{\Delta_H(h)} d\mu_s(x) d_r(h), \tag{6.4}$$

$$\frac{d\mu_s(xg)}{d\mu_s(x)} = \frac{\Delta_H(h(x,g))}{\Delta_G(h(x,g))}, \tag{6.5}$$

where $h(x, g) \in H$ is defined by the relation

$$s(x)g = h(x,g)s(xg). \tag{6.6}$$

If the group G is *unimodular*, i.e. if $\Delta_G \equiv 1$, and if it is possible to select a subgroup K that is complementary to H in the sense that almost every element of G can be uniquely written in the form

$$g = h \cdot k, \ h \in H, \ k \in K,$$

then it is natural to identify $X = H\backslash G$ with K and to choose s as the embedding of K in G. In such a case, the formula (6.4) assumes the form

$$dg = \Delta_H(h)^{-1} d_r h d_r k = d_l h d_r k. \tag{6.4'}$$

If both G and H are unimodular (or, more generally, if $\Delta_G(h)$ and $\Delta_H(h)$ coincide for $h \in H$), then there exists a G-invariant measure on $X = H\backslash G$. If it is possible to extend Δ_H to a multiplicative function on the group G, then there exists a *relatively invariant measure* on X which is multiplied by the factor $\frac{\Delta_H(g)}{\Delta_G(g)}$ under translation by g.

2.2. The Algebras $L_1(G)$ and $C^*(G)$. If G is a locally compact group, then it is possible to define the structure of an algebra on the space $L_1(G, d_l g)$, where $d_l g$ is the left Haar measure. The product in the algebra is defined by the convolution

$$(f_1 * f_2)(g) = \int_G f_1(h) f_2(h^{-1}g) d_l h.$$

It is also possible to define an *involution* $*$:

$$f^*(g) = \Delta_G(g)\overline{f(g^{-1})}.$$

For every unitary representation T of the group G, there is the corresponding representation of the algebra with involution $L_1(G, d_l g)$, denoted by the same letter, and given by the formula

$$T(f) = \int_G f(g)T(g)d_l g.$$

Conversely, if a representation T of the algebra with involution $L_1(G, d_l g)$ in a Hilbert space H is given and if the representation is nonsingular in the sense that the set $\{T(f)\xi \mid f \in L_1(G, d_l g), \xi \in H\}$ is dense in H, then the representation corresponds to a unitary representation of the group G.

The algebra $L_1(G, d_l g)$ is called the *group algebra of the locally compact group* G. It has the unity if and only if the group G is compact (hence its Haar measure has finite integral).

Let $C^*(G)$ denote the completion of the group algebra with respect to the norm

$$\|f\| = \sup_T \|T(f)\|,$$

where the supremum is taken over all unitary representations T of the group G. The algebra $C^*(G)$ is the enveloping C^*-algebra of the group algebra (see 1.1.) and is called the C^*-*algebra of the group* G. The category of unitary representations of the group G is naturally isomorphic to the category of nonsingular $C^*(G)$-modules.

One of the important applications of C^*-algebras of groups is the proof of the existence and completeness of irreducible unitary representations of locally compact groups.

Theorem 6.9. (Gel'fand-Rajkov) *Let G be a locally compact group and let g be a nontrivial element in G. Then there exists a unitary irreducible representation T of the group G such that $T(g)$ is a nontrivial operator.*

The statement of Theorem 6.9 can be expressed in the language of C^*-algebras as follows: for a given nontrivial element $a \in C^*(G)$, there exists an irreducible representation T such that $T(a) \neq 0$. Let f be any state on $C^*(G)$ for which $f(a) \neq 0$ (see 1.1). The Krejn-Mil'man theorem, applied to the compact convex set E of all states of the algebra $C^*(G)$, implies that f can be approximated by a convex linear combination of extremal points of E. Hence there exists a pure state which is different from zero on a. The representation T corresponding to this state is irreducible and $T(a) \neq 0$.

The original formulation of the theorem can be recovered as follows.

Let $\varphi \in L_1(G, d_l g)$ be supported in a neighbourhood V of the identity such that $V \cap gV = \emptyset$. Set $a = \varphi - L(g)\varphi$, where $L(g)$ is left translation by the element g in $L_1(G, d_l g)$. Then a is a nontrivial element of $C^*(G)$ because its image under the regular representation of G in $L_2(G, d_l g)$ is different from zero.

Theorem 6.10. (Gel'fand-Rajkov) *Every unitary representation T of a locally compact group G can be decomposed into a direct integral of irreducible representations.*

The proof of the theorem is based on the Choquet theorem on the representation of a point of a compact convex set in the form of an integral over extremal points.

2.3. Unitary Induction. The operation of induction can be carried over to locally compact groups due to the existence of a quasi-invariant measure on the factor space $X = H\backslash G$. Let S be a unitary representation of a subgroup H in a Hilbert space V and let μ be a measure on X satisfying the condition (6.5). Let $\mathcal{H}$ denote the space of all vector-valued functions f on X with values in V such that

$$\|f\|^2 := \int_X \|f(x)\|_V^2 d\mu(x) < \infty.$$

Let us consider the representation T given by the formula

$$[T(g)f](x) = A(x,g)f(xg), \qquad (6.7)$$

where

$$A(x,g) = \left[\frac{\Delta_H(h)}{\Delta_G(h)}\right]^{1/2} S(h) \qquad (6.8)$$

and where the element $h = h(x,g)$ is defined by (6.6). The representation T is called the *unitary induced* representation and denoted by $\mathrm{Ind}_H^G S$. There exists another realization of the unitary induced representation in the space of vector-valued functions on the group with prescribed transformation properties under left translations by elements $h \in H$. The operators of the unitary induced representation act on the space by right translations.

Example 6.8. Let $G = \mathrm{SL}(2,\mathbb{R})$, let H be the subgroup of upper triangular matrices of the form $h = \begin{pmatrix} \lambda & \mu \\ 0 & \lambda^{-1} \end{pmatrix}$ and let $S_{\rho,\varepsilon}(h) = |\lambda|^{i\rho}(\mathrm{sgn}\lambda)^\varepsilon$, $\rho \in \mathbb{R}$, $\varepsilon = 0,1$, be a one-dimensional representation of H. We also introduce the subgroup K of all lower triangular matrices of the form $\begin{pmatrix} 1 & 0 \\ x & 1 \end{pmatrix}$. In this case, the formula (6.4) is valid and equation (6.6) takes the form

$$\begin{pmatrix} 1 & 0 \\ x & 1 \end{pmatrix}\begin{pmatrix} \alpha & \beta \\ \gamma & \delta \end{pmatrix} = h(x,g) \cdot \begin{pmatrix} 1 & 0 \\ y & 1 \end{pmatrix};$$

hence

$$y = \frac{\alpha x + \gamma}{\beta x + \delta}, \quad h(x,g) = \begin{pmatrix} (\beta x + \delta)^{-1} & \beta \\ 0 & \beta x + \delta \end{pmatrix}.$$

Taking into account that $\Delta_G \equiv 1$, $\Delta_H \begin{pmatrix} \lambda & \mu \\ 0 & \lambda^{-1} \end{pmatrix} = \lambda^2$, the formula for the induced representation $T_{\rho,\varepsilon} = \mathrm{Ind}_H^G S_{\rho,\varepsilon}$ takes the form

$$[T_{\rho,\varepsilon}(g)f](x) = |\beta x + \delta|^{-1-i\rho}[\mathrm{sgn}(\beta x + \delta)]^\varepsilon f\left(\frac{\alpha x + \gamma}{\beta x + \delta}\right).$$

This is the so called *principal series of unitary representations* of the group $\mathrm{SL}(2,\mathbb{R})$ in the Hilbert space $L_2(\mathbb{R}, dx)$.

The formula (6.7) is characteristic for induced representations, as it was in the finite-dimensional case. There is also an analogue of the criterion of inducibility, due to Mackey, in terms of systems of imprimitivity. It looks similar to Theorem 3.6 but the proof is substantially more complicated.

If G is a Lie group, H a closed connected subgroup and if the induced representation $\mathrm{Ind}_H^G S$ is realized on vector-valued functions on the group, then the space of the representation can be described by differential equations of the form

$$L_X f + \rho(X)f = 0, \quad X \in \mathfrak{h}, \tag{6.9}$$

where L_X denotes the Lie derivative with respect to $X \in \mathfrak{h}$ and ρ is the representation of the Lie algebra $\mathfrak{h}$ corresponding to the representation of the group H.

The advantage of such a formulation is that it allows one to complexify the situation. Namely, the condition (6.9) has good meaning even if we take a complex subalgebra in $\mathfrak{g}^c$ instead of $\mathfrak{h}$.

In the case that $\mathfrak{h} + \overline{\mathfrak{h}}$ is a subalgebra in $\mathfrak{g}^c$, the space of solutions of the system (6.9) can be interpreted as the space of sections of a suitable bundle over $K \backslash G$ (where K is the Lie group corresponding to the Lie algebra $\mathfrak{h} \cap \mathfrak{g}$) which are holomorphic in certain directions.

In such a way, the notion of *holomorphically induced representations* arises.

Next, the space of all holomorphic sections of a fiber bundle L over X is just the zero term in the sequence of spaces $H^k(X, \mathcal{L})$ of cohomology of X with coefficients in the sheaf $\mathcal{L}$ of germs of holomorphic sections of L. This leads to the notion of a *representation in cohomology*. All these constructions have recently become a working apparatus in the theory of infinite-dimensional representations.

Example 6.9. Let G be a simply connected compact Lie group and T a maximal commutative subgroup.

The space $M = G/T$ is called a flag manifold (see Sect. 4 of Chap. 7). The reader is advised to make the theory described below more explicit by considering the special case $G = \mathrm{SU}(n)$ (even the case $G = \mathrm{SU}(2)$ is already important enough and instructive).

It is possible to define a G-invariant complex structure on M in the following way. Let $\mathfrak{g}$ be the Lie algebra of the group G, $\mathfrak{g}^c$ its complexification. Let G^c be the connected and simply connected Lie group with Lie algebra $\mathfrak{g}^c$ and B the Borel subgroup (maximal connected solvable subgroup) containing T.

It is possible to show that $B \cap G = T$, hence the space $M = G/T$ can be identified with the complex homogeneous manifold G^c/B.

Every linear G-fibre bundle L over M can be endowed with a unique structure of holomorphic G^c-fibre bundle, because every character χ of the group T can be extended uniquely to a holomorphic character χ^c of the group B.

The space $\Gamma(M, L_\chi)$ of sections of the fibre bundle L_χ (given by the character χ) can be identified with the space of functions φ on G having the property

$$\varphi(tg) = \chi(t)\varphi(g),\ t \in T,\ g \in G.$$

The subspace $\Gamma_{\text{hol}}(M, L_\chi)$ of holomorphic sections can be identified with the space of holomorphic functions on G^c having the property

$$\varphi(bg) = \chi^c(b)\varphi(g),\ b \in B,\ g \in G^c.$$

The space $\Gamma_{\text{hol}}(M, L_\chi)$ is finite-dimensional (because M is compact) and is nontrivial only if the character χ^c of the group B can be extended to a holomorphic function on G^c. The corresponding character χ of the group T is called *dominant*.

Let the group G have rank (i.e. the dimension of T) equal to l. The group $\hat{T}$ of characters of the group T is isomorphic to an l-dimensional lattice $\mathbf{Z}^l$. The group operation in $\hat{T}$ is usually written additively, so $\chi_1 + \chi_2$ denotes the character of T the value of which at a point $t \in T$ is equal to $\chi_1(t)\chi_2(t)$.

It is known that there exists a basis $\chi_1, \ldots, \chi_l$ of $\hat{T}$ such that the character $\chi = \sum_{i=1}^l n_i \chi_i$ is dominant if and only if all the n_i are non-negative integers. Let us set $\rho = \sum_{i=1}^l \chi_i$. It is possible to check that the character $(2\rho)^c$ coincides with the determinant of the adjoint representation of the group B.

Theorem.   (Borel-Weil)

a)   *The space $\Gamma_{\text{hol}}(M, L_\chi)$ is nontrivial only for dominant characters χ;*
b)   *the representation π_χ of the group G in the space $\Gamma_{\text{hol}}(M, L_\chi)$ is irreducible;*
c)   *every irreducible representation of the group G is equivalent to exactly one representation π_χ.*

The proof of the theorem uses the facts described above and the Cartan's theorems on the highest weights (see Sect. 6 of Chap. 5).

The Weyl group $W = N_G(T)/T$ acts naturally on T and $\hat{T}$ by automorphisms. It is generated by l involutive generators $s_1, \ldots, s_l$, acting on the basic characters $\chi_1, \ldots, \chi_l$ by the formula

$$s_i(\chi_j) = \chi_j - \delta_{ij} \sum_{k=1}^l a_{jk}\chi_k,$$

where $\|a_{jk}\|$ is the *Cartan matrix* of the group G.

A character χ is called *regular* if its stabilizer in W is $\{e\}$, otherwise it is called *singular* . It is known that every regular character can be written uniquely in the form $\chi = w(\lambda + \rho) - \rho$, where $w \in W$ and λ is a dominant character. We denote the length of the reduced expression of w in terms of the generators $s_1, \ldots, s_l$ by $l(w)$.

Let $\mathcal{L}_\chi$ denote the sheaf of germs of holomorphic sections of the bundle L_χ over M.

Theorem.   (Bott)

a)   *If the character $\chi + \rho$ is singular, then all cohomology groups $H^k(M, \mathcal{L}_\chi)$*
     are trivial;
b)   *if $\chi + \rho$ is a regular character and if $\chi + \rho = w(\lambda + \rho)$, where λ is a dominant*
     character, then $H^k(M, \mathcal{L}_\chi) = 0$ for $k \neq l(w)$ and the representation π_λ of
     the group G is realized in the space $H^{l(w)}(M, \mathcal{L}_\chi)$.

The proof of the Bott theorem can be deduced from the following fact.

Lemma 6.1.   (Demazure) *If $\chi = \sum_{i=1}^{l} n_i \chi_i$ and if there is an index j such*
that $n_j \leq 0$, then

$$H^k(M, \mathcal{L}_\chi) \cong H^{k-1}(M, \mathcal{L}_{s_j(\chi+\rho)-\rho}).$$

The techniques of cohomology theory with coefficients in sheaves allows one
to reduce the general case to the special case in which $k = 1$ and $G = \mathrm{SU}(2)$.
The discussion of this case is left to the reader as a useful exercise.

§3. Duality Theory

3.1. Topology on the Set of Irreducible Unitary Representations.
An analogue of the notion of a dual group in the case of a non-commutative
group G is the set $\hat{G}$ of equivalence classes of irreducible unitary representations of G. There are several natural ways of how to introduce a topology on
this set. We shall describe three methods in the case of locally compact groups
leading to equivalent definitions.

1.   A base for the open neighborhoods of a point $[T] \in \hat{G}$ consists of the
     sets $U(K; \xi_1, \ldots, \xi_n)$, where K is a compact subset of G and $\xi_1, \ldots, \xi_n$
     is a set of vectors from the space H of the representation T. By def-
     inition, the class $[S]$ of an irreducible representation S belongs to the
     set $U(K; \xi_1, \ldots, \xi_n)$ if there exist vectors $\eta_1, \ldots, \eta_n$ in the space of the
     representation S such that

$$|(T(g)\xi_i, \xi_j) - (S(g)\eta_i, \eta_j)| < 1 \quad \text{for all} \ \ g \in K, \ 1 \leq i, j \leq n.$$

2.   Let $BC(G)$ be the space of all bounded continuous functions on the
     group G with the topology of uniform convergence on compact sets. For
     every family A of unitary representations of G, a subspace $V(A) \subset BC(G)$
     is defined as the closure of the span of all matrix elements of the repre-
     sentations in A. We say that the family A_1 *is weakly contained* in A_2
     if $V(A_1) \subset V(A_2)$. We can now define the closure of a set $A \subset \hat{G}$ as the
     family of all representations $T \in \hat{G}$ weakly contained in the set A.
3.   For every representation $T \in \hat{G}$, there exists a two-sided ideal $I(T)$
     in $C^*(G)$, namely the kernel of the corresponding representation of $C^*(G)$.
     There is the so called *Jacobson topology* on the set of ideals; the closure

of a family A of ideals consists of all ideals containing the intersection of all ideals in A. The topology can be carried over to $\hat{G}$.

The topological space $\hat{G}$ is, as a rule, *non-Hausdorff;* points are not necessarily closed.

Theorem 6.11. *For a separable locally compact group G, the following properties are equivalent:*

a) *the group G is of type I;*
b) *For any $T \in \hat{G}$, the image of $C^*(G)$ under the representation T contains all compact operators in the representation space;*
c) *two irreducible representations with the same kernel in $C^*(G)$ are equivalent (in other words, the space $\hat{G}$ is semiseparated.)*

Deep connections between the topology of the space $\hat{G}$ and the structure of the group have been discovered. For example, if $\hat{G}$ is discrete, then G is compact. The group G is amenable if and only if the trivial one-dimensional representation is weakly contained in the regular representation. If the group G is discrete and if the trivial representation is an isolated point in $\hat{G}$, then G has a finite numbers of generators and the quotient group $G/[G,G]$ is finite. The last result, due to Kazhdan, opened access to new substantial information on the structure of discrete subgroups of semisimple Lie groups.

For groups which are not of type I, it is natural to consider the space $\tilde{G}$ consisting of all quasi-equivalence classes of unitary factor representations instead of $\hat{G}$. (Representations T_1 and T_2 are called *quasi-equivalent* if for any subrepresentations T_1' and T_2', the intertwining number $c(T_1', T_2')$ is not zero. In the case of representations of type I, the representations T_1 and T_2 are quasi-equivalent if they differ only by multiplicities of irreducible components in their decompositions into direct integrals. Hence, for groups of type I, the space $\tilde{G}$ coincides with $\hat{G}$.)

3.2. Abstract Plancherel Theorem. For a locally compact group G, it is possible to define the *Fourier transform.* If $f \in L_1(G, d_l g)$, then the operator-valued function $\hat{f}$ is defined by the formula

$$\hat{f}(\lambda) = \int_G f(g) T_\lambda(g)^* d_l g, \qquad (6.10)$$

where T_λ belongs to the class $\lambda \in \hat{G}$.

Theorem 6.12. *Let G be a locally compact unimodular group of type I. There exists a measure μ called the* Plancherel measure *on $\hat{G}$ such that*

$$\int_G |f(g)|^2 dg = \int_{\hat{G}} \mathrm{tr}\,[\hat{f}(\lambda)\hat{f}(\lambda)^*] d\mu(\lambda) \qquad (6.11)$$

for all $f \in L_1(G, dg) \cap L_2(G, dg)$.

The Fourier transform can be extended to an isomorphism between the spaces $L_2(G, dg)$ and $L_2(\hat{G}, \mu)$ (the last one consisting of square integrable operator-valued functions on $\hat{G}$ such that the value at each point $\lambda \in \hat{G}$ belongs to the space of Hilbert-Schmidt operators on H_λ).

The described theorem contains the results from Chapters 3 and 4, but it can still be generalized. One of the possible generalizations is connected with the study of groups which are not unimodular. In such a case, it is necessary to substitute for the Haar measure on the left hand side of the formula (6.11) the left or right Haar measure and a suitable operator-valued density should be inserted in the expression inside the trace on the right hand side of (6.11).

Another generalization is possible for groups which are not of type I. In this case, the integral on the right hand side of the formula is computed over the larger space $\tilde{G}$ and the ordinary trace is replaced by the trace in the sense of the corresponding factor.

The Fourier transform (6.10) has, as in the case of finite or compact groups, the following characteristic property: the left and right translations by $g \in G$ are carried over to the left and right multiplications by the operator-valued function $\lambda \mapsto T_\lambda(g)$. A similar property holds for the operations of left or right convolutions and, in the case of Lie groups, for differential operators generated by the Lie operations of left and right translations. This makes it possible, in particular, to establish a connection between the smoothness of the function f and the behaviour of its Fourier transform $\hat{f}$ at infinity.

To describe the image of one or another class of functions on G under the Fourier transform is, as a rule, a very difficult problem. For some types of groups, the description of the image of $C^*(G)$ as a subalgebra in the algebra of operator-valued functions on $\hat{G}$ with values in the algebra $\mathcal{K}$ of compact operators is known. The image of the space $\mathcal{D}(G)$ of smooth functions with compact support on G is known if G is a semisimple group. It is interesting that, apart from analytic conditions (holomorphicity or the behaviour at infinity), there are algebraic conditions on the image as well. More detailed discussion will appear in future volumes of the series.

3.3. Ring Groups and the Duality. The beauty and simplicity of the Pontryagin duality principle has inspired many attempts to generalize it to the case of non-commutative groups. One such possible generalization is the Tannaka-Krejn duality for compact groups. Instead of the original version, we shall describe here basic facts of this theory in more natural (in our view) terms.

As the object dual to a compact group G, we shall introduce the category $\Pi(G)$ of all finite-dimensional representations of G (the objects are representations and the morphisms are intertwining operators). Apart from the usual categorical operations in $\Pi(G)$, the tensor product of two objects and an involution sending a representation T to the contragredient representation T^* are defined. Note that the category $\Pi(G)$ contains more information than the set $\hat{G}$. (It is necessary to add an involution on the set $\hat{G}$ and to define, for every pair $\lambda, \mu \in \hat{G}$, an isomorphism of the space $V_\lambda \otimes V_\mu$ onto a direct sum of the form $\sum n^\nu_{\lambda,\mu} V_\nu$, where V_λ is the space of the irreducible representation T_λ of the class λ.) A *representation of the category* $\Pi(G)$ is a function φ on the set of objects of $\Pi(G)$, the value of which in a point T belongs to $\operatorname{Aut} V$, having the following properties:

1) $A\varphi(T_1) = \varphi(T_2)A$ for all $A \in \mathcal{C}(T_1, T_2)$,
2) $\varphi(T_1 \otimes T_2) = \varphi(T_1) \otimes \varphi(T_2)$.

It is possible to show that φ will then have the following additional properties:

3) $\varphi(T_1 + T_2) = \varphi(T_1) + \varphi(T_2)$,
4) $\varphi(T^*) = [\varphi(T)^*]^{-1}$.

A product and a topology can be defined on the set $\Gamma(\Pi)$ of all representations of the category Π by $(\varphi_1\varphi_2)(T) := \varphi_1(T)\varphi_2(T)$ and $\varphi_n \to \varphi$ if $\varphi_n(T) \to \varphi(T)$ for all T.

Theorem 6.13.

a) *Let G be a compact group and let φ_g be the representation of the category $\Pi(G)$ given by the formula $\varphi_g(T) = T(g)$. Then the mapping $g \mapsto \varphi_g$ is an isomorphism of the topological groups G and $\Gamma(\Pi(G))$.*

b) *Let Π be a category of linear spaces endowed with the operations of tensor product and involution. It is dual object to a compact group G if and only if the following conditions hold.*

  1) *There exists a unique object L_0 (up to isomorphism) having the property*

$$L_0 \otimes L \cong L \text{ for all } L.$$

  2) *Every object L can be decomposed into a sum of minimal objects.*
  3) *If L_1 and L_2 are minimal objects, then $\operatorname{Hom}(L_1, L_2)$ is either one-dimensional (for L_1 isomorphic to L_2) or trivial.*

If the conditions are satisfied, then $\Pi = \Pi(G)$, where G is the group of representations of the category Π.

There is a basic difference between the Tannaka-Krejn theory and the Pontryagin theory: the object dual to a group is no longer a group.

There is another version of the duality theorem, where more complicated objects are involved, which includes as a special case both groups and objects dual to them. We will describe this approach in the simplest case to expose its algebraic structure without hiding it behind analytic details.

Let L be a linear space over a field K. We say that a structure of *coalgebra* is given on L if there is a K-linear mapping

$$L \to L \otimes L$$

(comultiplication), for which the following diagram is commutative:

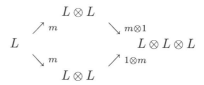

It is easy to see that the definition of a coalgebra can be obtained from the definition of an associative algebra simply by reversing the directions of arrows. The same is true for the definitions of *coalgebra with counit* and *cocommutative coalgebra*.

A linear space is called a *Hopf algebra* if it has the structure of an algebra and a coalgebra, and if comultiplication $m : L \to L \otimes L$ is an algebra homomorphism and multiplication $L \otimes L \to L$ defines a coalgebra homomorphism. In fact, the latter two conditions are equivalent; they both can be written using the same (quite complicated) commutative diagram.

Example 6.10. Let G be a finite group. We shall define a multiplication and a comultiplication in $\mathbb{C}[G]$ by the formulas

$$(f_1 f_2)(g) = f_1(g) f_2(g)$$

and

$$m(f)(g_1, g_2) = f(g_1 g_2).$$

(We identify $\mathbb{C}[G] \otimes \mathbb{C}[G]$ with $\mathbb{C}[G \times G]$.) It is easy to verify that it is a Hopf algebra with commutative multiplication and non-commutative (if G is non-commutative) comultiplication.

We can define a multiplication in the same space $\mathbb{C}[G]$ as the convolution

$$(f_1 * f_2)(g) = \sum_{h \in G} f_1(h) f_2(h^{-1} g)$$

and a comultiplication by the formula

$$m(f)(g_1, g_2) = \begin{cases} f(g_1), & \text{if } g_1 = g_2, \\ 0, & \text{if } g_1 \neq g_2. \end{cases}$$

It also defines a Hopf algebra but now the comultiplication is commutative and the multiplication is non-commutative (if G is non-commutative).

If L is a Hopf algebra, then the *dual Hopf algebra* is the dual space L^* equipped with the operations of multiplication and comultiplication adjoint to the corresponding operations in L. The Hopf algebras constructed in Ex. 6.10 are dual to each other. The construction of the second one (in Ex. 6.10) can be easily generalized so that it can be applied to any category $\Pi(G)$. In this way, for any finite group G, there exists a corresponding commutative Hopf algebra and for any category of the form $\Pi(G)$ there exists a cocommutative Hopf algebra. A class of Hopf algebras including both the cases described can be defined axiomatically. The elements of this class are called *ring groups* or *G.I.Kac's algebras*.

The interest in ring groups has increased recently in connection with a possible use of them in quantum field theory (see 5.3).

§4. The Theory of Characters

4.1. Generalized Characters. Recall that an operator A in a Hilbert space H is called *nuclear* (or of the *trace-class*), if for any orthonormal basis $\{\xi_\alpha\}$ in H, the series $\sum_\alpha (A\xi_\alpha, \xi_\alpha)$ is convergent. It is possible to show that the sum does not depend on the choice of basis. The corresponding number is called the *trace of the operator* A and is denoted by $\operatorname{tr}A$. The characteristic properties of the trace are

1) additivity: $\operatorname{tr}(A + B) = \operatorname{tr}A + \operatorname{tr}B$;
2) positivity: if $A \geq 0$, then $\operatorname{tr}A \geq 0$;
3) unitary invariance: $\operatorname{tr}(UAU^{-1}) = \operatorname{tr}A$ for any unitary operator U.

Now let $\mathfrak{A}$ be a von Neumann factor (see Sect. 1) and let $\mathfrak{A}_+$ be the set of all non-negative operators in $\mathfrak{A}$. The *trace* on $\mathfrak{A}$ is a map from $\mathfrak{A}_+$ into $\mathbb{R}_+ \cup \infty$ having all three properties described above (in the third one, we suppose that U is a unitary operator from $\mathfrak{A}$). It is known that the trace is uniquely defined (up to a positive multiple) on any factor. The factor is called *finite* if the trace of the identity operator is finite. In such a case, the trace is usually normalized by the condition $\operatorname{tr}\mathbb{1} = 1$. Finite factors are of type I_n (then the algebra $\mathfrak{A}$ is isomorphic to $\operatorname{Mat}_n(\mathbb{C})$ and the trace coincides with the ordinary trace multiplied by n^{-1}) or of type II_1 (then the trace of a projection operator on $\mathfrak{A}$ can take any value in the interval $[0, 1]$).

Example 6.11. Let G be a countable discrete group such that every non-trivial conjugacy class is infinite. Let us denote the von Neumann algebra generated by all operators of the left regular representation of the group G on the space $L_2(G)$ by $\mathfrak{A}$. It is possible to verify that the algebra $\mathfrak{A}^!$ is generated by operators of the right regular representation and that $\mathfrak{A} \cap \mathfrak{A}^! = \mathbb{C} \cdot \mathbb{1}$. So $\mathfrak{A}$ is a factor. Every element $A \in \mathfrak{A}^!$ defines a bounded function $a(g) = (A\delta_g, \delta_e)$, where $\delta_g \in L_2(G)$ is the function whose value is 1 at the point g and 0 at all other points. Define $\operatorname{tr}A = a(e)$; it is a normalized trace on $\mathfrak{A}$. Hence $\mathfrak{A}$ is a factor of type II_1.

A factor $\mathfrak{A}$ is called *semi-finite* if the condition $\operatorname{tr}A = \sup_B \operatorname{tr}B$ holds for every positive operator $A \in \mathfrak{A}$ (the supremum is taken over all operators B with finite trace satisfying the condition $0 \leq B \leq A$). Semi-finite factors are usually of type I_∞ (the algebra $\mathfrak{A}$ is isomorphic to the algebra $\mathcal{L}(H)$ of all bounded linear operators on a Hilbert space H and the trace coincides with the standard trace) and of type II_∞ (the algebra $\mathfrak{A}$ is isomorphic to the tensor product of factors of types II_1 and I_∞).

Now let T be a unitary representation of a locally compact group G such that the von Neumann algebra $\mathfrak{A}(T)$ is a semi-finite factor. Suppose that there is a subspace $\mathcal{D}_T(G)$ in $L_1(G, d_l g)$ such that the operators $T(a)$, $a \in \mathcal{D}_T(G)$, have a finite trace. Then the linear functional χ_T on $\mathcal{D}_T(G)$ defined by the formula

$$\chi_T(a) = \operatorname{tr} T(a),$$

is called the *generalized character of the representation* T.

Example 6.12. Let $T_{\rho,\varepsilon}$ be a representation which belongs to the principal series of the group $G = \mathrm{SL}(2, \mathbb{R})$, described in Ex 6.8. Take the space $\mathcal{D}(G)$ of smooth functions with compact support as the space $\mathcal{D}_T(G)$. A direct computation shows that for any $\varphi \in \mathcal{D}(G)$, the operator $T_{\rho,\varepsilon}(\varphi)$ is an integral operator with kernel

$$K_\varphi(x, y) = \int_H \varphi(s(x)^{-1} h s(y)) \pi_{\rho,\varepsilon}(h) d_l h,$$

$$\text{where} \quad s(x) = \begin{pmatrix} 1 & 0 \\ x & 1 \end{pmatrix}, \quad h = \begin{pmatrix} \lambda & \mu \\ 0 & \lambda^{-1} \end{pmatrix},$$

$$\pi_{\rho,\varepsilon}(h) = |\lambda|^{1+i\rho} (\operatorname{sgn}\lambda)^\varepsilon, \quad d_l h = \frac{d\lambda d\mu}{\lambda^2}.$$

If we consider x and y as points of the projective line, it is easy to see that $T_{\rho,\varepsilon}(\varphi)$ is an integral operator with smooth kernel on the compact manifold $H \backslash G$. Hence it has a finite trace given by the formula

$$\operatorname{tr} T_{\rho,\varepsilon}(\varphi) = \int_{H \backslash G} K_\varphi(x, x) dx.$$

There is a quite remarkable fact (common to all irreducible unitary representations of semisimple Lie groups) that the right hand side of the last formula can be written in the form

$$\int_G \varphi(g) \chi_{\rho,\varepsilon}(g) dg,$$

where $\chi_{\rho,\varepsilon}$ is a locally integrable function on G. So characters of representations of principal series are regular distributions on the group. They have the following explicit form:

$$\chi_{\rho,\varepsilon}(g) = \begin{cases} \frac{|\lambda|^{i\rho} + |\lambda|^{-i\rho}}{|\lambda - \lambda^{-1}|} (\operatorname{sgn}\lambda)^\varepsilon, & \text{if } g \text{ is conjugate to a matrix} \\ & \qquad\qquad \text{of the form } \begin{pmatrix} \lambda & \mu \\ 0 & \lambda^{-1} \end{pmatrix}; \\ 0, & \text{in the opposite case.} \end{cases}$$

We recommend to the reader to compare the above formula with the formula of H. Weyl for characters of finite-dimensional representations of $\mathrm{SL}(2, \mathbb{R})$. The method of orbits will make clear why the formulas coincide (see Chap. 7).

Generalized characters, similarly as ordinary ones, are invariant with respect to inner automorphisms of the group G: $\chi(ghg^{-1}) = \chi(h)$. For regular distributions, this means that they are constant on conjugacy classes.

Generalized characters of Lie groups are solutions of certain differential equations. This expresses the fact that they are generalized eigenfunctions of Laplace operators on the group (see 6.4.2 below). In some cases, the information is sufficient for explicit computation of characters and therefore also for the classification of its representations. In the case of groups of type I, generalized characters determine, as in the case of ordinary characters, representations up to equivalence (in the case of groups of type II, generalized characters only determine quasi-equivalence classes).

There is a formula, similar to the formula of Frobenius (3.13), for the character of a unitary induced representation. A generalization of this formula to the case of holomorphically induced representations and to representations on cohomology groups leads to an interesting analogy with the Atiyah-Singer-Lefschetz formula for the trace of an operator acting in sections of holomorphic fibre bundles.

4.2. Infinitesimal Characters. The definition of infinitesimal characters given in Chap. 5 in the case of finite-dimensional representations used the fact that matrix elements of representations are smooth.

Example 6.13. Let T be the regular representation of the group $\mathbb{R}$ in the Hilbert space $H = L_2(\mathbb{R}, dx)$. Then the matrix element corresponding to a pair of elements $\xi, \nu \in H$ has the form

$$t_{\xi,\nu}(s) = \int_{\mathbb{R}} \xi(x+s)\overline{\nu(x)}dx.$$

The function $t_{\xi,\nu}(s)$ is continuous in s for any choice of ξ and ν, but it need not be necessarily smooth. If, for example, ξ and ν are the characteristic functions of intervals, then $t_{\xi,\nu}(s)$ is piecewise-linear function, the graph of which is not smooth. At the same time, if ξ and ν are smooth functions, then $t_{\xi,\nu}$ is also a smooth function of s.

The above example leads to the following definition. The vector ξ in the space H of a representation T of a group G is called *differentiable* (resp. *analytic*) if the vector-valued function $g \mapsto T(g)\xi$ is a differentiable (resp. analytic) function on the group. (It is worth remarking here that if f is a continuous vector-valued function on a manifold M with values in a complete locally convex space L, then f is differentiable (analytic) if and only if it is weakly differentiable (weakly analytic), i.e. if the scalar functions $x \mapsto \langle F, f(x)\rangle$ are differentiable (analytic) for all $F \in L^*$.)

The set of all infinitely differentiable vectors in the space H of the representation T is called the Gårding subspace; we shall denote it by H^∞. The Gel'fand-Gårding theorem says that if T is a representation of a Lie group G in a complete locally convex space V, then the space V^∞ is dense in V.

The proof of the theorem is based on the following useful smoothing procedure. If φ is a smooth function with compact support on G, then the operator $T(\varphi) = \int_G \varphi(g)T(g)d_l g$ maps the space V into V^∞. If φ has its support in a small neighborhood of the identity, then the operator $T(\varphi)$ is close to the identity operator.

The space V^ω of analytic vectors is sometimes considered instead of the Gårding subspace V^∞. In the case of representations in Banach spaces, it is possible to show, using the smoothing technique, that the space V^ω is dense in V. Analytic functions almost supported in a neighborhood of the identity are used in place of the function φ. The existence of such functions can be proved using the heat equation on the group G with the Dirac function δ as the initial condition.

Let T be a representation of a Lie group G in the space V. It induces a representation of the Lie algebra $\mathfrak{g}$ of the group G and its enveloping algebra $U(\mathfrak{g})$ in the Gårding space V^∞. It is possible to show that if T is a unitary irreducible representation, then elements of $Z(\mathfrak{g})$ carry over to scalar operators. This makes it possible to define the infinitesimal character λ_T of the representation T as was done in Chap. 5:

$$T(z) = \lambda_T(z) \cdot \mathbb{1}, \; z \in Z(\mathfrak{g}).$$

The question whether the representation T in V is determined by representations of $\mathfrak{g}$ and $U(\mathfrak{g})$ in V^∞ is far from being trivial. For unitary representations, there are useful criteria of Nelson and of FS^3 = Flato, Simon, Snellman and Sternheimer discussing when a representation of the Lie algebra $\mathfrak{g}$ can be "integrated" to a unitary representation of the group G. Apart from the natural condition that the operators X_k corresponding to the generators of the Lie algebra should be skew-symmetric, it is necessary to suppose either that the operator $\Delta = \sum_k X_k^2$ is essentially selfadjoint on the space V^∞ or that there exists a common dense invariant subspace V^ω in V^∞ consisting of analytic vectors for the operators X_k.

Chapter 7
The Method of Orbits in the Representation Theory

§1. Symplectic Geometry of Homogeneous Spaces

1.1. Local Lie Algebras. A *symplectic manifold* is an even-dimensional smooth manifold M with a given nonsingular closed differential 2-form ω. A systematic explanation of the theory of symplectic geometry is given in the paper by V.I.Arnol'd and A.B.Givental' in the fourth volume of this series. The role of symplectic geometry in the construction of the mathematical model of classical mechanics is also discussed there. We shall describe here the way in which symplectic and contact geometries arise naturally in connection with the notion of a local Lie algebra.

Let E be a vector bundle over a manifold M. We denote the space of all smooth sections of E by $\Gamma(E)$. We say that E is equipped with the structure of a *local Lie algebra* if for any two sections $s_1, s_2 \in \Gamma(E)$, a bracket $[s_1, s_2] \in \Gamma(E)$ is defined satisfying the conditions:

1) the bracket operation is bilinear, skew-symmetric and satisfies the Jacobi identity (i.e.the space $\Gamma(E)$ has the structure of a Lie algebra);
2) the bracket operation $[s_1, s_2]$ is continuous in the C^∞-topology simultaneously in both variables s_1 and s_2;
3) the locality condition

$$\text{supp}\,[s_1, s_2] \subset \text{supp}\,s_1 \cap \text{supp}\,s_2$$

holds, where supp s denotes the support of the section s.

Example 7.1. If $E = TM$ is the tangent bundle over M, then the commutator of vector fields defines the structure of a local Lie algebra on E.

This example can be generalized in a natural way.

Example 7.2. Let E_k denote the $(k+1)$-st exterior power of the tangent bundle and set $E = \oplus_{k=-1}^{dim\,M-1} E_k$. In particular, $E_{-1} = M \times \mathbb{R}$, $E_0 = TM$. There is the unique structure of a graded Lie superalgebra on E compatible in a natural way with the structure of the exterior product such that its restriction to E_0 coincides with the structure of the local Lie algebra described in Ex. 7.1 (more details on superalgebras can be found in Manin (1984)). The commutator in E is called the *Schouten bracket*. If one of the terms in the commutator is of degree 0, i.e. if it is a vector field on M, then the bracket coincides (up to sign, see Kirillov (1980)) with the Lie derivative L_ξ acting on the second term.

If E is the trivial line bundle over M, then it is possible to identify $\Gamma(E)$ with $C^\infty(M)$ and the structure of the local Lie algebra on E defines the structure of a Lie algebra in $C^\infty(M)$.

It is possible to show that in this case the bracket always has the form

$$[f_1, f_2] = \Lambda(df_1, df_2) + f_1 a(df_2) - f_2 a(df_1), \tag{7.1}$$

where Λ is a bivector field and a is a vector field on M satisfying the conditions

$$a \wedge \Lambda = \frac{1}{2}[\Lambda, \Lambda], \quad [a, \Lambda] = 0, \tag{7.2}$$

where $[\ ,\ ]$ is the Schouten bracket described in Ex. 7.2. A pair (Λ, a) having the property (7.2) is also called a *Jacobi structure* on M. A special case of Jacobi structure (when $a \equiv 0$) is the well-known *Poisson structure* (see the quoted paper by V.I.Arnol'd and A.B.Giventaĺ').

A Jacobi structure given by a pair (A, a) is called *transitive* if the vector fields $A(df), f \in C^\infty(M)$, and a generate the tangent space at every point x in M.

Next, if M can be written as a union of submanifolds M_α (possibly of different dimensions) and if the structures of local Lie algebras on $C^\infty(M)$ and $C^\infty(M_\alpha)$ are compatible so that the bracket operation commutes with the restriction to M_α (i.e. if the value of $[f, g]$ at a point $x \in M_\alpha$ depends only on the restrictions of f and g to M_α), then we shall say that the local Lie algebra $C^\infty(M)$ (resp. the corresponding Jacobi structure (A, a)) *decomposes* into local Lie algebras $C^\infty(M_\alpha)$ (resp. Jacobi structures (A_α, a_α)).

Lemma 7.1. *Any Jacobi structure decomposes into transitive ones.*

Let us introduce an equivalence relation among Jacobi structures: the pair (A, a) is equivalent to a pair (A', a') if there exists a function ρ such that $A' = \rho A$, $a' = \rho a + A(d\rho)$. Then the corresponding local Lie algebras $C^\infty(M)$ are isomorphic and the isomorphism is given by the operator of multiplication by ρ :

$$[f, g]' = \rho^{-1}[\rho f, \rho g].$$

There exists a canonical Poisson structure on every symplectic manifold (M, ω) : the bivector A is defined to be the bivector inverse to the form ω. It is possible to verify that the conditions (7.2) imply that the form ω is closed.

If the manifold M is *strictly contact manifold*, i.e. if a 1-form α is defined on M such that $\alpha \wedge (d\alpha)^k \neq 0$, where $k = \frac{1}{2}(\dim M - 1)$, then there exists a canonical Jacobi structure on M, defined by the conditions

$$\alpha(a) \equiv 1, \ A(\alpha) = 0, \ d\alpha \circ A = 1 - \alpha \circ a.$$

(Here $d\alpha, A, \alpha$ and a are considered as operators on forms.)

Lemma 7.2. *Any transitive Jacobi structure is equivalent either to a symplectic or to a contact structure (depending on the parity of the dimension of the manifold M).*

This is the reason, in our opinion, behind the fact that symplectic and contact structures appear in many problems in mathematics and mechanics.

Example 7.3. Let $M = \mathfrak{g}^*$ be the dual space to a Lie algebra $\mathfrak{g}$. Elements of $\mathfrak{g}^*$ are called *momenta* *of the group* G corresponding to the Lie algebra $\mathfrak{g}$. If $X_1, \ldots, X_n$ is a basis in $\mathfrak{g}$ with the commutation relations

$$[X_i, X_j] = \sum_k c_{ij}^k X_k,$$

then the formula

$$[f_1, f_2] = \sum_{i,j,k} c_{ij}^k X_k \frac{\partial f_1}{\partial X_i} \frac{\partial f_2}{\partial X_j}$$

defines a Poisson structure on the space $C^\infty(M)$. Elements $X_i \in \mathfrak{g}$ are considered here as linear functionals on $\mathfrak{g}^*$ giving a coordinate system on M.

It was clarified recently that the above formula was known already to Sophus Lie. Being out of use for a long time, it was rediscovered by F.A.Berezin 20 years ago (after the discovery of the method of orbits).

Lemma 7.3. *The decomposition of the local Lie algebra described in Ex. 7.3, into transitive components coincides with the decomposition of the phase space $\mathfrak{g}^*$ into orbits of the coadjoint representation of a connected Lie group G corresponding to the Lie algebra $\mathfrak{g}$.*

In the above situation, all components are even-dimensional, so they are of symplectic type. Let us define the vector field $K(X_i)$ on $\mathfrak{g}^*$ by the formula

$$K(X_i) = \sum_{j,k} c_{ij}^k X_k \frac{\partial}{\partial X_j}. \tag{7.3}$$

Then the symplectic structure on the orbit containing the point with coordinates $X_1, \ldots, X_n$ has the form

$$B(K(X_i), K(X_j)) = \sum_k c_{ij}^k X_k. \tag{7.4}$$

This special form B on orbits of the coadjoint representation was found in the paper Kirillov (1962) which was the starting point of the method of orbits. The form B will be called the *canonical form* on the orbit.

1.2. Homogenenous Symplectic Manifolds. An action of a Lie group G on a symplectic manifold (M, ω) is called *symplectic* if it preserves the form ω. In such a case, the vector fields ξ_X on M corresponding to generators $X \in \mathfrak{g}$ will be Hamiltonian vector fields. i.e.

$$L_{\xi_X} \omega = 0, \ X \in \mathfrak{g}.$$

An action of the group G on M is called a *Poisson* action if there exists a linear map $X \mapsto h_X$ of the Lie algebra $\mathfrak{g}$ into $C^\infty(M)$ such that h_X is the generating function of the Hamiltonian vector field ξ_X for all $X \in \mathfrak{g}$ and moreover $h_{[X,Y]} = [h_X, h_Y]$ for all $X, Y \in \mathfrak{g}$ (a function f is the generating function of a vector field ξ if $\omega(\xi, .) = df$).

It is clear that every Poisson action of a connected Lie group G is symplectic. The converse is not true in general.

Nevertheless, every symplectic action can be reduced to a Poisson action if the group G and the manifold M are slightly modified.

Lemma 7.4. *If (M, ω) is a connected symplectic manifold and if a symplectic action of a connected group G on (M, ω) is given, then there exist*

a) *a covering $\tilde{M} \xrightarrow{p} M$ of the manifold M,*

b)   a central extension $\tilde{G}$ of the group G by $\mathbb{R}$,

c)   a Poisson action $\tilde{G}$ on $(\tilde{M}, \tilde{\omega})$, where $\tilde{\omega} = p^*\omega$, such that the diagram

$$\tilde{G} \quad \times \quad \tilde{M} \quad \to \quad \tilde{M}$$

$$\downarrow \qquad \qquad \downarrow \qquad \qquad \downarrow$$

$$G \quad \times \quad M \quad \to \quad M$$

is commutative (the horizontal arrows describe the actions of $\tilde{G}$ on $\tilde{M}$ and G on M, the vertical ones are the natural projections.

We shall now describe the main lines of the proof because it clarifies the content of Lemma 7.4.

Let Ham (M) denote the Lie algebra of Hamiltonian vector fields on the symplectic manifold M. There is the following exact sequence of Lie algebras

$$0 \to \mathbb{R} \xrightarrow{\alpha} C^\infty(M) \xrightarrow{\beta} \text{Ham } (M) \xrightarrow{\gamma} H^1(M, \mathbb{R}) \to 0.$$

The map α is the imbedding of $\mathbb{R}$ into $C^\infty(M)$ as the subspace of constants; the map β sends a function f into the Hamiltonian vector field ξ such that f is its generating function; finally, the element $\gamma(\xi)$ is given by the integral of the closed 1-form $\omega(\xi, \cdot)$ over a 1-cycle C.

The symplectic action of G determines the homomorphism $X \mapsto \xi_X$ of the Lie algebra $\mathfrak{g}$ into Ham (M) (for connected groups G, the action is determined by this homomorphism).

Consider now a simply connected group G. Then the action of G on M lifts to action on any connected covering $\tilde{M} \xrightarrow{p} M$. Let us choose $\tilde{M}$ in such a way that the generating functions of vector fields $\xi_X, X \in \mathfrak{g}$, are single-valued. The map $X \mapsto \xi_X$ lifts to a homomorphism of $\mathfrak{g}$ into $C^\infty(\tilde{M})/\mathbb{R}$. In other words, there exists a linear map $X \mapsto h_X$ of the Lie algebra $\mathfrak{g}$ into $C^\infty(M)$ such that

$$h_{[X,Y]} = [h_X, h_Y] + c(X, Y).$$

It is clear that c is a 1-cocycle on the Lie algebra $\mathfrak{g}$ with values in $\mathbb{R}$.

Let $\tilde{\mathfrak{g}}$ be the central extension of $\mathfrak{g}$ corresponding to the cocycle c. It consists of pairs (X, a), $X \in \mathfrak{g}$, $a \in \mathbb{R}$ with multiplication law given by

$$[(X, a), (Y, b)] = ([X, Y], c(X, Y)).$$

Let us set

$$h_{(X,a)} = h_X + a.$$

The correspondence $(X, a) \mapsto h_{(X,a)}$ induces a homomorphism of the algebra $\tilde{\mathfrak{g}}$ into $C^\infty(\tilde{M})$, hence also a Poisson action of the simply connected group $\tilde{G}$, corresponding to the Lie algebra $\tilde{\mathfrak{g}}$, on $\tilde{M}$. The described construction implies the commutativity of the diagram described in Lemma 7.4.

Let us now note that a Poisson action of the group G on the manifold M induces the map from M to $\mathfrak{g}^*$ called the momentum map:

$$M \ni m \mapsto (X \mapsto h_X(m)).$$

The momentum map is equivariant (commutes with the action of G) and a local homeomorphism (because the kernel of its tangent map coincides with the Lie algebra of the stabilizer of the point m).

Hence we have

Lemma 7.5. *Every symplectic manifold with a transitive Poisson action of a group G is a covering space of a G-orbit in $\mathfrak{g}^*$.*

Putting together Lemmas 7.3, 7.4 and 7.5, we obtain the following classification of G-homogeneous symplectic manifolds.

Theorem 7.1. *Every symplectic manifold with a transitive and symplectic action of a connected Lie group G is locally isomorphic to an orbit of the coadjoint representation of the group G itself or of its central extension $\tilde{G}$.*

Example 7.4. Let $M = \mathbb{R}^{2n}$ be the standard symplectic linear space with coordinates $p_1, \ldots, p_n, q_1, \ldots, q_n$ and with the form

$$\omega = \sum_{k=1}^{n} dp_k \wedge dq_k.$$

Let us consider the vector group $G = \mathbb{R}^{2n}$ acting on M by translations. Then the group $\tilde{G}$ is the so called *Heisenberg group* H_n of dimension $2n + 1$. We shall now describe two matrix realizations of this group. The first one consists of all $(n + 2) \times (n + 2)$-matrices of the form

$$\begin{pmatrix} 1 & \xi & t \\ 0 & \mathbb{1}_n & \eta \\ 0 & 0 & 1 \end{pmatrix},$$

where ξ is a row-vector; η is a column-vector and $\mathbb{1}_n$ is the identity $n \times n$ matrix.

The second realization describes the group H_n as a subgroup of the symplectic group $\mathrm{Sp}(2n + 2, \mathbb{R})$ consisting of matrices of the form

$$\begin{pmatrix} 1 & \xi' & \eta' & t \\ 0 & \mathbb{1}_n & 0 & \eta \\ 0 & 0 & \mathbb{1}_n & -\xi \\ 0 & 0 & 0 & 0 \end{pmatrix},$$

where ξ and η are column vectors and prime denotes the transposition.

We shall not compute the action of the coadjoint representation of H_n explicitly here (see Sect. 2); the identification of $\mathbb{R}^{2n}$ with an orbit of the action will be left to the reader as an exercise.

Example 7.5. Let us consider the space $M = \mathbb{R}^{2n} \setminus \{0\}$ with the same form ω as in Ex. 7.4, let $G = \mathrm{Sp}(2n, \mathbb{R})$ be the group of symplectic linear transformations. Then the basic vector fields on M have the form:

$$\xi_{ij} = -p_i \frac{\partial}{\partial p_j} + q_j \frac{\partial}{\partial q_i}, \quad \eta_{ij} = p_i \frac{\partial}{\partial q_j} + p_j \frac{\partial}{\partial q_i}, \quad \zeta_{ij} = -q_i \frac{\partial}{\partial p_j} - q_j \frac{\partial}{\partial p_i}.$$

The generating functions for these fields are given by the following functions on M :

$$f_{ij} = p_i q_j, \; g_{ij} = p_i p_j, \; h_{ij} = q_i q_j.$$

They determine mapping of M into $\mathfrak{g}^* \approx \mathfrak{g}$, the image of which is a $2n$-dimensional orbit in the Lie algebra $\mathrm{sp}(2n, \mathbb{R})$. The space $\mathrm{sp}(2n, \mathbb{R})$ can be naturally identified with the space $\mathrm{Sym}(2n, \mathbb{R})$ of all symmetric matrices of order $2n$: if J is the matrix of the symplectic form ω, then $\mathrm{sp}(2n, \mathbb{R})$ is characterized by the equation

$$X'J + JX = 0.$$

Taking into account that J is skew-symmetric, the matrix $Y = JX$ is symmetric. Let us consider the subset of $\mathrm{Sym}(2n, \mathbb{R})$ consisting of matrices of rank less than or equal to 1. It is a cone with vertex in the origin. It splits into the vertex and the two half-cones under the coadjoint action of the group $\mathrm{Sp}(2n, \mathbb{R})$ (coinciding with the standard action of the linear group on quadratic forms). It is easy to see that the image of the momentum map is the two-fold covering of one of the half-cones.

1.3. Orbits in the Coadjoint Representation. Let us consider now the problem of the classification of orbits of a linear group G. The orbits are often given by a system of equations

$$f_k(x) = c_k, \; k = 1, \dots, m,$$

where the f_k are G-invariant functions on $\mathfrak{g}^*$.

In particular, if G is an algebraic group (i.e. an algebraic submanifold of the general linear group), then generic orbits are almost distinguished by invariant rational functions (i.e. level sets consist of a finite number of orbits). The following Lemma gives useful information.

Lemma 7.6. *If a rational function f is invariant with respect to a linear action of a connected Lie group G, then f can be written in the form $f = P/Q$, where P and Q are eigen-polynomials for the group G, i.e. polynomials with the property*

$$T(g)P = \lambda(g) \cdot P, \; T(g)Q = \lambda(g) \cdot Q,$$

where $T(g)$ is the operator of translation by the element $g \in G$ and λ is a homomorphism of G into the multiplicative group of non-zero numbers, called the weight of P, resp. Q.

Example 7.6. Let G be the group of diagonal matrices acting on the space M of all $n \times n$ matrices by the formula

$$X \mapsto DXD^{-1}, \; X \in M, \; D \in G.$$

Every monomial in the coefficients of the matrix X

$$P_{IJ}(X) = X_{i_1 j_1} \ldots X_{i_k j_k}$$

is an eigen-polynomial with weight

$$\lambda(D) = \frac{\delta_{i_1} \ldots \delta_{i_k}}{\delta_{j_1} \ldots \delta_{j_k}} = \frac{\lambda_I(D)}{\lambda_J(D)},$$

where δ_i are the eigenvalues of the matrix D.

This implies that the field of all G-invariant rational functions is generated by expressions of the form

$$P_{IJ}/P_{I'J'}, \text{ where } I + J' = J + I'.$$

The most convenient way of computing invariants of the group G in a space V explicitly is the following one. Let Ω be a G-orbit of maximal dimension in V and let v_0 be a point of Ω. Consider a surface S containing v_0 and transversal to Ω (this means that the sum of the tangent spaces $T_{v_0}S$ and $T_{v_0}\Omega$ is equal to V). Then orbits close to Ω have the same dimension and they intersect S in points close to v_0. If $u_1, \ldots, u_m$ is a local coordinate system on S in a neighbourhood of v_0, then the set of G-invariants will be given by functions $I_1, \ldots, I_m$ defined in a neighbourhood of Ω as follows: the value $I_k(v)$ is equal to the k-th coordinate of the intersection of S with the G-orbit passing through the point v. It is easy to verify that any G-invariant defined on the same neighbourhood is a function of $I_1, \ldots, I_m$.

Example 7.7. Let us again consider the same situation as in Ex. 7.6. Let Ω be the orbit of an element X_0 such that all elements in the first column are equal to 1. Let the surface S be given by equations $x_{i1} = 1, i = 2, 3, \ldots, n$. The matrix coefficients x_{11} and $x_{ij}, j \geq 2$ are local coordinates on S.
Then the invariants have the form

$$I_{11}(X) = x_{11}, \quad I_{ij}(X) = \frac{x_{ij} x_{j1}}{x_{i1}}.$$

It is not difficult to show that these $n^2 - n + 1$ invariants are independent and that they generate the whole field of rational invariants.

Polynomial invariants can be described in a similar way. In the case $n = 2$ the answer is that all polynomial invariants are polynomials in three independent invariants:

$$I_1 = x_{11}, \ I_2 = x_{22}, \quad I_3 = x_{12} x_{21}.$$

But the ring of invariants is already not free for $n = 3$. It has 8 generators $I_1 = x_{11}, I_2 = x_{22}, I_3 = x_{33}, I_4 = x_{12} x_{21}, I_5 = x_{23} x_{32}, I_6 = x_{31} x_{13}, I_7 = x_{12} x_{23} x_{31}, I_8 = x_{13} x_{32} x_{21}$ connected by one relation $I_4 I_5 I_6 = I_7 I_8$.

The polynomial and rational invariants of the coadjoint representation play an important rôle in the representation theory due to their connections with infinitesimal characters (see Sect. 5 of Chap. 4). It seems that non-rational invariants can also be used in the construction of intertwining operators, but this has not yet been done.

Example 7.8. Let us show now that the classification of orbits of the coadjoint representation leads for certain Lie groups to the situation discussed in Exs. 7.6 and 7.7.

Let G be the Lie group consisting of all real matrices of order $2n$ of the form

$$g = \begin{pmatrix} D & A \\ 0 & D \end{pmatrix},$$

where D is a diagonal $n \times n$ matrix with positive entries on the diagonal and A is any real $n \times n$ matrix. The Lie algebra $\mathfrak{g}$ consists of all matrices X of the same form but without the positivity condition. We shall use the following useful notation for elements of $\mathfrak{g}^*$. Every linear functional $F \in \mathfrak{g}^*$ can be extended to the space $\mathrm{Mat}_{2n}(\mathbb{R})$ of all matrices of order $2n$, hence it can be written in the form

$$F_Y(X) = \mathrm{tr}\,(XY),\ Y \in \mathrm{Mat}_{2n}(\mathbb{R}).$$

Denote the set of all elements $Y \in \mathrm{Mat}_{2n}(\mathbb{R})$ such that $F_Y = 0$ by $\mathfrak{g}^\perp$ and suppose that V is a subspace complementary to $\mathfrak{g}^\perp$ in $\mathrm{Mat}_{2n}(\mathbb{R})$. The space $\mathfrak{g}^*$ can be identified with $\mathrm{Mat}_{2n}(\mathbb{R})/\mathfrak{g}^\perp$, hence also with V. Let us choose the space V as the space of all matrices of the form

$$Y = \begin{pmatrix} B & 0 \\ C & 0 \end{pmatrix},$$

where B is a diagonal matrix and C is an arbitrary matrix. Let P be the projection of $\mathrm{Mat}_{2n}(\mathbb{R})$ onto V along $\mathfrak{g}^\perp$. Then the coadjoint representation of G on $\mathfrak{g}^*$ can be written as:

$$g : F_Y \mapsto F_{Y_1}, \text{ where } Y_1 = P(gYg^{-1}). \tag{7.5}$$

This general formula has, in our special case, the form:

$$\begin{pmatrix} B & 0 \\ C & 0 \end{pmatrix} \mapsto P\left(\begin{pmatrix} D & A \\ 0 & D \end{pmatrix} \begin{pmatrix} B & 0 \\ C & 0 \end{pmatrix} \begin{pmatrix} D & A \\ 0 & D \end{pmatrix}^{-1} \right) =$$

$$= \begin{pmatrix} B + \delta[A, CD^{-1}] & 0 \\ DCD^{-1} & 0 \end{pmatrix},$$

where δA denotes the diagonal part of the matrix A. Thus the classification of G-orbits in $\mathfrak{g}^*$ is reduced to the classification of pairs (B, C) up to the transformations

$$(B, C) \mapsto (B + \delta[A, CD^{-1}],\ DCD^{-1}).$$

Let us consider first transformations with $D = 1$:

$$(B, C) \mapsto (B + \delta[A, C], C).$$

Fixing a matrix C, the space $W(C)$ of all matrices of the form $\delta[A, C]$ is a subspace of the space W of all diagonal matrices. Let us find its orthogonal complement with respect to the bilinear form

$$\langle A, B \rangle = \mathrm{tr}\,(AB).$$

We have

$$A \in W(C)^{\perp} \iff \mathrm{tr}\,(A[B, C]) \overset{\mathrm{B}}{=} 0 \iff \mathrm{tr}\,(B[C, A]) \overset{\mathrm{B}}{=} 0 \iff [C, A] = 0.$$

So the space $W(C)^{\perp}$ consists of all diagonal matrices commuting with C; this is the space of scalar matrices for almost all C. (For example, it is true if all elements of the first column of the matrix C are nonzero.) So for almost all C, the space $W(C)$ consists of all traceless matrices. Noting that $\mathrm{tr}\,B$ is an invariant of the coadjoint representation it follows that, for almost all C, it is the only invariant depending on B. The pair (B, C) can be reduced to the form $(b.1, C)$, where $b = \frac{1}{n}\mathrm{tr}\,B$ for almost all C. So we are left with transformations such that $A = 0$:

$$(b \cdot 1, C) \mapsto (b \cdot 1, DCD^{-1}).$$

This is the problem considered in Ex. 7.6 and Ex. 7.7. The dimension of a generic orbit is $2n - 2$ and its codimension (i.e. the number of functionally independent invariants) is equal to $n^2 - n + 2$.

§2. Representations of Nilpotent Lie Groups

2.1. The Formulation of the Basic Result. A Lie group G is called nilpotent if there exists a sequence of normal subgroups

$$\{e\} = G_0 \subset G_1 \subset \ldots \subset G_{n-1} \subset G_n = G$$

such that G_{k+1}/G_k belongs to the center of G/G_k. We shall suppose in what follows that, as a rule, all considered nilpotent Lie groups are connected and simply connected. Then the exponential map is a one-to-one map between the group G and its Lie algebra $\mathfrak{g}$. Lie subalgebras of $\mathfrak{g}$ correspond under this map to closed connected and simply connected Lie subgroups in G and ideals in $\mathfrak{g}$ correspond to normal subgroups in G. Every nilpotent Lie group can be realized as a subgroup of the group $N_+(n, \mathbb{R})$ of all upper triangular real matrices of order n with ones on the main diagonal. (The converse is also true; all subgroups of $N_+(n, \mathbb{R})$ are nilpotent.) Even if nilpotent groups have a relatively simple structure, their classification is not known up to now and it seems that it is a hopelessly difficult problem. (In lower dimensions, i.e. for $n \leq 6$, there are only a finite number of non-isomorphic nilpotent Lie groups. Starting with $n = 7$ continuous parameters appear.)

The basic result of the method of orbits, applied to nilpotent Lie groups, is the description of a one-to-one correspondence between two sets:

a) The set $\hat{G}$ of all equivalence classes of irreducible unitary representations
    of a connected and simply connected nilpotent Lie group G;
b) The set $\mathcal{O}(G)$ of all orbits of the group G in the space $\mathfrak{g}^*$ dual to the Lie
    algebra $\mathfrak{g}$ with respect to the coadjoint representation.

To construct this correspondence, we introduce the following definition. A
subalgebra $\mathfrak{h} \subset \mathfrak{g}$ is *subordinate* to a functional $f \in \mathfrak{g}^*$ if

$$\langle f, [x, y] \rangle = 0 \text{ for all } x, y \in \mathfrak{h},$$

i.e. if $\mathfrak{g}$ is an isotropic subspace with respect to the bilinear form defined by
$B_f(x, y) = \langle f, [x, y] \rangle$ on $\mathfrak{g}$.

An equivalent description of this property can be given using the language
of symplectic geometry and in terms of representation theory.

Lemma 7.7.   *The following conditions are equivalent.*

a) *A subalgebra $\mathfrak{h}$ is subordinate to the functional f.*
b) *The image of $\mathfrak{h}$ in the tangent space $T_f \Omega$ to the orbit Ω in the point f is
    an isotropic subspace.*
c) *The map*

$$x \mapsto \langle f, x \rangle$$

is a one-dimensional real representation of the Lie algebra $\mathfrak{h}$.

If the conditions of Lemma 7.7 are satisfied, we define the one-dimensional
unitary representation $U_{f,H}$ of the group $H = \exp \mathfrak{h}$ by the formula

$$U_{f,H}(\exp x) = e^{2\pi i \langle f, x \rangle}.$$

Theorem 7.2.

a) *Every irreducible unitary representation T of a connected and simply con-
    nected nilpotent Lie group G has the form*

$$T = \operatorname{Ind}_H^G U_{f,H},$$

where $H \subset G$ is a connected subgroup and $f \in \mathfrak{g}^*$.
b) *The representation $T_{f,H} = \operatorname{Ind}_H^G U_{f,H}$ is irreducible if and only if the Lie
    algebra $\mathfrak{h}$ of the group H is a subalgebra of $\mathfrak{g}$ subordinate to the functional f
    with maximal possible dimension.*
c) *Irreducible representations T_{f_1, H_1} and T_{f_2, H_2} are equivalent if and only if
    the functionals f_1 and f_2 belong to the same orbit of $\mathfrak{g}^*$.*

Theorem 7.2, like many other theorems of this chapter, can be proved by
induction with respect to the dimension of the group G. The following cases
are to be considered along the way:

1. The dimension of the center of the group G is bigger than 1. Irreducible
    representations of the group G cannot be faithful in this case; they can

be reduced to a representation of a quotient group G/C_0, where C_0 is a subgroup of the center contained in the kernel of the representation.

2. The center C of the group G has dimension equal to 1 and it belongs to the kernel of the representation. This case is similar to the previous one.

3. The center C of the group G has dimension 1 and it does not belong to the kernel of the representation. Then the structure theory of nilpotent Lie groups with one-dimensional center should be used.

Lemma 7.8. *Every nilpotent Lie group $\mathfrak{g}$ with one-dimensional center $\mathfrak{c}$ admits a basis $x, y, z, w_1, \ldots, w_k$ with the following commutation relations:*

$$[x, y] = z; \ [y, w_i] = 0, \ 1 \le i \le k, \ z \in \mathfrak{c}.$$

The induction step in case 3 is based on the following fact.

Lemma 7.9. *If $\mathfrak{g}$ is a nilpotent Lie algebra with one-dimensional center and if G is the corresponding connected and simply connected Lie group, then every locally faithful representation T of the group G is induced from the subgroup $G_0 = \exp \mathfrak{g}_0$, where $\mathfrak{g}_0$ is the subalgebra spanned by the vectors $y, z, w_1, \ldots, w_k$ described in Lemma 7.8.*

Theorem 7.2 implies that the classification of irreducible unitary representations of the group G can be reduced to the purely algebraic problem of the classification of G-orbits in $\mathfrak{g}^*$. For a given functional $f \in \mathfrak{g}^*$, the explicit realization of these representations is then reduced to the problem of finding a subalgebra subordinate to f of maximal possible dimension.

The last problem can be solved by the standard induction described above. An alternative method was presented by M. Vergne.

Lemma 7.10. *Let V be a linear space and let B be a skew-symmetric bilinear form on V. A filtration of V, i.e. a family of subspaces satisfying the condition*

$$0 = V_0 \subset V_1 \subset \ldots \subset V_n = V, \ \dim V_k = k,$$

will be denoted briefly by s. Let us define the space

$$W(s, B) = \sum_{i=0}^{n} V_i \cap V_i^{\perp}.$$

Then:

a) *The space $W(s, B)$ is a maximal isotropic subspace of V with respect to B.*
b) *If V is a Lie algebra, if B has the form B_f for a functional $f \in V^*$, where*

$$B_f(x, y) = \langle f, [x, y] \rangle$$

and if all spaces V_i are ideals in V, then $W(s, B)$ is a subalgebra in V.

Example 7.9. Let $G = N_+(n, \mathbb{R})$ be the group of real upper triangular matrices with ones along the main diagonal. The Lie algebra $\mathfrak{g} = \mathfrak{n}_+(n, \mathbb{R})$ of this group consists of all upper triangular matrices with zeros along the main diagonal. The space $\mathfrak{g}^*$ can be identified with the space $\mathfrak{n}_-(n, \mathbb{R})$ of all lower triangular matrices X using

$$f_X(A) = \mathrm{tr}\,(AX), \ A \in \mathfrak{n}_+(n, \mathbb{R}), \ X \in \mathfrak{n}_-(n, \mathbb{R}).$$

The action of G on $\mathfrak{g}^*$ (the coadjoint representation) has the form (compare with Ex. 7.8. in 1.2 and the formula (7.5))

$$K(a) : X \mapsto (aXa^{-1})_{\mathrm{low}},$$

where the index "low" means that the matrix is replaced by its lower triangular part. It is easy to see that the action preserves the element in the left bottom corner of the matrix X. (The fact that $a \in \mathfrak{n}_+(n, \mathbb{R})$ implies that the map $K(a)$ acts as follows: to a given column of X, a linear combination of the previous columns is added and to a given row of X, a linear combination of the following rows is added.) More generally, the minors $\Delta_k, k = 1, 2, \ldots, [\frac{n}{2}]$, consisting of the last k rows and first k columns of X are invariants of the action. It is possible to show that if all the numbers c_k are different from zero, then the manifold given by the equations

$$\Delta_k = c_k, 1 \leq k \leq \left[\frac{n}{2}\right]$$

is a G-orbit in $\mathfrak{g}^*$. Hence generic orbits have codimension equal to $[\frac{n}{2}]$ and dimension equal to $\frac{n(n-1)}{2} - [\frac{n}{2}]$.

To obtain a representative for such an orbit, we can take a matrix X of the form

$$X = \begin{pmatrix} 0 & 0 \\ \Lambda & 0 \end{pmatrix},$$

where Λ is a matrix of order $[\frac{n}{2}]$ such that all nonzero elements are contained in the anti-diagonal. It is easy to find a subalgebra of dimension $[\frac{n}{2}] \cdot [\frac{n+1}{2}]$ subordinate to the functional f_X. It consists of all matrices of the form

$$\begin{pmatrix} 0 & B \\ 0 & 0 \end{pmatrix},$$

where B is an $[\frac{n}{2}] \times [\frac{n+1}{2}]$ matrix.

Let us note that the classification of orbits of lower dimensions is a difficult problem. The solution is known only for $n \leq 6$. It was shown recently that it is so called wild classification problem (containing the classification of pairs of matrices with respect to simultaneous similarity transformation as a subproblem). At the same time, every individual orbit has a simple structure.

Lemma 7.11. *Let G be a connected nilpotent Lie group. Every G-orbit in $\mathfrak{g}^*$ can be described in a suitable coordinate system $X_1, \ldots, X_{2r}, Y_1, \ldots, Y_l$ by a system of equations*

$$Y_i = P_i(X_1, \ldots, X_{2r}), \ 1 \le i \le l,$$

where the P_i are polynomials.

Corollary 7.1.

a) *Every orbit is diffeomorphic to an even-dimensional Euclidean space.*
b) *The stabilizer G_f of the point f of an orbit is a connected subgroup in G coinciding with $\exp \mathfrak{g}_f$, where $\mathfrak{g}_f$ is the kernel of the form B_f.*

Example 7.10. Let N_n be a Lie group consisting of coordinate transformations in a neighbourhood of the point $x = 0$ having the form

$$x \mapsto x + a_1 x^2 + \ldots + a_n x^{n+1} + o(x^{n+1})$$

together with the operation of composition of maps. It is an n-dimensional subgroup of the group G_n, studied in Ex. 5.1. Its Lie algebra $\mathfrak{n}_n$ can be identified with the space of all vector fields on $\mathbb{R}$ defined on a neighbourhood of the origin and having the form

$$L = \left(\sum_{k=1}^{n} \alpha_k x^{k+1} + o(x^{n+1}) \right) \frac{d}{dx}.$$

The vector fields

$$L_k = x^{k+1} \frac{d}{dx}, \ k = 1, 2, \ldots, n,$$

form a basis for $\mathfrak{n}_n$.

A functional $f \in \mathfrak{n}_n^*$ is determined by the coordinates

$$c_k = \langle f, L_k \rangle, \ 1 \le k \le n$$

and is equivalent, as a geometrical object, to the expression

$$\sum_{k=1}^{n} \frac{c_k}{(k+1)!} \delta^{(k+1)}(x) dx,$$

where $\delta(x)$ is the Dirac δ-function. The action of the field L_i on $\mathfrak{n}_n^*$ in the coordinates c_k has the form (see (7.3)):

$$K(L_i) = \sum_{k=1}^{n} (k - i) c_{k+i} \frac{\partial}{\partial c_k},$$

where $c_m = 0$ for $m > n$.

Let $n = 2m+1$. Using the action of the maps $\exp t L_k$, $k = 1, 2, \ldots, n-1$, it is possible to trivialize the coordinates $c_{n-1}, c_{n-2}, \ldots, c_1$. The coordinate c_n

is clearly an invariant of the coadjoint representation. So if $n = 2m + 1$, then a generic orbit is given by the equation $c_n = \text{const} \neq 0$.

If $n = 2m$, then it is possible to trivialize in the same way all coordinates except c_{2m} and c_m. The coefficient c_{2m} and a polynomial

$$P_m(c_m, c_{m+1}, \ldots, c_{2m})$$

of order m and weight $m(2m - 1)$ (the weight of c_k is k) are invariants in this case. There exists a generating function for the polynomials P_m, namely:

$$\left(\sum_{k=0}^{\infty} y_k y_0^{k-1} \right)^{\frac{1}{2}} = \sum_{m=0}^{\infty} P_m(y_m, y_{m-1}, \ldots, y_0).$$

Let $\mathcal{O}_0(N_n)$ denote the set of all N_n-orbits in $\mathfrak{n}_n^*$ such that $c_n \neq 0$; then the set $\mathcal{O}(N_n)$ will be the disjoint union of subsets of the form $\mathcal{O}_0(N_k)$, $k = n, n - 1, \ldots, 1, 0$. If k is odd, then $\mathcal{O}_0(N_k)$ coincides with $\mathbb{R} \setminus \{0\}$; if $k > 0$ is even, then $\mathcal{O}_0(N_k)$ coincides with $\mathbb{R}^2 \setminus \mathbb{R} = (\mathbb{R} \setminus \{0\}) \times \mathbb{R}$, and $\mathcal{O}_0(N_0) = \mathcal{O}(N_0) \setminus \{\text{point}\}$.

2.2. The Topology of $\hat{G}$ in Terms of Orbits.

Let us now describe $\hat{G}$ as a topological space.

Theorem 7.3. *Consider the space $\hat{G}$ with the topology described in Chap. 6 and the space $\mathcal{O}(G)$ with the natural quotient topology. Then the correspondence between $\hat{G}$ and $\mathcal{O}(G)$ is a homeomorphism.*

Example 7.11. Let us recall the group N_n from Ex. 7.10. It coincides, in the special case $n = 3$, with the Heisenberg group H_1 (see Ex. 7.4). The topological space $\hat{N}_3 = \hat{H}_1$ is shown in Fig. 6. To construct it, we take the vertical line (the z-axis, see Fig. 6), we remove the origin ($x = y = z = 0$) and we glue in the horizontal plane $z = 0$ instead of it. The topology of the line with the origin removed and the topology of the plane are the standard ones. But if a sequence of points of the line approaches the removed origin, then all points of the plane are limits of the sequence. As an intuitive illustration, points of the line in Fig. 7 are "thick", while the horizontal plane is "small".

Let us explain the topological structure of $\hat{H}_1$ in the language of representation theory. For a given number $\lambda \neq 0$, the corresponding irreducible infinite-dimensional representation T_λ of the group H_1 acts on $L_2(\mathbb{R}, dx)$ by the formula

$$\left[T_\lambda \begin{pmatrix} 1 & a & c \\ 0 & 1 & b \\ 0 & 0 & 1 \end{pmatrix} f \right] (x) = e^{2\pi i \lambda (bx+c)} f(x + a). \tag{7.6}$$

(The realization of the group H_1 by triangular matrices described in Ex. 7.4 is used here.)

If $(\mu, \nu) \in \mathbb{R}^2$, the corresponding one-dimensional representation $S_{\mu, \nu}$ of the group H_1 is given by

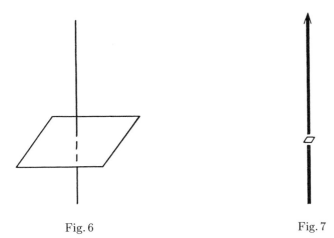

Fig. 6   Fig. 7

$$S_{\mu,\nu} \begin{pmatrix} 1 & a & c \\ 0 & 1 & b \\ 0 & 0 & 1 \end{pmatrix} = e^{2\pi i(\mu a + \nu b)}.$$

It follows from (7.6) that if $\lambda \to 0$, then the representations T_λ approaches the reducible representation T_0 acting in the space $L_2(\mathbb{R}, dx)$ by the formula

$$\left[T_0 \begin{pmatrix} 1 & a & c \\ 0 & 1 & b \\ 0 & 0 & 1 \end{pmatrix} f \right] (x) = f(x + a).$$

It is clear that T_0 is the direct integral of the representations $S_{\mu,0}$:

$$T_0 = \int_{-\infty}^{\infty} S_{\mu,0} d\mu.$$

At first sight, this leads to a contradiction with the topology of $\hat{H}_1$ described above. It seems that the set of all limits of the representations T_λ for $\lambda \to 0$ is not the set of all representations $S_{\mu,\nu}$, but only the set $S_{\mu,0}$. The "contradiction" is explained in the following way. The representation T_λ is equivalent to the representation

$$\left[\tilde{T}_\lambda \begin{pmatrix} 1 & a & c \\ 0 & 1 & b \\ 0 & 0 & 1 \end{pmatrix} g \right] (y) = e^{2\pi i \lambda(c + ay)} g(y - b). \tag{7.7}$$

(To prove the equivalence, we can use the Fourier transform in the space $L_2(\mathbb{R}) : g(y) = \frac{1}{\sqrt{2\pi}} \int_{-\infty}^{\infty} f(x) e^{-ixy} dx$.)

Passing to the limit $\lambda \to 0$ in (7.7), we obtain that the representation $\tilde{T}_0$ is equivalent to the direct integral $\int_{-\infty}^{\infty} S_{0,\nu} d\nu$. It is possible to show that we can obtain any direct integral of the form $\int_{-\infty}^{\infty} S_{\alpha t, \beta t} dt$, where (α, β) is any nonzero vector, as the limit representation for a suitable choice of the realizations of the representations T_λ. In this way, the limit $T_\lambda, \lambda \to 0$, really contains all one-dimensional representations $S_{\mu,\nu}$.

2.3. The Functors Res **and** Ind. The next important problem in the theory of representation – the description of the functors Res_H^G and Ind_H^G – can also be solved in an intuitive way using the language of the theory of orbits.

Theorem 7.4. *Let G be a connected and simply connected nilpotent Lie group and let H be a connected subgroup. Denote the natural projection of $\mathfrak{g}^*$ onto $\mathfrak{h}^*$ by p; it maps a functional f on $\mathfrak{g}$ to its restriction to $\mathfrak{h}$.*

a) *Let an irreducible representation T of the group G corresponds to an orbit $\Omega \subset \mathfrak{g}^*$. Then the representation $\text{Res}_H^G T$ of the group H decomposes into the direct integral of irreducible representations of H corresponding to the orbits $\omega \subset \mathfrak{h}^*$ such that $\omega \subset p(\Omega)$.*
b) *Suppose that an irreducible representation S of the group H corresponds to an orbit $\omega \subset \mathfrak{h}^*$. Then the representation $\text{Ind}_H^G S$ of the group G decomposes into the direct integral of irreducible representations of G corresponding to orbits $\Omega \subset \mathfrak{g}^*$ such that $p(\Omega) \supset \omega$.*
c) *Let T_1 and T_2 be irreducible representations of the group G corresponding to orbits Ω_1 and Ω_2 in $\mathfrak{g}^*$. Then the tensor product $T_1 \otimes T_2$ decomposes into the direct integral of irreducible representations of G corresponding to orbits $\Omega \subset \mathfrak{g}^*$ such that Ω is included in the sum $\Omega_1 + \Omega_2$.*

Note that the assertions a) and b) of Theorem 7.4 lead to the following analogue of Frobenius duality for nilpotent Lie groups.

Lemma 7.12. *A representation $S \in \hat{H}$ is weakly contained in $\text{Res}_H^G T$ for $T \in \hat{G}$ if and only if T is weakly contained in $\text{Ind}_H^G S$.*

Part c) of the Theorem 7.4 follows from part a) applied to the representation $T_1 \times T_2$ of the group $G \times G$ restricted to the diagonal subgroup.

Note also that part b) implies the following property of orbits.

Lemma 7.13. *Let Ω be an orbit in $\mathfrak{g}^*$, let $f \in \Omega$ and let $\mathfrak{h}$ be a subalgebra of maximal dimension subordinate to the functional f. Then Ω contains the affine manifold $f + (\mathfrak{h})^\perp$ of dimension $\frac{1}{2}\dim \Omega$.*

In this way, orbits of the coadjoint representation of a nilpotent Lie group are always fibred by flat submanifolds of half the dimension. It is possible to check that the corresponding fibration is a polarization, i.e. that its fibers are Lagrangian submanifolds of the symplectic manifold Ω (see Lemma 7.7 b)). The condition $f + (\mathfrak{h})^\perp \subset \Omega$ of Lemma 7.13 is called the *Pukanszky condition*. The condition is always satisfied for nilpotent Lie groups, but it is not generally true already for solvable groups (see Sect. 3).

2.4. Computation of Characters by Orbits. Let us now compute characters of irreducible representations. For this purpose, we shall define the *Fourier transform* for a function φ on a nilpotent Lie group G as the function $\tilde{\varphi}$ on the space $\mathfrak{g}^*$ given by the formula

$$\tilde{\varphi} = \int_{\mathfrak{g}} \varphi(\exp X) e^{2\pi i f(X)} dX. \tag{7.8}$$

The Fourier transform is well-defined by (7.8) for all functions φ in the space $\mathcal{S}(G)$ (we say that $\varphi \in \mathcal{S}(G)$ if the function $X \mapsto \varphi(\exp X)$ belongs to the Schwartz space $\mathcal{S}$ of rapidly decreasing functions).

We shall define, as usual, the Fourier transform of distributions $\chi \in \mathcal{S}'(G)$ by the formula

$$\langle \tilde{\chi}, \tilde{\varphi} \rangle = \langle \chi, \varphi \rangle \tag{7.9}$$

Theorem 7.5. *Let T be an irreducible unitary representation of a connected and simply connected nilpotent Lie group G corresponding to an orbit $\Omega \subset \mathfrak{g}^*$. The character of the representation T is a distribution in $\mathcal{S}'(G)$ and its Fourier transform coincides with the canonical measure on Ω given by the symplectic structure.*

Using Theorem 7.5, it is possible to give an explicit formula for the character. If $\varphi \in \mathcal{S}(G)$ and if $T(\varphi) = \int_{\mathfrak{g}} \varphi(\exp X) T(\exp X) dX$, then the operator $T(\varphi)$ is a nuclear operator and

$$\operatorname{tr} T(\varphi) = \int_{\Omega} \tilde{\varphi}(f) d_{\Omega} f, \tag{7.10}$$

where $d_{\Omega} f$ is the measure on the orbit Ω given by the form $\frac{1}{k!} B_{\Omega} \wedge \ldots \wedge B_{\Omega}$ (k factors) and

$$B_{\Omega}(f)(K(X), K(Y)) = \langle f, [X, Y] \rangle = B_f(X, Y) \tag{7.11}$$

is the symplectic form on the orbit (see (7.4)).

Example 7.12. Let us now consider the representation T_λ of the Heisenberg group given by the formula (7.6) from Ex. 7.11. The exponential map then has the following form:

$$\exp \begin{pmatrix} 0 & \alpha & \gamma \\ 0 & 0 & \beta \\ 0 & 0 & 0 \end{pmatrix} = \begin{pmatrix} 1 & \alpha & \gamma + \frac{\alpha\beta}{2} \\ 0 & 1 & \beta \\ 0 & 0 & 1 \end{pmatrix}.$$

The operator $T_\lambda(\varphi)$ is an integral operator in $L_2(\mathbb{R})$ with the kernel

$$K_\varphi^\lambda(x, y) = \int_{\mathbb{R}} \int_{\mathbb{R}} \varphi(y - x, \beta, \gamma + \frac{\beta(y - x)}{2}) e^{2\pi i \lambda(\gamma + \beta \frac{x+y}{2})} d\beta d\gamma.$$

Hence

$$\operatorname{tr} T_\lambda(\varphi) = \int_{\mathbb{R}} K_\varphi^\lambda(x,x) = \int_{\mathbb{R}} \int_{\mathbb{R}} \int_{\mathbb{R}} \varphi(0,\beta,\gamma)e^{2\pi i\lambda(\gamma+\beta x)}d\beta d\gamma dx =$$

$$= \frac{1}{\lambda}\int_{\mathbb{R}}\int_{\mathbb{R}} \tilde{\varphi}(x,y,\lambda)dxdy$$

The symbol $\tilde{\varphi}$ here denotes the Fourier transform (7.8). The group H_1 acts on the orbit Ω_λ going through the point $(0,0,\lambda) \in \mathfrak{g}^*$ by the formula

$$K\begin{pmatrix} 1 & a & c \\ 0 & 1 & b \\ 0 & 0 & 1 \end{pmatrix} : (x,y,\lambda) \mapsto (x+\lambda a, y+\lambda b, \lambda).$$

The form B_Ω given by (7.11) has the form:

$$B_\Omega(0,0,\lambda)(\lambda\frac{\partial}{\partial x}, \lambda\frac{\partial}{\partial y}) = \lambda.$$

Hence $B_\Omega = \frac{1}{\lambda}dx \wedge dy$, $d_\Omega f = \frac{dx\,dy}{\lambda}$ and the expression for $\operatorname{tr} T_\lambda(\phi)$ coincides with the right hand side of the formula (7.10).

The described formula for characters makes it possible to give an explicit formula for the Plancherel measure on $\hat{G}$. Let us first note that it is possible, for a given nilpotent group G, to choose a linear subspace $Q \subset \mathfrak{g}^*$ in such a way that generic G-orbits intersect Q in exactly one point.

Let us choose a basis $X_1,\ldots,X_l,Y_1,\ldots,Y_{n-l}$ in $\mathfrak{g}$ in such a way that elements $Y_1,\ldots,Y_{n-l}$, considered as linear functions on $\mathfrak{g}^*$, are constant on Q. Then $X_1,\ldots,X_l$ are coordinates on Q, hence also on an open dense subset of $\mathcal{O}(G)$. For every point $f \in Q$ with coordinates $X_1,\ldots,X_l$, we consider the skew-symmetric matrix A with elements

$$a_{ij} = \langle f, [Y_i, Y_j]\rangle.$$

Denote the Pfaffian of the matrix A by $P(X_1,\ldots,X_l)$. It is a homogeneous polynomial of order $(n-l)/2$.

Theorem 7.6. *The Plancherel measure on $\hat{G} \approx \mathcal{O}(G)$ is concentrated on the set of generic orbits and it has the form*

$$\mu = P(X_1,\ldots,X_l)dX_1 \wedge \ldots \wedge dX_l. \tag{7.12}$$

in the coordinates $X_1,\ldots,X_l$.

The proof is based on the well-known interpretation of the Plancherel theorem as the decomposition of the δ-function on the group G into characters of irreducible unitary representations of G. Applying the Fourier transform (in the canonical coordinates on G), the δ-function is changed into the Lebesgue measure on $\mathfrak{g}^*$ and characters of irreducible unitary representations are transformed into canonical measures on G-orbits. The Plancherel measure coincides in this way with the measure induced on the set of orbits by the factorization and it is easy to compute it.

The values of infinitesimal characters of representations are often used as coordinates on $\hat{G}$. We shall discuss this in the next section.

2.5. Infinitesimal Characters and Orbits. Infinitesimal characters of irreducible unitary representations can be described by orbits very intuitively.

Let us recall that we have described, using the symmetrization map, a linear isomorphism between the enveloping algebra $U(\mathfrak{g})$ and the symmetric algebra $S(\mathfrak{g})$ such that the center $Z(\mathfrak{g})$ of the algebra $U(\mathfrak{g})$ goes over to $S(\mathfrak{g})^G$. Let us now identify the algebra $S(\mathfrak{g})$ with the algebra $P(\mathfrak{g}^*)$ of polynomial functions on $\mathfrak{g}^*$ in such a way that an element $X \in \mathfrak{g}$ is mapped to the linear function

$$f \mapsto 2\pi i \langle f, X \rangle$$

on $\mathfrak{g}^*$. Then G-invariant polynomials on $\mathfrak{g}^*$ correspond to elements of $Z(\mathfrak{g})$. If $A \in Z(\mathfrak{g})$ and if $p_A \in P(\mathfrak{g}^*)^G$ is the corresponding polynomial, then p_A is constant on every G-orbit $\Omega \subset \mathfrak{g}^*$. We shall denote the corresponding value by $p_A(\Omega)$.

Theorem 7.7. *If T is an irreducible unitary representation of a nilpotent Lie group G corresponding to an orbit $\Omega \subset \mathfrak{g}^*$, then the infinitesimal character of T has the form*

$$\lambda_T(A) = p_A(\Omega). \tag{7.13}$$

Corollary 7.2. *For a nilpotent Lie group G, the identifications of $Z(\mathfrak{g})$ with $S(\mathfrak{g})^G$ and with $P(\mathfrak{g}^*)^G$ are isomorphisms of algebras.*

Corollary 7.3. *A representation T is determined by its infinitesimal character if and only if the corresponding orbit is determined by the values of G-invariant polynomials on it.*

For nilpotent groups, generic orbits satisfy the condition of Corollary 7.3. More precisely, there exist G-invariant polynomials $p_0, p_1, \ldots, p_l$ on $\mathfrak{g}^*$ such that the numbers $c_k = p_k(f)$, $1 \le k \le l$, can be taken as coordinates on the set $V = \{f \in \mathfrak{g}^* | p_0(f) \ne 0\}$. In these coordinates, the Plancherel measure has the form

$$\mu = R(c_1, \ldots, c_l) dc_1 \wedge \ldots \wedge dc_l, \tag{7.14}$$

where R is a rational function. For a specific group G and a given set $p_1, \ldots p_l$, it is possible to compute the function R explicitly (using Theorem 7.6) but a simple description of general case is not known.

Example 7.13. Let us consider the group $N_+(n, \mathbb{R})$ described in Ex. 7.9. Using the notation $p_k = \Delta_k, 1 \le k \le m, m := [\frac{n}{2}]$, the function R in (7.14) for the Plancherel measure is given explicitly by

$$R(c_1, \ldots, c_m) = c_m^{n-2}(c_1 c_2 \ldots c_{m-1})^{-3}.$$

If, in particular, $n = 3$, then we have the Plancherel measure for the Heisenberg group H_1. The parameter c_1 then coincides with the parameter λ used in Ex. 7.12 and the Plancherel measure has the form $\mu = \lambda d\lambda$.

§3. Representations of Solvable Lie Groups

The class of solvable Lie groups plays an important rôle in the general theory because every simply connected Lie group G decomposes into a semidirect product of a semisimple subgroup S and a solvable normal factor R :

$$G = S \ltimes R.$$

3.1. Exponential Groups. There exists a wide class of solvable Lie groups such that the theory described in Sect. 2 can be transferred to them with only minor alterations.

Definition 7.1. A connected Lie group G is called *exponential* if the map $\exp : \mathfrak{g} \to G$ is a diffeomorphism of manifolds.

A simply connected Lie group G is exponential if and only if for every $X \in \mathfrak{g}$, the operator $\operatorname{ad} X$ has no nonzero purely imaginary eigenvalues. It follows that all exponential groups are solvable.

Theorem 7.2, describing the correspondence between $\hat{G}$ and $\mathcal{O}(G)$, can be transferred to exponential groups with the following changes. It is necessary to suppose in part b) of Theorem 7.2 that the algebra $\mathfrak{h}$ subordinate to the functional f satisfies the *Pukanszky condition*: the orbit Ω is contained in the affine space $f + h^{\perp} \subset \mathfrak{g}^*$. Given an exponential group G and $f \in \mathfrak{g}^*$, it can be shown that among all subalgebras of $\mathfrak{h}$ of maximal dimension subordinate to the functional f, there exists at least one subalgebra satisfying the Pukanszky condition (see Busyatskaya (1973), Bernat et al. (1972)).

Example 7.14. The simplest connected solvable Lie group is the group

$$G = \operatorname{Aff}(1, \mathbb{R})$$

of affine transformations of the real line preserving orientation:

$$x \mapsto ax + b, a > 0.$$

(This group is sometimes called the "$ax + b$"-group.) The group G is also isomorphic to the group of all matrices of the form $g_{a,b} = \begin{pmatrix} a & b \\ 0 & 1 \end{pmatrix}$. Let us identify $\mathfrak{g}$ with the Lie algebra of matrices of the form $X_{\alpha,\beta} = \begin{pmatrix} \alpha & \beta \\ 0 & 0 \end{pmatrix}$ and $\mathfrak{g}^*$ with the space of matrices of the form $f_{x,y} = \begin{pmatrix} x & 0 \\ y & 0 \end{pmatrix}$. The coadjoint representation has the form

$$K(g_{a,b}^{-1})f_{x,y} = f_{x-by,ay}.$$

The orbits of the representation are open half-planes $\Omega_{\pm} = \{f_{x,y}|\pm y > 0\}$ and the points $\omega_c = \{f_{c,0}\}$. Let us consider the points $f_{\pm} := f_{0,\pm 1} \in \Omega_{\pm}$. All one-dimensional subalgebras $\mathfrak{h} \subset \mathfrak{g}$ are clearly subordinate to f and there exists only one, namely the normal subalgebra characterized by the equation $\alpha = 0$, satisfying the Pukanszky condition. The corresponding irreducible representations $T_{\pm}$ act on the space $L_2(\mathbb{R}_+, \frac{dt}{t})$ by

$$T_{\pm}(g_{a,b})\varphi(t) = e^{\pm 2\pi i b t}\varphi(at).$$

The one-point orbits ω_c correspond to the one-dimensional representations

$$S_c(g_{a,b}) = a^{2\pi i c}.$$

As far as is known to the author, Theorem 7.3 for exponential Lie groups cannot be found in the published literature, even though it apparently holds. Theorem 7.4 holds without changes (see Busyatskaya (1973)). It contains, in particular, part b) of Theorem 7.2 in a new version (i.e. with the Pukanszky condition).

The formula for characters of irreducible representations (an analogue of Theorem 7.7) is basically true, but it needs some comments.

Firstly, for exponential groups, canonical measures on G-orbits in $\mathfrak{g}^*$ are not always distributions in $\mathcal{S}'(\mathfrak{g}^*)$. It is certainly not true for orbits which are not closed. Hence the integral on the right hand side of (7.10) is not defined for all test functions on G. Supplementary conditions are needed on smooth and rapidly decreasing functions to guarantee the existence of the integral. The exact formulation of these conditions is not known in general. It seems that they will be connected with the structure of the closure of the orbit.

Secondly, the left and right Haar measures do not necessarily coincide for exponential groups and they are both different from the image of the Lebesgue measure on $\mathfrak{g}$ under the exponential map (see Ex. 6.7).

This implies that it is necessary to introduce a correction term into the definition (7.8) of the Fourier transform to guarantee its unitarity. The left and right Haar measures on a Lie group have the following form in canonical coordinates:

$$d_l(\exp X) = \det\left(\frac{1 - e^{\text{ad} X}}{\text{ad} X}\right) dX,$$

$$d_r(\exp X) = \det\left(\frac{e^{\text{ad} X} - 1}{\text{ad} X}\right) dX.$$

Let us define the function $q(X)$ as the density of the "geometrical mean" of the left and right Haar measures:

$$q(X) = \sqrt{\frac{d_l(\exp X)}{dX} \frac{d_r(\exp X)}{dX}} = \det\left(\frac{\sinh(\text{ad} X/2)}{\text{ad} X/2}\right). \tag{7.15}$$

It is an analytic function on $\mathfrak{g}$ and its zeros are the points where the Jacobian of the exponential map vanishes. Thus, for exponential groups, the function $p(X) = \sqrt{q(X)}, p(0) = 1$ is well defined and analytic.

Let us change (7.8) and define the *Fourier transform for exponential Lie groups* by the formula

$$\tilde{\varphi}(f) = \int_{\mathfrak{g}} \varphi(\exp X) e^{2\pi i f(X)} \frac{dX}{p(X)}. \tag{7.8'}$$

Then (7.10) is true, at least for closed orbits of maximal dimension, as was shown in Duflo (1970) (compare also Duflo and Moore (1976), where an analogue of the Plancherel formula for exponential Lie groups was shown). A discussion of the validity of (7.10) for other classes of groups can be found below.

Example 7.15. Let us consider the group $G = \text{Aff}(1, \mathbb{R})$ (see Ex. 7.14). The exponential map has the form

$$\exp \begin{pmatrix} \alpha & \beta \\ 0 & 0 \end{pmatrix} = \begin{pmatrix} a & b \\ 0 & 1 \end{pmatrix}, \text{ where } a = e^\alpha, \ b = \frac{e^\alpha - 1}{\alpha} \beta.$$

The operator $T_\pm(\varphi) = \int_{\mathfrak{g}} \varphi(\exp X) T_\pm(\exp X) dX$ can be realized as an integral operator in $L_2(\mathbb{R})$ with kernel

$$K_\varphi^\pm(x, y) = \frac{y - x}{2 \sinh(\frac{y-x}{2})} \int_{-\infty}^\infty \tilde{\varphi}\left(\tau, \frac{e^y - e^x}{y - x}\right) e^{2\pi i \tau(x-y)} d\tau.$$

If $\varphi \in \mathcal{D}(G)$, then the operator $T_\pm(\varphi)$ is nuclear if and only if the function $\tilde{\varphi}$ is equal to zero on the common boundary of the orbits $\Omega_\pm$. Note that the condition holds if and only if the operators $S_c(\varphi)$ are trivial for all one-dimensional unitary representations S_c of the group G. If the condition is satisfied, then the trace of the operators is given by the formula (7.10).

As concerns infinitesimal characters, exponential Lie groups do not differ from solvable ones (and even from all non-nilpotent Lie groups). Hence we refer for further details to Sect. 5 below, where the method of orbits is discussed for general Lie groups.

3.2. General Solvable Groups. It is well-known that there are solvable groups which are not of type I. The first such example was found by Mautner and was rediscovered many times.

Example 7.16. Let G_α be the group of all matrices of the form

$$\begin{pmatrix} e^{it} & 0 & z \\ 0 & e^{i\alpha t} & w \\ 0 & 0 & 1 \end{pmatrix}, \ t \in \mathbb{R}, \ z, w \in \mathbb{C},$$

where α is a fixed real irrational number. Identifying the dual $\mathfrak{g}_\alpha^*$ with the space of all matrices of the form

$$\begin{pmatrix} i\tau & 0 & 0 \\ 0 & i\alpha\tau & 0 \\ a & b & 0 \end{pmatrix}, \ \tau \in \mathbb{R}, \ a, b \in \mathbb{C},$$

we get the following explicit formulas for the coadjoint representation (see Exs. 7.8 and 7.9):

$$K(t, z, w) : (a, b, \tau) \mapsto (ae^{-it}, be^{-i\alpha t}, \tau + \mathrm{Im}(az + bw)).$$

It is easy to see that generic G_α-orbits (with $(a, b) \neq (0, 0)$) are two-dimensional cylinders with their axis parallel to the τ-axis, the base of which is formed (for $ab \neq 0$) by a dense subset of the torus $|a| = r_1$, $|b| = r_2$. Two substantially different decompositions of the regular representation of the group G_α into a direct integral of irreducible representations are described in Kirillov (1978). The representations corresponding to G_α-orbits in $\mathfrak{g}_\alpha^*$ are used in one of them, while the representations corresponding to the so called *virtual orbits*, i.e. to ergodic measures in $\mathfrak{g}_\alpha^*$ with respect to G_α, are used in the other one.

A criterion describing when a solvable group is of type I is found in Auslander, Kostant (1971). They described the set $\hat{G}$ for all such groups. We shall present these results here together with some complements, discussed in Shchepochkina (1977).

Recall that a topological space is called *semiseparated* or a T_0 *space* if one of each pair of distinct points admits a neighbourhood not containing the other. The space of orbits $\mathcal{O}(G_\alpha)$ for the group G_α in Ex. 7.16 is not semiseparated.

Theorem 7.8. *A connected and simply connected solvable Lie group G belongs to type I if and only if the space $\mathcal{O}(G)$ is semiseparated and the canonical forms B_Ω on all orbits Ω are exact.*

An example of a seven-dimensional solvable Lie group such that the second condition of Theorem 7.8 is not satisfied can be found in Kirillov (1978).

Even for solvable Lie groups of type I, the correspondence between orbits and representations need not be one-to-one. To describe it better, we shall introduce the notion of a *rigged orbit*.

Let G be any simply connected Lie group and Ω an orbit in $\mathfrak{g}^*$. Let G_f be the stabilizer of the point $f \in \Omega$ and $\mathfrak{g}_f = \ker B_f$ its Lie algebra. Unlike the case for nilpotent and exponential groups, the orbit Ω need not be simply connected. Simple topological considerations show that the fundamental group $\pi_1(\Omega)$ of the orbit is isomorphic to G_f/G_f^0, where G_f^0 is the connected component of the identity in G_f. A pair (f, χ), where $f \in \mathfrak{g}^*$ and χ is a unitary character of the group G_f whose differential at the identity is equal to $2\pi i f|_{\mathfrak{g}_f}$, will be called a *rigged momentum* of the group G. The set of all rigged momenta of G will be denoted by $\mathfrak{g}_{\mathrm{rig}}^*$. The group G acts on $\mathfrak{g}_{\mathrm{rig}}^*$ in a natural way:

$$g : (f, \chi) \mapsto (f \circ \operatorname{Ad} g, \ \chi \circ A(g)),$$

where $A(g) : x \mapsto gxg^{-1}$ is an inner automorphism of G. The canonical projection $\pi : \mathfrak{g}^*_{\mathrm{rig}} \to \mathfrak{g}^* : (f, \chi) \mapsto f$ commutes with the action of G.

It turns out that the structure of the set $\mathfrak{g}^*_{\mathrm{rig}}$ is determined by topological properties of G-orbits in $\mathfrak{g}^*$. We say that an orbit Ω is *integral* if the two-dimensional cohomology class defined by the canonical 2-form B_Ω belongs to $H^2(\Omega, \mathbf{Z})$. (In other words, the integral of the form B_Ω over any two-dimensional cycle in Ω is an integer.) Also, let us denote the fundamental group of the orbit Ω by $\pi_1(\Omega)$ and the group of its unitary characters by Γ_Ω.

Lemma 7.14. *Let G be a connected and simply connected Lie group and Ω an orbit in $\mathfrak{g}^*$.*

a)   *The set $\pi^{-1}(\Omega) \subset \mathfrak{g}^*_{\mathrm{rig}}$ is nonempty if and only if the orbit Ω is integral.*

b)   *If Ω is an integral orbit, then $\pi^{-1}(\Omega)$ is a principal fiber bundle with base Ω and group Γ_Ω.*

In particular, Theorem 7.8 implies that, for solvable Lie groups of type I, the class of the form B_Ω is trivial, hence every momentum $f \in \mathfrak{g}^*$ admits an extension to $(f, \chi) \in \mathfrak{g}^*_{rig}$. If the orbit Ω containing f is, moreover, simply connected, then the extension is unique.

The proof of Lemma 7.14 is based on the study of the exact homotopy sequence of the fibration $G_f \to G \to \Omega$:

$$\ldots \to \pi_2(G) \to \pi_2(\Omega) \to \pi_1(G_f) \to \pi_1(G) \to$$
$$\to \pi_1(\Omega) \to \pi_0(G_f) \to \pi_0(G) \to \ldots,$$

where the middle and the last terms are trivial by assumption. It is a well-known fact from the theory of Lie groups that the first term above is trivial. As a consequence, it is possible to identify $\pi_1(\Omega)$ with $\pi_0(G_f) = G_f / G_f^0$ and to define the action of Γ_Ω on the set $\pi^{-1}(\Omega)$ as well as to identify $H_2(\Omega, \mathbf{Z}) \simeq \pi_2(\Omega)$ with $\pi_1(G_f) \simeq H_1(G_f, \mathbf{Z})$ and to associate an element of $H^1(G_f, \mathbf{Z}) \simeq H^2(\Omega, \mathbf{Z})$ to any character χ of the group G_f. It is possible to check that this element coincides with the class of the form B_Ω and with the first Chern class of the line bundle over Ω given by the character χ. (The space of all sections of the bundle is the space of the representation $\operatorname{Ind}_{G_f}^G \chi$.)

Let G be a connected and simply connected Lie group. The fundamental result of Kostant and Auslander says that there is a bijection between the set $\hat{G}$ and the set $\mathcal{O}_{\mathrm{rig}}(G)$ of all orbits of the group in the space $\mathfrak{g}^*_{\mathrm{rig}}$.

The usual induction procedure is not sufficient for the reconstruction of the unitary representation of the group G from a given orbit $\Omega_{\mathrm{rig}} \subset \mathfrak{g}^*_{\mathrm{rig}}$. This is a consequence of the fact that, for a given rigged momentum $(f, \chi) \in \Omega_{\mathrm{rig}}$, it is not always possible to find a subalgebra of $\mathfrak{g}$ of the required dimension subordinate to the functional f. On the other hand, it is always possible to find a complex subalgebra $\mathfrak{h}$ of dimension $\frac{1}{2}(\dim \mathfrak{g}+\dim \mathfrak{g}_f)$ in the complexification $\mathfrak{g}^c$ subordinate to f. Such algebras are called *polarizations*.

A polarization $\mathfrak{h}$ is called *admissible* if the following conditions are satisfied:

1) $\mathfrak{h}$ is invariant with respect to $\mathrm{Ad}\,G_f$,
2) $\mathfrak{h}+\bar{\mathfrak{h}}$ is a subalgebra of $\mathfrak{g}^c$.

For every admissible polarization $\mathfrak{h}$, we shall consider two subalgebras in $\mathfrak{g}$: $\mathfrak{d} = \mathfrak{h}\cap\mathfrak{g}$ and $\mathfrak{e} = (\mathfrak{h}+\bar{\mathfrak{h}})\cap\mathfrak{g}$. Let D_0 and E_0 be the corresponding subgroups in G and let $D = D_0G_f$, $E = E_0G_f$. The form B_f is nonsingular on $\mathfrak{e}/\mathfrak{d}$ and it induces a symplectic structure on $X = E/D$. On the other hand, $(\mathfrak{e}/\mathfrak{d})^c \simeq (\mathfrak{h}+\bar{\mathfrak{h}})/\mathfrak{h}\cap\bar{\mathfrak{h}} = \mathfrak{h}/\mathfrak{h}\cap\bar{\mathfrak{h}}\oplus\bar{\mathfrak{h}}/\mathfrak{h}\cap\bar{\mathfrak{h}}$. So we can introduce an almost complex structure J on X given by the multiplication by i on $\mathfrak{h}/\mathfrak{h}\cap\bar{\mathfrak{h}}$ and by $-\mathrm{i}$ on $\bar{\mathfrak{h}}/\mathfrak{h}\cap\bar{\mathfrak{h}}$. The hermitean structure on X is given by the canonical form $S(u,v) = B_f(Ju,v)$.

The polarization $\mathfrak{h}$ is called *positive* if the form S is positive; then X is a Kähler manifold.

The polarization $\mathfrak{h}$ satisfies the Pukanszky condition (see Sect. 2) if the condition $f+\mathfrak{e}^\perp \subset \Omega_f$ holds (note that $\mathfrak{e}^\perp = \mathfrak{h}^\perp\cap\mathfrak{g}$, where $\mathfrak{h}^\perp$ is the annihilator of $\mathfrak{h}$ in $(\mathfrak{g}^c)^*$).

Theorem 7.9.   *Let G be a connected and simply connected Lie group of type I, let $(f,\chi) \in \mathfrak{g}_{\mathrm{rig}}^*$ be a rigged momentum of the group.*

a) *There exists an admissible positive polarization $\mathfrak{h}$ for (f,χ) satisfying the Pukanszky's condition.*
b) *The holomorphically induced representation $T_{\mathfrak{h},f,\chi}$ of the group G (constructed using the momentum (f,χ) and the polarization $\mathfrak{h}$) is irreducible and its equivalence class depends only on the orbit $\Omega_{\mathrm{rig}} \subset \mathfrak{g}_{\mathrm{rig}}^*$ containing (f,χ).*
c) *All irreducible unitary representations of the group G can be obtained in this way.*

It is a quite interesting problem to clarify whether the conditions concerning the polarization $\mathfrak{h}$ in Theorem 7.9 are necessary. The experience with real semisimple groups shows that the lack of positivity can be compensated by the replacement of sections of the holomorphic bundle by higher cohomology spaces. Up to now, it is not clear whether the theorem is true without the second assumption of admissibility: $\mathfrak{h}+\bar{\mathfrak{h}}$ is a subalgebra of $\mathfrak{g}^c$. The conditions specifying the type of functions of the representation space cannot be reduced to partial holomorphicity in such a case.

There is no doubt that an analogue of Theorem 7.3 holds for solvable Lie groups of type I but it is not yet proved.

The universal formula (7.10) for the generalized characters of solvable groups was studied by Pukanszky, Duflo and their students. In particular, it turned out that an analogue of the formula holds for groups which do not belong to type I if the generalized character is defined using the trace on a factor of type II and if ergodic measures are considered instead of G-orbits in $\mathfrak{g}^*$.

Example 7.17. Let $\mathfrak{g}$ be the so called *oscillator Lie algebra* with generators T, X, Y, Z and relations

$$[T, X] = Y, \quad [T, Y] = -X, \quad [X, Y] = Z, \quad [Z, \mathfrak{g}] = 0.$$

A matrix realization of the algebra has the form:

$$xX + yY + zZ + tT = \begin{pmatrix} 0 & x - iy & 2iz \\ 0 & it & x + iy \\ 0 & 0 & 0 \end{pmatrix}.$$

The invariants of the coadjoint representation are

$$I_1 = Z, \quad I_2 = X^2 + Y^2 + 2TZ.$$

The level sets $I_1 = c_1$, $I_2 = c_2$ are two-dimensional orbits, with the exception of the line $c_1 = c_2 = 0$ which decomposes into one-point orbits.

For $c_1 \neq 0$, the orbit is simply connected. If $c_1 = 0, c_2 \neq 0$, then the orbit is a cylinder. It is the projection of a set of rigged orbits parametrized by points of the circle. Hence the set $\hat{G} = \mathcal{O}_{\mathrm{rig}}(G)$ looks like the plane with coordinates c_1, c_2 with circles glued in instead of the points $(0, c), c \neq 0$, and a line instead of the origin $(0, 0)$. Note that the described result refers to the simply connected group G with Lie algebra $\mathfrak{g}$. If we restrict ourselves to representations of the group G_1 generated by the Lie algebra $\mathfrak{g}$ in the matrix realization above, then there is only a point instead of every glued-in circle and a lattice of integer points instead of the glued-in line.

§4. The Method of Orbits for Other Classes of Groups

4.1. Semisimple Groups. The theory of unitary representations of compact semisimple groups was constructed a long time before the discovery of the method of orbits. But the explicit geometrical realization of irreducible representations, given by the Borel-Weil-Bott theorem, can be naturally included into the method of orbits. Let us describe it here. Let G be a connected and simply connected compact Lie group. Then the G-orbits in $\mathfrak{g}^*$ are simply connected and they have G-invariant complex structure, a Riemannian metric and also a Kähler structure. These manifolds are called *flag manifolds* because their elements are realized in terms of "flags" for classical Lie groups G. A flag is a family of subspaces

$$0 = V_0 \subset V_{i_1} \subset \ldots \subset V_{i_k} \subset V,$$

where V is the complex vector space of the standard representation of the group G, $(i_1, \ldots, i_k)$ is a subset of $(1, 2, \ldots, n)$ and $\dim V_i = i$. (For orthogonal and symplectic groups there are supplementary conditions of "selfduality" of the flag: $V_k = V_{n-k}^{\perp}$, $n = \dim V$.) Orbits of maximal dimension consist of nonsingular flags such that $k = n$.

The integrality condition (see 3.2) distinguishes a countable family of orbits in $\mathcal{O}(G)$.

Let Ω be an integral orbit of maximal dimension, $f \in \mathfrak{g}^*$ and $\mathfrak{h}$ a positive admissible polarization for f. If R_X denotes the left-invariant complex vector field on G corresponding to $X \in \mathfrak{h}$, then the space of functions φ on G having the property

$$R_X \varphi = 2\pi i f(X) \varphi, \; X \in \mathfrak{h}$$

can be identified with the space of holomorphic sections of the line bundle L over Ω corresponding to the character

$$\chi(\exp X) = e^{2\pi i f(X)}$$

of the (connected) group G_f. The representation of the group G in this space is irreducible and does not depend on the choice of a positive polarization on $\mathfrak{h}$. All irreducible representations can be constructed in this way. A generalization of this result, due to A.Borel and A.Weil, was found by R.Bott. He showed that the positivity condition for $\mathfrak{h}$ can be dropped if the space of holomorphic sections of L (which is trivial for non-positive polarizations) is replaced by the cohomology spaces $H^k(\Omega, \mathcal{L})$, where k is the number of negative squares in the Kähler form and $\mathcal{L}$ is the sheaf of germs of holomorphic sections (see Sect. 2 of Chap.6).

The formula (7.10) for characters of irreducible representations (they consist of ordinary functions in our case) is valid for compact groups and gives a simple integral representation of the character:

$$\chi(\exp X) = \frac{1}{p(X)} \int_{\Omega} e^{2\pi i f(X)} d_{\Omega} f.$$

It implies

Lemma 7.15. *The dimension of an irreducible representation of a compact Lie group is equal to the volume of the corresponding orbit.*

Note that this agrees with the interpretation of the method of orbits in quantum mechanics (see Kirillov (1980)) and it reflects the fact that the number of degrees of freedom in quantum mechanics is equal to the volume of the phase space counted in Planck units.

The described relation also admits a topological interpretation and makes it possible to calculate explicitly the structure of the cohomology ring of the orbit Ω (see Atiyah and Hirzebruch (1961)). More precisely, all orbits of maximal dimension have the same structure as G-spaces: they can be identified with the so called flag space $M = G/T$, where T is a maximal torus of the group G. Let $l = \dim T$ and let $t_1, \ldots, t_l$ be a basis of the lattice in the Lie algebra $\mathfrak{t}$ of the torus T which is the inverse image of the identity under the exponential map. Then the cycles $C_i = \exp([0,1] \cdot t_i)$ form a basis in $H_1(T, \mathbf{Z})$. We know that $\pi_1(G) = \pi_2(G) = 0$, hence there is a unique (up to homotopy) two-dimensional surface in G the boundary of which is the cycle C_i. The projection of the surface to $M = G/T$ will be denoted by $\tilde{C}_i$. The two-dimensional cycles $\tilde{C}_i$ form a basis of $H_2(M, \mathbf{Z})$.

Let us now consider an element $f \in \mathfrak{t}^* \subset \mathfrak{g}^*$ (The spaces $\mathfrak{t}$ and $\mathfrak{t}^*$, resp. $\mathfrak{g}$ and $\mathfrak{g}^*$ are identified using the Killing metric). The symplectic form B_Ω on the orbit Ω passing through the point f gives a two-dimensional cohomology class $[B_\Omega]$ in $H^2(\Omega, \mathbb{R})$. If we identify Ω with M, this class goes over to $\sum_{i=1}^{l} k_i \omega_i$, where $\{\omega_i\}$ is the basis of $H^2(M, \mathbf{Z})$ dual to the basis $\{\tilde{C}_i\}$ in $H_2(M, \mathbf{Z})$ and $k_i = f(t_i)$. The orbit Ω is integral if and only if all the numbers k_i are integers. Now let $P(k_1, \ldots, k_l)$ be the polynomial of Weyl of degree $m = \frac{1}{2}\dim M$, expressing the dimension of an irreducible representation of the group G using the coordinates of its highest weight. Then the following formula is a consequence of Lemma 7.15:

$$\frac{1}{m!}\left(\sum_i k_i \omega_i\right)^m [\Omega] = P(k_1 - 1, \ldots, k_l - 1). \tag{7.17}$$

Using it, we can reconstruct the multiplication in $H^*(M)$.

Example 7.18. Consider the group $G = \mathrm{SU}(3)$. Then $l = 2$, $m = 3$, and $P(k_1 - 1, k_2 - 1) = \frac{1}{2}k_1 k_2 (k_1 + k_2)$. It follows from (7.17) that

$$\omega_1^3[\Omega] = \omega_2^3[\Omega] = \omega_1\omega_2(\omega_1 - \omega_2)[\Omega] = 0,$$
$$\omega_1\omega_2^2[\Omega] = \omega_1^2\omega_2[\Omega] = 1.$$

The Poincaré duality shows that ω_1^2 and ω_2^2 form a basis of $H^4(M, \mathbf{Z})$ with the relation $\omega_1\omega_2 = \omega_1^2 + \omega_2^2$. The element $\omega_1\omega_2^2 = \omega_1^2\omega_2$ forms the basis of $H^6(M, \mathbf{Z})$.

In the case of noncompact semisimple groups, at present the contribution of the method of orbits consists of an intuitive interpretation of known results in terms of orbits. Different series of representations (the principal, degenerate, discrete and complementary series) correspond to different types of G-orbits in $\mathfrak{g}^*$ (for the complementary series – in $(\mathfrak{g}^*)^c$). Formula (7.10) for the characters remains true for representations of the principal and discrete series (in the last case, for elliptic elements of $\mathfrak{g}$). The recent treatment of the formula using the theory of equivariant cohomology and the theory of reduction in symplectic geometry is quite interesting.

4.2. General Lie Groups. It is worthwhle describing two results here. The first one describes an isomorphism between the algebras $Z(\mathfrak{g})$ and $S(\mathfrak{g})^G$ for any Lie algebra $\mathfrak{g}$. Note that the isomorphism σ (the symmetrization), described in Chap. 5, does not preserve multiplication. Nevertheless, it was noted, even before the discovery of the method of orbits, that the algebras

$Z(\mathfrak{g})$ and $S(\mathfrak{g})^G$ are isomorphic in all known examples. Theorem 7.17 on infinitesimal characters (together with the evident isomorphism between $S(\mathfrak{g})^G$ and $P(\mathfrak{g}^*)^G$) proves the hypothesis for nilpotent Lie groups and indicates the method of proof in general case. The first proof was given by Duflo in Gel'fand (1975). A more general result was proved in Ginzburg (1981). The main idea behind it is the fact that for any complex Lie group G, it is possible to choose real coordinates $p_1, \ldots, p_n, q_1, \ldots, q_n, z_1, \ldots, z_k$ in an open subset $V \subset \mathfrak{g}^*$ such that

a) the orbits are given by the equations $z_i = \text{const}$, $1 \leq i \leq k$,
b) the functions p_i, q_j give canonical coordinates on generic orbits;
c) the conditions $z_i = \text{const}$, $q_j = \text{const}$ give a compatible system of polarizations on generic orbits.

The Fourier transform (7.10) maps the universal enveloping algebra $U(\mathfrak{g})$ into the space $P(\mathfrak{g}^*)$ of all polynomials on $\mathfrak{g}^*$.

Let A denote the subalgebra of $P(\mathfrak{g}^*)$ consisting of all polynomials which are constant on the fibres of the polarization described above (i.e. independent of p_i, $1 \leq i \leq n$). Let B be the inverse image of A in $U(\mathfrak{g})$ under the Fourier transform (7.10).

Theorem 7.10. *The set B is a commutative subalgebra in $U(\mathfrak{g})$ and the Fourier transform gives an isomorphism between the algebras A and B.*

The second important result is a general construction, due to Duflo, of a representation of the Lie group G starting from a G-orbit Ω in $\mathfrak{g}^*$. The point of departure is a point $f \in \Omega$ and a solvable polarization $\mathfrak{h} \subset \mathfrak{g}^c$ satisfying the Pukanszky condition. The stabilizer of the point f in G will be denoted by G_f. Let $\tilde{G}_f$ be the two-fold covering of G_f corresponding to the symplectic representation of G_f in the tangent space $T_f\Omega$.

A functional f will be called admissible if there exists an irreducible representation τ of the group $\tilde{G}_f$ with the properties:
a) the differential τ_* of the representation τ is a multiple of $2\pi i f|_{\mathfrak{g}_f}$;
b) $\tau(\varepsilon) = -1$, where ε is the nontrivial inverse image of the unit $e \in G_f$ in $\tilde{G}_f$.

The pair (f, τ) is an analogue of the rigged momentum introduced in Sect. 3. For every pair (f, τ), Duflo constructs (in a complicated way, including induction with respect to the dimension of the group) an irreducible unitary representation $T_{(f,\tau)}$ of the group G. It is shown that the family of representations constructed in this way is rich enough to decompose the regular representation of a complex unimodular algebraic Lie group.

4.3. Infinite-Dimensional Lie Groups. A large number of results concerning the theory of representations of infinite-dimensional Lie groups is already available. Three classes of such groups are especially often met in different applications: 1) groups of linear operators in infinite-dimensional linear spaces; 2) groups of diffeomorphisms of smooth manifolds; 3) gauge groups (resp. current groups) consisting of functions on a set (resp. manifold) X with values in an ordinary Lie group G. The latter two types of groups are sometimes unified inside authomorphism group of vector bundles.

The method of orbits was applied to groups of the first type in Kirillov (1973). It initiated a long series of papers on the theory of representations of the infinite-dimensional classical groups GL(∞), $U(\infty)$, O(∞), Sp(∞) and their real forms.

Using the method of orbits for groups of the second and third types, it is possible to construct a large number of unitary representations with a finite functional dimension connected with finite-dimensional G-orbits in $\mathfrak{g}^*$.

The study of infinite-dimensional orbits has just begun but it promises to be very interesting. Only the case of the group $G = \text{Diff } S^1$ and its central extension has been studied up to now but there are interesting unsolved questions connected with the geometry of orbits even in this case.

More about representations of infinite-dimensional Lie groups can be found in future volumes of Encyclopaedia.

4.4. Representations of Lie Supergroups and Superalgebras. An intuitive understanding offered by the method of orbits is again quite important but many specific problems are still not solved. These questions will be also discussed in future volumes (see also Kac (1977).

Comments on the References

The overall number of all papers and books on the representation theory and harmonic analysis is nowadays counted by thousands. An attempt to collect a detailed bibliography can be found in Barut, Rcaczka (1977), Hewitt, Ross (1970), and Zhelobenko, Shtern (1983). Our list is much more modest, it containts basically classical and contemporary papers having a substantial influence on the evolution of the subject, textbooks, review papers and proceedings of conferences on harmonic analysis and its applications. For beginners or non-specialists, we recommend to start with Kirillov (1978), Kirillov (1985a), Gel'fand, Najmark (1947), Najmark (1976), Serre (1967), Zhelobenko (1970). A historical perspective can be found in reviews Dieudonne (1977), Mackey (1963), Mackey (1980). For more special questions of the representation theory of finite groups, see Brauer (1963), Curtis, Reiner (1962), Deligne, Lusztig (1976), Feit (1982), James (1978), Kazhdan, Lusztig (1979), Lusztig (1977). Finite-dimensional representations are discussed in Hewitt, Ross (1963), Ismagilov (1983), Weil (1940), Weyl (1939), Zhelobenko (1970).

A general introduction to the contemporary representation theory is Kirillov (1978). Infinite-dimensional representations of semisimple groups are studied in Berezin (1957), Berezin (1963), Gel'fand et al. (1962), Gel'fand, Najmark (1950), Graev (1958), Najmark (1976), Knapp, Zuckerman (1982), Operator algebras and group representation I,II (1984), Schmid (1975, 1976), Stein (1967), Vogan (1981), Warner (1972), Wolf et al. (1980), Zhelobenko (1974). Recent results on the classification of unitary irreducible representations of semisimple groups are described in Carmona, Vergne (1975), Carmona, Vergne (1976), Herb et al. (1984), Knapp, Zuckerman (1982), Tadič (1985). The Langlands's programme and related problems of the representations theory of p-adic and adèle groups can be found in Gel'fand, Kirillov (1968), Herb et al. (1984), Langlands (1966), Operator algebras and group representations I,II (1984). The method of orbits and its connections with geometrical quantization are treated in Auslander, Kostant (1971), Bernat et al. (1972), Bleuler et al. (1977), Brezin (1968), Brown (1973), Busyatzkaya (1973), Carmona, Vergne (1975), Carmona, Vergne (1976), Carmona, Vergne (1981), Duflo (1970), Garcia et al. (1980), Ginzburg (1981), Gulick, Lipsman (1972), Kirillov (1967), Kirillov (1980), Kirillov (1985b), Kostant (1970), Pukanszky (1972), Rossman (1978), Shchepochkina (1977), Vergne (1979). The algebraic aspects of the method of orbits connected with the study of the structure of the universal enveloping algebra and its field of quotients were developed in Dixmier (1974), Duflo (1977), Gel'fand (1975), Gel'fand, Kirillov (1966), Gel'fand et al. (1958), Ginzburg (1981), Chow (1969), Josèph (1974), McConnel (1974), Rentschler, Vergne (1973). Infinite-dimensional Lie groups and algebras are discussed in Vershik et al. (1973), Kirillov (1962), Kirillov (1973), Kirillov (1982), Kirillov (1984), Najmark (1976), Kac (1990). See also the two issues from Advances in Soviet Mathematics, AMS edition, Vol. 2: Topics in Representation Theory, A.A. Kirillov (ed.) (1991) and Vol. 16 Gel'fand's Seminar, S.G. Gindikin (ed.) (1993). The survey of my last results is in Contemporary Mathematics, Vol. 145, AMS edition (1993).

References*

Arsene G., Stratila S., Verona A., Voiculescu D. (eds.) (1984): Operator algebras and group representations. Proc. of the Int. Conference held in Neptun (Romania), Boston, Pitman, Vol. I, 277 pp., Vol. II, 250 pp., Zbl. 515.00017

Atiyah M.F., Hirzebruch F. (1961): Vector bundles and homogeneous spaces in differential geometry. Proc. Symp. Pure Math. *3*, 7–38, Zbl. 108,177

Auslander L., Kostant B. (1971): Polarization and unitary representations of solvable Lie groups. Invent. Math. *14*, 255–354, Zbl. 233.22005

Bargmann V. (1947): Irreducible unitary representations of the Lorentz group. Ann. Math., II. Ser. *48*, 568–640, Zbl. 41,362

Bargmann V. (1954): On unitary ray representations of continuous groups. Ann. Math., II. Ser. *59*, No. 1, 1–46, Zbl. 55,103

Barut A.O., Rcaczka R. (1977): Theory of Group Representations and Applications I, II. Warszawa, PWN, Zbl. 471.22021

Berezin F.A. (1957): Laplace operators on semisimple Lie groups. Tr. Mosk. Mat. O.-va *6*, 371–463. English transl.: Transl., II. Ser., Am. Math. Soc. *21*, 239–339 (1962), Zbl. 91,282

Berezin F.A. (1963): Letter to the editor. Tr. Mosk. Mat. O.-va *12*, 453–466; English transl.: Trans. Mosc. Math. Soc. 1963, 510–524 (1965), Zbl. 154,391

---

*For the convenience of the reader, references to reviews in Zentralblatt für Mathematik (Zbl.), compiled using the MATH database, and Jahrbuch über die Fortschritte in der Mathematik (FdM) have, as far as possible, been included in this bibliography.

Berezin F.A. (1967): Some remarks on the associative envelope of a Lie algebra. Funkts. Anal. Prilozh. *1*, No. 2, 1–14; English transl.: Funct. Anal. Appl. *1*, 91–102 (1968), Zbl. 227.22020

Bernat P., Conze N., Duflo M. et al. (1972): Représentations des Groupes de Lie Résolubles. Paris, Dunod, 272 pp., Zbl. 248.22012

Blattner R. (1961): On induced representations I, II. Am. J. Math. *83*, 79–98; 499–512, Zbl. 122,284; 139,77

Bleuler K. et al. (eds.) (1977): Differential Geometrical Methods in Mathematical Physics I,II. Heidelberg, Springer-Verlag, Lect. Notes Math. *570* (1977), 576 pp., Zbl. 336.00008; *676* (1978), 526 pp., Zbl. 378.00011; *836* (1980), 538 pp., Zbl. 436.00018

Brauer R. (1963): Representations of Finite Groups. Lect. Modern Math. *1*, 133–175, Zbl. 124,265

Bredon G.E. (1972): Introduction to Compact Transformation Groups. New York, Acad. Press, XIII+459 pp., Zbl. 246.57017

Brezin J. (1968): Theory of Unitary Representations of Solvable Lie Groups. Mem. Am. Math. Soc. *79*, 122 pp., Zbl. 157,366

Brown I.D. (1973): Dual topology of a nilpotent Lie group. Ann. Sci. Ec. Norm. Supér., IV Sér. *6*, No. 3, 407–411, Zbl. 284.57026

Busyatskaya N.K. (1973): Representations of exponential Lie groups. Funkts. Anal. Prilozh. *7*, No. 2, 79–80. English transl.: Funct. Anal. Appl. *7*, 151–152 (1973), Zbl. 286.22013

Carmona J., Vergne M. (eds.) (1975): Non-commutative Harmonic Analysis. Heidelberg, Springer-Verlag, Lect. Notes Math. *466*, 231 pp., Zbl. 299.00023

Carmona J., Vergne M. (eds.) (1976): Non-commutative Harmonic Analysis. Heidelberg, Springer-Verlag, Lect. Notes Math. *587*, 240 pp., Zbl. 342.00009

Carmona J., Vergne M. (eds.) (1981): Non-commutative Harmonic Analysis and Lie Groups. Heidelberg, Springer-Verlag, Lect. Notes Math. *880* (1981), 553 pp., Zbl. 453.00017; *1020* (1983), 187 pp., Zbl. 511.00012

Cartan E. (1984): Oeuvres Complétes, Vol. 1. CNRS, Paris, 1356 pp.

Chow Y. (1969): Gelfand-Kirillov conjecture on the Lie field of an algebraic Lie algebra. J. Math. Phys. *10*, 975–992, Zbl. 179,56

Curtis C.W., Reiner I. (1962): Representation Theory of Finite Groups and Associative Algebras. New York, Interscience Publ., XIV, 685 pp., Zbl. 131,256

Deligne P., Lusztig G. (1976): Representations of reductive groups over finite fields. Ann. Math. *103*, 103–161, Zbl. 336.20029

Dieudonne J. (1977): Panorama des Mathematiques Pures. Le Choix Bourbachique. Paris, Gauthier-Villars, XVII, 302 pp., Zbl. 482.00002

Dixmier J. (1969): Les C^*-algèbres et leur représentations. Paris, Gauthier-Villars, 390 pp., Zbl. 174,186

Dixmier J. (1974): Algèbres Enveloppantes. Paris, Gauthier-Villars, 350 pp., Zbl. 308.17007

Drinfel'd V.G. (1986): Quantum groups. Zap. Nauchn. Semin. Leningr. Otd. Mat. Inst. Stekolova *155*, 18–49. English transl.: J. Sov. Math. *41*, No. 2, 898–915 (1988), Zbl. 617.16004

Duflo M. (1970): Characteres des groupes et des algèbres de Lie resolubles. Ann. Sci. Ec. Norm. Supér., IV Sér. *3*, 23–74, Zbl. 223.22016

Duflo M. (1977): Operateurs differentiels bi-invariants sur un group de Lie. Ann. Sci. Ec. Norm. Supér., IV Sér. *10*, 265–283, Zbl. 353.22009

Duflo M., Moore C.C. (1976): On the regular representation of a nonunimodular locally compact group. J. Funct. Anal. *21*, No. 7, 209–243, Zbl. 317.43013

Eymard P. et al. (eds.) (1976): Analyse Harmonique sur les Groupes de Lie. Heidelberg, Springer-Verlag, Lect. Notes Math. *497*, 710 pp., Zbl. 309.00022

Feit W. (1982): The Representation Theory of Finite Groups. Amsterdam, North-Holland, 502 pp., Zbl. 493.20007

Frobenius G. (1968): Gesammelte Abhandlungen I–III. Heidelberg, Springer-Verlag, 650; 733; 740 pp., Zbl. 169,289

Garcia P. et al. (eds.) (1980): Differential Geometrical Methods in Mathematical Physics. Heidelberg, Springer-Verlag, Lect. Notes Math. *836*, 538 pp., Zbl. 436.00018

Gel'fand I.M. (1950): The centre of an infinitesimal group ring. Mat. Sb., Nov. Ser. *26*, No. 1, 103–112 (Russian), Zbl. 35,300

Gel'fand I.M. (ed.) (1975): Lie Groups and their Representations. Budapest, Akad. Kiadó, 726 pp., Zbl. 297.00009

Gel'fand I.M., Graev M.I., Piatetski-Shapiro N.N. (1966): Generalized Functions, Vol. 6: The Theory of Representations and Automorphic Functions. Moscow, Izdat. Nauka, 512 pp. English transl.: Philadelphia, Saunders (1969), 512 pp., Zbl. 138,72

Gel'fand I.M., Graev M.I., Vilenkin N.Ya. (1962): Generalized Functions, Vol. 5: Integral Geometry and Representation Theory. Moscow, Fizmatgiz., 656 pp. English transl.: New York, Academic Press (1966), 499 pp., Zbl. 115,167

Gel'fand I.M., Kirillov A.A. (1966): Sur les corps liés aux algèbres enveloppantes des algèbres de Lie. Publ. Math., Inst. Hautes Etud. Sci. *31*, 509–523, Zbl. 144,21

Gel'fand I.M., Kirillov A.A. (1969): Structure of the Lie skew-field connected with a semisimple decomposable Lie algebra. Funkts. Anal. Prilozh. *3*, No. 1, 7–26. English transl.: Funct. Anal. Appl. *3*, 6–21 (1969), Zbl. 244.17007

Gel'fand I.M., Minlos R.A., Shapiro Z.Ya. (1958): Representations of the rotation group and of the Lorentz group. Moscow, Fizmatgiz., 368 pp. English transl.: New York, Pergamon Press (1963), Zbl. 108,220

Gel'fand I.M., Najmark M.A. (1947a): Unitary representations of the Lorentz group. Izv. Akad. Nauk SSSR, Ser. Mat. *11*, No. 5, 411–504 (Russian), Zbl. 37,153

Gel'fand I.M., Najmark M.A. (1947b): Unitary representations of the group of linear transformations of the line. Dokl. Akad. Nauk SSSR *55*, No. 7, 567–570 (Russian), Zbl. 29,5

Gel'fand I.M., Najmark M.A. (1950): Unitary representations of the classical groups. Tr. Mat. Inst. Steklova *36*, 288 pp. German transl.: Unitäre Darstellungen der klassischen Gruppen. Berlin, Akademie Verlag (1957), 333 pp., Zbl. 41,362

Gel'fand I.M., Ponomarev V.A. (1969): Remarks on the classification of a pair of commuting linear transformations in a finite-dimensional space. Funkts. Anal. Prilozh. *3*, No. 4, 81–82. English transl.: Funct. Anal. Appl. *3*, 325–326 (1970), Zbl. 204,453

Gel'fand I.M., Rajkov D.A. (1943): Irreducible unitary representations of locally bicompact groups. Mat. Sb., Nov. Ser *13*, 301–316 (Russian)

Ginzburg V.A. (1981): The method of orbits in the representation theory of complex Lie groups. Funkts. Anal. Prilozh. *15*, No. 1, 23–37. English transl.: Funct. Anal. Appl. *15*, 18–28 (1981), Zbl. 457.22004

Graev M.I. (1958): Unitary representations of real simple Lie groups. Tr. Mosk. Mat. O.-va *7*, 335–389, Zbl. 87,24

Greenleaf F. (1969): Invariant Means on Topological Groups and their Applications. New York, Van Nostrand, Zbl. 174,190

Gulick D., Lipsman R.L. (eds.) (1972): Conference on Harmonic Analysis. Heidelberg, Springer-Verlag, Lect. Notes Math. *266*, 323 pp., Zbl. 228.00005

Harish-Chandra (1984): Collected papers, Vol. 1–4. Heidelberg, Springer-Verlag, 566 pp.; 613 pp.; 669 pp.; 641 pp., Zbl. 546.01013

Heckman G.J. (1982): Projections of orbits and asymptotic behavior of multiplicities for compact connected Lie groups. Invent. Math. *67*, 333–356, Zbl. 497.22006

Herb R. et al. (eds.) (1984): Lie Group Representations I,II. Heidelberg, Springer-Verlag, Lect. Notes Math. *1024*, 369 pp., Zbl. 511.00011; *1041*, 340 pp., Zbl. 521.00012

Hewitt E., Ross K.A. (1963): Abstract Harmonic Analysis I, II. Heidelberg, Springer-Verlag, 519 pp., Zbl. 115,106; (1970), 771 pp., Zbl. 213,401

Ismagilov R.S. (1983): Infinite-dimensional groups and their representations. In: Proc. Int. Congr. Math., Warszaw 1983, Vol. 2, 861–875 (Russian), Zbl. 575.22003

James G.D. (1978): The Representation Theory of the Symmetric Groups. Heidelberg, Springer-Verlag, Lect. Notes Math. *682*, 156 pp., Zbl. 393.20009

Josèph A. (1974): Proof of the Gelfand-Kirillov conjecture for solvable Lie algebras. Proc. Am. Math. Soc. *45*, No. 1, 1–10, Zbl. 293.17006

Kac V.G. (1977): Lie superalgebras. Adv. in Math. *26*, 8–96

Kac V.G. (1990): Infinite dimensional Lie algebras. 3rd ed., Cambridge University Press (1990)

Kazhdan D., Lusztig G. (1979): Representations of Coxeter groups and Hecke algebras. Invent. Math. *53*, 165–184, Zbl. 499.20035

Kirillov A.A. (1962): Unitary representations of nilpotent Lie groups. Usp. Mat. Nauk. *17*, No. 4, 57–110. English transl.: Russ. Math. Surv. *17*, No. 4, 53–104 (1962), Zbl. 106,250

Kirillov A.A. (1967): On the Plancherel measure for nilpotent Lie groups. Funkts. Anal. Prilozh. *1*, No. 4, 84–85. English transl.: Funct. Anal. Appl. *1*, 330–331 (1968), Zbl. 176,303

Kirillov A.A. (1973): Representations of infinite-dimensional unitary groups. Dokl. Akad. Nauk SSSR *212*, No. 2, 288–290. English transl.: Sov. Math. Dokl. *14*, 1355–1358 (1974), Zbl. 288.22020

Kirillov A.A. (1976): Local Lie algebras. Usp. Mat. Nauk *31*, No. 4, 57–76. English transl.: Russ. Math. Surv. *31*, No. 4, 55–75 (1976), Zbl. 352.58014

Kirillov A.A. (1978): Elements of the Theory of Representations. 2nd ed., Moscow, Izdat. Nauka, 344 pp. English translation of the first edition (Moscow, Izdat. Nauka (1972): Heidelberg, Springer-Verlag (1975), 310 pp., Zbl. 342.22001

Kirillov A.A. (1980): Invariant Operators on Geometrical Quantities. Itogi Nauki Tekh., Ser. Sovrem. Probl. Mat. *16*, 3–29. English transl.: J. Sov. Math. *18*, 1–21 (1982), Zbl. 453.53008

Kirillov A.A. (1982): Infinite-dimensional Lie groups: their orbits, invariants and representations. Geometry of moments. In: Doebner H.D., Palev T.D. (eds.): Twistor Geometry and Non-Linear Systems. Heidelberg, Springer-Verlag, Lect. Notes Math. *970*, 101–123, Zbl. 498.22017

Kirillov A.A.(1984): Unitary representations of the group of diffeomorphisms and of some of its subgroups. Sel. Math. Sov. *1*, No. 4, 351–372, Zbl. 515.58009

Kirillov A.A. (ed.) (1985a): Representations of Lie Groups and Lie Algebras. Budapest, Akad. Kiadó, 225 pp., Zbl. 581.00006

Kirillov A.A. (1985b): Geometrical Quantization. Itogi Nauki Tekh., Ser. Sovrem. Probl. Mat., Fundam. Napravleniya *4*, 141–178 (Russian), Zbl. 591.58014

Kirillov A.A., Yur'ev D.V. (1987): Kähler geometry of the infinite-dimensional homogeneous space Diff$_+S^1$/Rot S^1. Funkts. Anal. Prilozh. *21*, No. 4, 35–46. English transl.: Funct. Anal. Appl. *21*, No. 4, 284–294 (1987), Zbl. 671.58007

Knapp A. (1986): Representation Theory of Semisimple Groups. An Overview Based on Examples. Princeton, Princeton Univ. Press, 790 pp., Zbl. 604.22001

Knapp A., Zuckerman G. (1982): Classification of irreducible tempered representations of semi-simple groups I, II. Ann. Math., II. Ser. *116*, No. 2–3, 389–455, Zbl. 516.22011; 457–501. Correction: Ann. Math., II. Ser. (1984), *119*, No. 3, 639, Zbl. 539.22012

Kostant B. (1959): A formula for the multiplicity of a weight. Trans. Am. Math. Soc. *93*, No. 1, 53–73, Zbl. 131,272

Kostant B. (1970): Quantization and unitary representations. In: Lectures in Modern Analysis and Applications III, Heidelberg, Springer-Verlag, Lect. Notes Math. *170*, 87–208, Zbl. 223.53028

Lang S. (1975): $SL_2(R)$. Reading, Addison-Wesley, 428 pp., Zbl. 311.22001

Langlands R.P. (1966): Eisenstein series. In: Algebraic groups and discontinuous subgroups. Providence, Proc. Symp. Pure Math. *9*, 235–252, Zbl. 204,96

Leznov A.N., Savel'ev M.V. (1985): Group Methods for the Integration of Nonlinear Dynamical Systems. Moscow, Izdat. Nauka, 280 pp. English transl.: Basel, Birkhäuser (1992), Zbl. 667.58020

Lusztig G. (1977): Irreducible representations of finite classical groups. Invent. Math. *43*, 125–175, Zbl. 372.20033

Macdonald I.G. (1979): Symmetric Functions and Hall Polynomials. Oxford, Clarendon Press, 180 pp., Zbl. 487.20007

Mackey G.W. (1952): Induced representation of locally compact groups I. Ann. Math., II. Ser. *55*, No. 1, 101–139, Zbl. 46,116

Mackey G.W. (1953): Induced representations of locally compact groups II. Ann. Math., II. Ser. *58*, No. 2, 193–221, Zbl. 51,19

Mackey G.W. (1963): Infinite-dimensional group representations. Bull. Am. Math. Soc. *69*, No. 5, 628–686, Zbl. 136,115

Mackey G.W. (1980): Harmonic analysis as the exploitation of symmetry – a historical survey. Bull. Am. Math. Soc., New Ser. *3*, No. 1, 543–698, Zbl. 437.43001

Manin Yu.I. (1984): Gauge Fields and Complex Geometry. Moscow, Izdat. Nauka, 336 pp. English transl.: Berlin Heidelberg New York, Springer-Verlag (1988), Zbl. 641.53001

McConnell J.C. (1974): Representation of solvable Lie algebras and the Gel'fand-Kirillov conjecture. Proc. Lond. Math. Soc., III. Ser. *29*, No. 3, 453–484, Zbl. 323.17005

Moore C.C. (ed.) (1973): Harmonic Analysis on Homogeneous Spaces. Proc. Symp. Pure Math. *26*, Zbl. 272.00009

Najmark M.A. (1976): Theory of Group Representations. Moscow, Izdat. Nauka, 560 pp. English transl.: Berlin Heidelberg New York, Springer-Verlag (1982), Zbl. 327.20001

Ol'shanskij G.I. (1969): On the Frobenuis duality theorem. Funkts. Anal. Prilozh. *3*, No. 4, 49–58. English transl.: Funct. Anal. Appl. *3*, 295–302 (1970), Zbl. 234.22012

Ol'shanskij G.I. (1984): Infinite-dimensional classical groups of finite R-rank: the description of representations and asymptotic theory. Funkts. Anal. Prilozh. *18*, No. 1, 28–42. English transl.: Funct. Anal. Appl. *18*, 22–34 (1984), Zbl. 545.22020

Pontryagin L.S. (1977): Topological Groups. 3rd ed., Moscow, Izdat. Nauka, 515 pp. English transl.: New York, Gordon and Breach (1960), 543 pp., Zbl. 265.22001

Pressley A., Segal G. (1986): Loop Groups. Oxford, Clarendon Press, 318 pp., Zbl. 618.22011

Pukanszky L. (1972): Unitary representations of solvable Lie groups. Ann. Sci. Ec. Norm. Supér., IV. Ser. *4*, 457–608, Zbl. 238.22010

Rentschler R., Vergne M. (1973): Sur le semi-centre du corps enveloppant d'une algèbre de Lie. Ann. Sci. Ec. Norm. Supér., IV. Ser. *6*, No. 3, 389–405, Zbl. 293.17007

Ricci F., Weiss G. (eds.) (1982): Harmonic Analysis. Heidelberg, Springer-Verlag, Lect. Notes Math. *908*, 325 pp., Zbl. 471.00014

Rossmann W. (1978): Kirillov's character formula for reductive Lie groups. Invent. Math. *48*, No. 3, 207–220, Zbl. 372.22011

Schmid W. (1975): Some properties of square integrable representations of semisimple Lie groups. Ann. Math., II. Ser. *102*, 535–564, Zbl. 347.22011

Schmid W. (1976): L^2-cohomology and the discrete series. Ann. Math., II. Ser. *103*, 375–394, Zbl. 333.22009

Schur I. (1973): Gesammelte Abhandlungen, Bd. I–III. Heidelberg, Springer, 1465 pp., Zbl. 274.01054

Shchepochkina I.M. (1977): On representations of solvable Lie groups. Funkts. Anal. Prilozh. *11*, No. 2, 93–94. English transl.: Funct. Anal. Appl. *11*, 159–161 (1977), Zbl. 353.22012

Segal I.E. (1950): An extension of Plancherel's formula to separable unimodular locally compact groups. Ann. Math., II. Ser. *52*, No. 2, 272–292, Zbl. 41,363

Serre J.-P. (1967): Représentation Linéaires des Groupes Finis. Paris, Hermann, Collection Méthodes, 170 pp.

Stein E.M. (1967): Analysis in matrix spaces and some new representations of $SL(n, C)$. Ann. Math. *86*, No. 3, 461–495

Stratila S., Voiculesu D. (1975): Representations of AF-algebras and of the Group $U(\infty)$. Heidelberg, Springer-Verlag, Lect. Notes Math. *486*, 169 pp., Zbl. 318.46069

Tadič M. (1985): Proof of a conjecture of Bernstein. Math. Ann. *272*, 11–16, Zbl. 547.22010

Vergne M. (1979): On Rossmann's character formula for discrete series. Invent. Math. *54*, 11–14, Zbl. 428.22010

Vershik A.M., Gel'fand I.M., Graev M.I. (1973): Representations of the group $SL(2, R)$, where R is a ring of functions. Usp. Mat. Nauk *28*, No. 3, 82–128. English transl.: Russ. Math. Surv. *28*, 87–132 (1973), Zbl. 288.22005

Vogan D. (1981): Representations of Real Reductive Groups. Boston, Birkhäuser, 754 pp., Zbl. 469.22012

Vogan D. (1987): Unitary Representations of Reductive Lie Groups. Princeton, Princeton Univ. Press, 700 pp., Zbl. 626.22011

Warner G. (1972): Harmonic Analysis on Semisimple Lie Groups I, II. Heidelberg, Springer-Verlag, 529 pp.; 491 pp., Zbl. 265.22020; Zbl. 265.22021

Weil A. (1940): L'intégration dans les Group Topologiques et ses Applications. Paris, Hermann

Weyl H. (1939): The Classical Groups, their Invariants and Representations. Princeton, Princeton Univ. Press, Zbl. 20,206

Wolf J.A. et al. (eds.) (1980): Harmonic Analysis and Representations of Semisimple Lie Groups. Dordrecht, D. Reidel, 495 pp., Zbl. 433.00001

Young A. (1901): Quantitative substantional analysis, I–VIII. Proc. Lond. Math. Soc., II. Ser. (1901), *33*, No. 1, 97–146, FdM 32,157; (1902), *34*, No. 1, 361–397, FdM 33,158; (1928), *28*, No. 2, 255–292 FdM 54,150; (1930), *31*, No. 2, 253–272; 273–288, FdM 56,135; (1932), *34*, No. 2, 196–230, Zbl. 5,97; (1933), *36*, No. 2, 304–368, Zbl. 8,49; (1934), *37*, No. 2, 441–495, Zbl. 9,301

Zhelobenko D.P. (1970): Compact Lie Groups and their Representations. Moscow, Izdat. Nauka, 664 pp. English transl.: Am. Math. Soc., Transl. Math. Monogr. *40*, (1973), Zbl. 228.22013

Zhelobenko D.P. (1974): Harmonic Analysis on Semisimple Complex Lie Groups. Moscow, Izdat. Nauka, 240 pp. (Russian), Zbl. 341.22001

Zhelobenko D.P., Shtern A.I. (1983): Representations of Lie Groups. Moscow, Izdat. Nauka, 360 pp. (Russian), Zbl. 521.22006

II. Representations of Virasoro and Affine Lie Algebras

Yu. A. Neretin

Translated from the Russian
by V. Souček

Contents

§0. Introduction

The first important results in the theory of representations of infinite-dimensional Lie groups were obtained in the book by Friedrichs "Mathematical problems of quantum field theory " (1953), which inspired whole series of papers on automorphisms of the commutation relations. A summary of this evolution was given in the book by F.A.Berezin "Methods of second quantization" (Berezin (1965)). The basic result cotained therein was the description of explicit formulas for spinor representations of the group $(O(2\infty, \mathbb{R}), U(\infty))$ and for the Weil representation of the group $(Sp(2\infty, \mathbb{R}), U(\infty))$, where the symbols $(O(2\infty, \mathbb{R}), U(\infty))$ (resp. $(Sp(2\infty, \mathbb{R}), U(\infty)))$ denote the group of all orthogonal (resp. symplectic) operators that can be written as a sum of a unitary and a Hilbert-Schmidt operator (for more details see Sect.3).

The group $\mathrm{Diff}\,(M^n)$ of all diffeomorphisms of an n-dimensional compact manifold M^n and the *current group* $C^\infty(M^n, G)$ consisting of all smooth functions on M^n with values in a Lie group G are important examples of infinite-dimensional Lie groups. The corresponding Lie algebras are the Lie algebra $\mathrm{Vect}(M^n)$ of vector fields on M^n, resp. the Lie algebra $C^\infty(M^n, \mathfrak{g})$ of vector fields on M^n with values in the Lie algebra $\mathfrak{g}$ of G. For the past ten years the main interest has been concentrated on the case where the manifold M^n is the circle S^1; this is the case studied here. On the one hand, the case $M^n = S^1$ offers a rich representation theory having various ties with other fields of mathematics, which cannot be, it seems, generalized to higher-dimensional cases. On the other hand, these groups and algebras have important applications in the so-called "string model" of quantum field theory, which offers good hopes for the future. There are two main reasons behind the connection between the representation theory of infinite-dimensional groups and quantum field theory. Firstly, infinite-dimensional groups arise naturally as symmetry groups in quantum field theory (for example, the string model studies objects concentrated along curves – "strings" – and the group $\mathrm{Diff}(S^1)$ appears as the group of all their reparametrizations). Then the quantization of the theory is akin to the problem of the construction of representations of the group. Secondly, both theories cope with difficulties in one common field, namely in the analysis of spaces of functions of an infinite number of variables (at present such an analysis is developed only in two cases – in bosonic and fermionic Fock spaces, see Sect.3).

Let us consider the algebra $\mathrm{Vect}(S^1)$ of vector fields on S^1 in more detail. We shall consider elements $A_n = e^{in\varphi}\frac{\partial}{i\partial\varphi}$ in its complexification $\mathrm{Vect}_C(S^1)$. It is easy to see that

$$[A_n, A_m] = (m - n)A_{n+m}. \qquad (0.1)$$

The Virasoro algebra $\mathcal{L}$ is the (unique) central extension of Vect_C. The basis A_n is extended by a central element z and the relation (0.1) is replaced in the case $m = -n$ by

$$[A_{-n}, A_n] = 2nA_0 + \frac{n^3 - n}{12} z.$$

Representations of the algebra $\mathcal{L}$ coincide with projective representations of the algebra Vect, which, in turn, correspond (up to a few comments) to projective representations of $\mathrm{Diff}(S^1)$. An irreducible module over $\mathcal{L}$ is called a highest weight module if it contains a common eigenvector for A_0 and z annihilated by all "raising operators" $A_1, A_2, A_3, \dots$. The theory of highest weight representations of the Virasoro algebra recalls the theory of highest weight modules of noncompact semisimple groups and is almost complete at present.

Now let $\mathfrak{g}$ be a simple complex Lie algebra with corresponding Killing form $\langle \cdot, \cdot \rangle$ and a Cartan subalgebra $\mathfrak{h}$. Let X_α be a basis such that

$$[X_\alpha, X_\beta] = \sum C_{\alpha\beta}^\gamma X_\gamma.$$

Then the vectors $X_\alpha^{(n)} = X_\alpha e^{in\varphi}$ form a basis in $C^\infty(S^1, \mathfrak{g})$ and their commutation relations are

$$[X_\alpha^{(n)}, X_\beta^{(k)}] = \sum C_{\alpha\beta}^\gamma X_\gamma^{(n+k)}. \tag{0.2}$$

The affine Lie algebra $\widehat{\mathfrak{g}}$ is a central extension of $C^\infty(S^1, \mathfrak{g})$ by an element z, the commutation relations (0.2) are modified by an additional term $n\langle X_\alpha, X_\beta \rangle z$ added in the case $n = -k$.

The elements $X_\alpha^{(n)}$, $n > 0$ and $X_\alpha^{(0)}$, where X_α is a raising operator of $\mathfrak{g}$, are called raising operators of $\widehat{\mathfrak{g}}$. Any eigenvector for $\mathfrak{h} \oplus \mathbb{C}z$ annihilated by all raising operators of $\widehat{\mathfrak{g}}$ is called a highest weight vector. There is a deep analogy between the theory of highest weight modules for $\widehat{\mathfrak{g}}$ and the representation theory of compact groups (they even share the property that there are strong general theorems available, while the answers to specific problems are usually confronted with serious combinatorial difficulties).

In this paper, we do not restrict ourselves only to highest weight representations and we try to reflect the existing diversity of the subject. Complementary facts, not fitting into the main line of reasoning, are described in appendices. Most of the constructions used in the paper fit into one of the following schemes.

1. The imbedding technique. The group G can be imbedded into one of the groups $(\mathrm{Sp}(2\infty, \mathbb{R}), \mathrm{U}(\infty))$ or $(\mathrm{O}(2\infty, \mathbb{R}), \mathrm{U}(\infty))$ and the Weil representation of $(\mathrm{Sp}(2\infty, \mathbb{R}), \mathrm{U}(\infty))$, respectively the spinor representation of $(\mathrm{O}(2\infty, \mathbb{R}), \mathrm{U}(\infty))$ can then be restricted to G (in other words, the group G is realized in the group of automorphisms of the commutation or anticommutations relations). As in the next two schemes, the methods of second quantization are used here and the groups always act on the Fock space (see Sections 3–5 and 8.1, 9.2–9.4).

2. Vertex operators. They are used for the construction of projective representations of the group $C^\infty(S^1, K)$, where K is a compact group. The re-

striction of the representation to a maximal abelian subgroup is a priori very
simple and can well be irreducible. Then there is the problem of constructing
an extension from the subgroup to the original group.

Next two types of constructions were intensively used in the theory of
representations of the groups $\text{Diff}(M^n)$ and $C^\infty(M^n, G)$ in the 1970s.

3. Araki's scheme. We construct an "affine action" of the group G, i.e. an
imbedding of G into the group Isom of affine isometries of a Hilbert space
and then we restrict a special representation of Isom to G. Formally, this is
a degenerate case of the imbedding technique, see A.3.1.

**4. The construction of representations using a quasi-invariant measure
on G.** It is an old and well-known generalization of the induction procedure,
see Sections 1.2–1.5, 3.2, 5.4.

5. (G, H)-pairs. A class of representations of G with the property that the
restriction to the subgroup H has given (a priori guessed) properties is studied.
The majority of attempts to classify representations use this approach, see
Section 3.6 and A.1.

In the present paper, analytic questions of the theory are explained with a
reasonable amount of detail. Applications to the theory of integrable systems
and some applications to physics will be discussed in one of the following
volumes of the Encyclopaedia. The theory of characters and combinatorial
questions related to it as well as the connection with finite sporadic groups
(Fischer-Griess monster group) is not discussed here.

We shall use the notation $G = A \ltimes B$ for the semi-direct product, where A
is a subgroup and B is a normal subgroup of the group G.

§1. The Group of Diffeomorphisms of the Circle

1.0. Notation. Let S^1 be the circle parametrized by $\varphi : 0 \leq \varphi \leq 2\pi$, and
let Diff be the group of all orientation-preserving C^∞-diffeomorphisms of S^1.
It is natural to consider the Lie algebra Vect of vector fields on S^1 as its Lie
algebra. The complexification of Vect will be denoted by Vect_C. The vector
fields

$$Y_n = e^{in\varphi} \frac{\partial}{i\partial\varphi}, \quad n \in \mathbb{Z}$$

form a basis of Vect_C and satisfy the following commutation relations:

$$[Y_n, Y_m] = (m - n)Y_{m+n}. \tag{1.1}$$

1.1. Finite-Dimensional Subgroups of Diff. We shall realize $\text{SL}_2(\mathbb{R})$ as
the group of all matrices of the form $\begin{pmatrix} \alpha & \beta \\ \overline{\beta} & \overline{\alpha} \end{pmatrix}$, where $|\alpha^2| - |\beta^2| = 1$ (it
corresponds to the isomorphism $\text{SL}_2(\mathbb{R}) \simeq \text{SU}(1,1)$, see Zhelobenko and

Shtern (1983)). It is easy to see that Möbius transformations $z \to \frac{\alpha z + \beta}{\bar{\beta} z + \bar{\alpha}}$ preserve the circle $|z| = 1$. The matrix $-\mathbb{1}$ acts as the identity, so we get an imbedding of $PSL_2(\mathbb{R})$ into Diff. The action of $PSL_2(\mathbb{R})$ on S^1 implies that the n-fold covering group of $PSL_2(\mathbb{R})$, denoted by $PSL_2^{(n)}(\mathbb{R})$, acts on the n-fold covering space of S^1 (recall that the fundamental group of $PSL_2(\mathbb{R})$ is isomorphic to $\mathbb{Z}$). However, all finite covering spaces over the circle are homeomorphic to the circle, so we have constructed a family of subgroups in Diff isomorphic to $PSL_2(\mathbb{R})$. The Lie algebra corresponding to the subgroup $PSL_2^{(n)}(\mathbb{R})$ is generated by the vector fields $(Y_n - Y_{-n})/2$, $(Y_n + Y_{-n})/2i$, Y_0.

The imbedding of $SL_2(\mathbb{R}) = PSL_2^{(2)}(\mathbb{R})$ into Diff can be described more easily: the group $SL_2(\mathbb{R})$ acts on the set M of all rays in $\mathbb{R}^2$ emanating from the origin O ($M \simeq S^1$).

1.2. The Covering $\mathrm{Diff}^{(\infty)}$ over Diff. Let us consider the *group* $\mathrm{Diff}^{(\infty)}$ of all diffeomorphisms q of the real line $\mathbb{R}$ satisfying the periodicity condition $q(x + 2\pi) = q(x) + 2\pi$. The center of the group $\mathrm{Diff}^{(\infty)}$ consists of all maps of the form $q_n(x) = x + 2\pi n$. It is clear that the center is isomorphic to $\mathbb{Z}$ and that $\mathrm{Diff}^{(\infty)}/\mathbb{Z} \simeq \mathrm{Diff}$. The inverse image of any subgroup $PSL_2^{(n)}(\mathbb{R})$ in $\mathrm{Diff}^{(\infty)}$ is isomorphic to $PSL_2^{(\infty)}(\mathbb{R})$, where $PSL_2^{(\infty)}(\mathbb{R})$ is the universal covering group of $PSL_2(\mathbb{R})$. Let $\mathrm{Diff}^{(k)}$ denote the k-fold covering group over Diff; clearly $\mathrm{Diff}^{(k)} = \mathrm{Diff}^{(\infty)}/k\mathbb{Z}$.

1.3. The Central Extension of $\mathrm{Diff}^{(\infty)}$. The *Bott cocycle* c (an element of $H^2(\mathrm{Diff}^{(\infty)}, \mathbb{R})$) is defined by the formula

$$c(p, q) = \int_h^{h+2\pi} \ln(p'(q(x))) d\ln(q'(x)),$$

where $p, q \in \mathrm{Diff}^{(\infty)}$. The periodicity of the functions $p'(x)$ and $q'(x)$ implies that the cocycle c does not depend on h. The following formula defines the group structure on $\mathrm{Diff}^{(\infty)} \times \mathbb{R}$:

$$(p_1(x), d_1)(p_2(x), d_2) = (p_2(p_1(x)), d_1 + d_2 + c(p_1, p_2)).$$

The *group* $\mathrm{Diff}^{\sim}$ defined in this way is a nontrivial central extension of the group $\mathrm{Diff}^{(\infty)}$.

The *Virasoro algebra* $\mathcal{L}$ is the algebra over $\mathbb{C}$ with basis L_n ($n \in \mathbb{Z}$) and z and with the commutation relations

$$[L_n, L_m] = (m - n)L_{m+n} + \frac{1}{12}(n^3 - n)\delta_{n,-m}z, \tag{1.2}$$

$$[L_n, z] = 0$$

(where $\delta_{k,l} = 0$ for $k \neq l$, $\delta_{k,k} = 1$). It is easy to see that the element z belongs to the center of $\mathcal{L}$ and $\mathcal{L}/\mathbb{C}z \simeq \mathrm{Vect}_\mathbb{C}$. The algebra $\mathcal{L}$ is the unique nontrivial central extension of Vect_C. It carries a natural grading $\deg L_k = k$, $\deg z = 0$; we shall consider only graded modules over $\mathcal{L}$.

The Lie algebra of the group $\mathrm{Diff}^{\sim}$ is the real subalgebra $\mathcal{L}_R$ of $\mathcal{L}$ generated by the vectors

$$\frac{1}{2i}(L_n + L_{-n}), \quad \frac{1}{2}(L_n - L_{-n}), \quad iz.$$

Let M be a module over $\mathcal{L}$. A symmetric bilinear (resp. hermitian) form $\langle \cdot, \cdot \rangle$ on M is called *invariant* if the condition $\langle L_n v, w \rangle = \langle v, L_{-n} w \rangle$ is satisfied for all n and for all $v, w \in M$. A hermitian form $\langle \cdot, \cdot \rangle$ is invariant if and only if $\langle Pv, w \rangle + \langle v, Pw \rangle = 0$ for all $v, w \in M$, $P \in \mathcal{L}_R$. If there exists an invariant positive definite hermitian form on a module M, then M is called *unitarizable*.

1.4. The Trivial Series of Representations. Let us choose two complex numbers $\alpha, \mu \in \mathbb{C}$. Suppose that the Lie algebra Vect_C acts on the vector space with the basis e_n, $n \in \mathbb{Z}$, by the formula

$$r_{\alpha,\mu}(Y_k)e_n = (n + \alpha + k\mu)e_{n+k}. \tag{1.3}$$

It is clear that the representations $r_{\alpha,\mu}$ and $r_{\alpha,\mu+1}$ are equivalent. Let H_α be the space of all smooth functions on $\mathbb{R}$ satisfying the condition

$$f(x + 2\pi) = e^{2\pi i \alpha} f(x).$$

Let $\mathrm{Diff}^{(\infty)}$ acts on H_α by the formula

$$R_{\alpha,\mu}(q)f(x) = f(q(x))q'(x)^\mu, \tag{1.4}$$

where $q \in \mathrm{Diff}^{(\infty)}$.

It is natural to suppose that the *representations* $R_{\alpha,\mu}$ of $\mathrm{Diff}^{(\infty)}$ correspond to the representations $r_{\alpha,\mu}$ of the algebra Vect and the vectors e_n correspond to functions $e^{i(n+\alpha)\varphi}$. A special role is played by the *representations* $P_{\alpha,s} = R_{\alpha,1+is/2}$ and $T_{\alpha,s} = R_{\alpha,(1+s)/2}$, where $\alpha \in \mathbb{R}$, $s \in \mathbb{R}$.

The representations $P_{\alpha,s}$ are unitary with respect to the scalar product given by the integration over a period:

$$\langle f, g \rangle = \int_0^{2\pi} f(x)g(x)dx.$$

Example. Let $\alpha = 0$. Then the representation $R_{0,\mu}$ is a singlevalued representation of the group Diff and is realized naturally on the space of functions on the circle:

$$R_{0,\mu}f(\varphi) = f(q(\varphi))q'(x)^\mu.$$

It is clear that this is just the natural action of Diff on tensor densities of weight μ.

Appendices

A.1.1. Problems of a Type of Invariant Theory.

The adjoint action. Let the group Diff act naturally on the space of all smooth vector fields on S^1. The orbit of the vector field $a(\varphi)\partial/\partial\varphi$ has a finite codimension if and only if $a(\varphi)$ has finite number of zeros and each of them has finite multiplicity. The following set of invariants distinguishes such orbits:

1. The number of zeros and their multiplicies.
2. The residues of the forms $d\varphi/a(\varphi)$ at the zeros of a (finite pieces of the Laurent series of $a(\varphi)^{-1}$ are well-defined for $a(\varphi) \in C^\infty(S^1)$).
3. The integral $I = \int_0^{2\pi} a(\varphi)^{-1}d\varphi$ regularized in the following sense. Suppose that $\varphi_j, j = 1,\ldots,p$, are the zeros of the function $a(\varphi)$ and k_j their multiplicities. Let

$$b_j(z) = \sum_{n=0}^{2k_j} \frac{a^{(n)}(\varphi_j)}{n!}(z - \varphi_j)^n,$$

where $z \in \mathbb{C}$. Then

$$I = \lim_{\varepsilon \to 0^+} \left(\int_{[0,2\pi]\setminus\cup_j[\varphi_j-\varepsilon,\varphi_j+\varepsilon]} a(\varphi)^{-1}d\varphi + \sum_j \int_{\gamma_j} b_j(z)^{-1}dz \right),$$

where γ_j is the curve parametrized by $\gamma_j(\theta) = \varphi_j - \varepsilon e^{i\theta}, \theta \in [0, \pi]$.

The coadjoint action. Let the group Diff act on the quadratic differentials of the form $p(\varphi)d\varphi^2$. Let $\varphi_1,\ldots,\varphi_n$ be the zeros of $p(\varphi)$. Then the following list of invariants distinguishes orbits of finite codimension:

1. The multiplicities of the zeros $\varphi_1, \varphi_2, \ldots, \varphi_n$,
2. $\int_{\varphi_k}^{\varphi_{k+1}} \sqrt{|p(\varphi)|}$ (the "length" of the segment $[\varphi_k, \varphi_{k+1}]$),
3. $\text{sgn}\,(p(\varphi))$ for $\varphi_1 \leq \varphi \leq \varphi_2$.

The structure of the group Diff. This structure is very complicated. In particular, the problem of characterization of conjugacy classes contains the classical problem of reduction of a diffeomorphism without fixed points to a rotation as a subproblem, see Arnol'd and Il'yashenko (1985). The image of Vect in Diff under the exponential map generates a nowhere dense normal subgroup of Diff.

A.1.2. Representations of Diff of Finite Functional Dimension. Let

$D(\varphi_1,\ldots,\varphi_k)$ be the stabilizer of the points $\varphi_1,\ldots,\varphi_k \in S^1$. Consider the group $A(\varphi_1,l_1,\ldots,\varphi_k,l_k)$ of all $q \in D(\varphi_1,\ldots,\varphi_k)$ having the same l_j-jet at the points φ_j as the identity map. Then the corresponding quotient group $D(\varphi_1,\ldots,\varphi_k)/A(\varphi_1,l_1\ldots,\varphi_k,l_k)$ is a finite-dimensional solvable Lie group. Let us now introduce a class K of unitary representations of the group Diff.

It consists of all representations induced from unitary representations of the subgroups $D(\varphi_1, \ldots, \varphi_k)$ which are trivial on $A(\varphi_1, l_1, \ldots, \varphi_k, l_k)$ (note that $\text{Diff}/A(\varphi_1, l_1 \ldots, \varphi_k, l_k)$ is a finite-dimensional manifold, so the induction procedure is well defined). The class K includes, in particular, all finite tensor products of representations of the trivial series. The other representations of the class K can clearly be obtained as limits of such tensor products.

Orbits and representations. (Kirillov (1974)) For finite-dimensional Lie groups, it is well-known that there is a close connection between unitary representations and orbits of the coadjoint representation. Let the group Diff act on the space of quadratic differentials of the form $p(\varphi)d\varphi^2$, where $p(\varphi)$ is a distribution. The orbit going through $q(\varphi)d\varphi$ is finite-dimensional if and only if $q(\varphi)$ is a finite sum of δ-functions and its derivatives. Such orbits correspond to representations of the class K similarly as was the case in the finite-dimensional theory.

It would be quite natural to consider infinite tensor products of representations of the trivial series, but up to now, only a few first steps has been made toward a probably substantial general scheme. Examples of countable tensor products can be found in A.1.4, some continuous tensor products are realized in 5.4.

A.1.3. Dynamical Systems and Group Representations. Let the group G act on a space Ω with measure μ in such a way that μ is quasi-invariant with respect to G. A series T_ω of unitary representations of G on $L^2(\Omega)$ is given by the following formula

$$T_\omega(g)f(x) = f(gx)[d\mu(gx)/d\mu(x)]^{\frac{1}{2}+i\omega}, \qquad (1.5)$$

where $g \in G$, $\omega \in \mathbb{R}$, $x \in \Omega$ and the derivative in the bracket means the Radon-Nikodým derivative of the transformation g.

A generalization. Let H be the space of L^2-functions on Ω with values in a Hilbert space H_0, and let $g \in G$. We shall consider the operators

$$T(g)f(x) = f(gx)a(g,x)[d\mu(gx)/d\mu(x)]^{\frac{1}{2}} \qquad (1.6)$$

on H, where $a(g,x)$ is a function on $G \times \Omega$ with values in the space of unitary operators, satisfying the relation

$$a(g_2 g_1, x) = a(g_1, x)a(g_2, g_1 x).$$

Then $T(g)$ is a unitary representation of G.

Remark. To make sure that the representations $T_\omega(g)$ and $T(g)$ are continuous, it is necessary to add some supplementary conditions concerning the action of G on Ω and properties of the function $a(g, x)$.

Remark. If Ω is a homogeneous space and if G is finite-dimensional (often even in the case of infinite-dimensional groups), then the described construction is equivalent to unitary induction.

Thus, to construct representations of G, it is useful to look for ergodic quasi-invariant measures; constructions of representations along these lines can be found in 1.4., A.1.2., A.1.4., A.3.2. and 5.4.

A.1.4. Quasi-Invariant Measures with Respect to $\mathrm{Diff}(M^n)$.

Poisson measures. (A.M.Vershik, I.M.Gel'fand, M.I.Graev, R.S.Ismagilov.) Let M^n be a noncompact smooth manifold with volume form ω and let μ be the corresponding Lebesgue measure on M^n (for U open, $\mu(U)$ is defined as the integral of ω over U). Suppose that $\mu(M^n) = \infty$.

Let Γ be the space of all discrete countable subsets of the manifold M^n. An "event" $A_n(U)$ (i.e. a subset of Γ) is defined by the property that a set $\kappa \in \Gamma$ belongs to $A_n(U)$ if $\kappa \cap U$ has exactly n points. The Poisson measure ν on γ is then defined by the condition

$$\nu(A_n(U)) = \mu(U)^n/n!$$

Let us consider the natural action of the group $\mathrm{Diff}_c(M^n)$ of diffeomorphisms of M^n with compact support on Γ. Then the Poisson measure ν is quasi-invariant and its Radon-Nikodým derivative is

$$d\nu(g\kappa)/d\nu(\kappa) = \prod_{x \in \kappa}(d\mu(gx)/d\mu(x)).$$

We can construct many unitary irreducible representations of $\mathrm{Diff}_c(M^n)$ using the scheme A.1.3; they are, in principle, obtained by the operation of induction.

Measures on the space of convergent sequences. (Ismagilov (1971)) Suppose that $M^n = S^1$ and that δ_k is a sequence of positive numbers such that $\sum \delta_k < \infty$. Let us consider a sequence x_k of independent random variables in $[-\pi, \pi]$ described by their distributions

$$p_k = \frac{\delta_k}{2\pi^{\delta_k}}\frac{dx}{|x|^{1-\delta_k}}, \quad p_0 = \frac{dx}{2\pi}.$$

The circle S^1 will be realized as $\mathbb{R}/2\pi\mathbb{Z}$ and we define a set of new random variables $y_k = \sum_{j \le k} x_j$. Then a probability measure arises on the set Ω of all sequences $\{y_k\}$ of points in S^1; we shall denote it by μ. The condition $\sum \delta_k < \infty$ implies that the series $\sum Dx_k$ is convergent, where Dx_k is the dispersion of the random variable x_k. The Kolmogorov-Khinchin theorem on series of random variables implies that almost all sequences $\{y_k\}$ (with respect to the measure μ) are convergent. It is not difficult to show that the measure μ is quasi-invariant with respect to Diff (roughly: in a small neighbourhood of the point $\lim y_n$, the diffeomorphism looks like a dilation and the density $p_k(x)$ is approximately equal to the density dx/x, which is invariant with respect to dilations).

Let $\widehat{\Omega}$ be the space of all countable unordered subsets of the circle S^1 and let $p : \Omega \to \widehat{\Omega}$ be the natural projection and $\widehat{\mu} = p(\mu)$. Then the measure $\widehat{\mu}$ is quasi-invariant and the representation constructed using formula (1.5) for $\omega = 0$ is irreducible.

Remark. For examples of Diff-quasi-invariant measures on the space of the closed noncountable subsets of the S^1 see Neretin (1993). This paper contains many other examples of quasi-invariant measures.

Shavgulidze's measure. Let us consider the group $\mathrm{Diff}_2'[0,1]$ of all diffeomorphisms of the interval $[0,1]$ that are twice continuously differentiable and satisfy the condition $\varphi''(0) = 0$. Let $C_0[0,1]$ denote the space of continuous functions on $[0,1]$ such that $f(0) = 0$. We shall consider the Wiener measure on $C_0[0,1]$ and we shall introduce the one-to-one map A from $\mathrm{Diff}_2'[0,1]$ to $C_0[0,1]$ given by the formula

$$A(\varphi) = \varphi''/\varphi' = (\ln \varphi')'.$$

The probability measure induced on $\mathrm{Diff}_2'[0,1]$ by the map A will be denoted by μ.

The measure μ on $\mathrm{Diff}_2'[0,1]$ is quasi-invariant with respect to the left translations $T_\alpha : \varphi \to \alpha(\varphi)$ by a three times continuously differentiable diffeomorphism.

Remark. This construction is the partial case of the multidimensional Shavgulidze construction, see Khafizov (1990), Shavgulidze (1988).

Another example of an invariant measure with respect to the group Diff is described in 5.4.

A.1.5. Representations of the Class I. Let $D[a,b]$ be the group of diffeomorphisms of S^1 which are the identity on the segment $[a,b]$, $a \neq b$. We say that a unitary irreducible representation ρ of the group Diff is of class I if the restriction of ρ to a subgroup $D[a,b]$ (hence to any such subgroup) contains an invariant vector.

All representations of Diff considered up to now belong to class I. For all representations considered below, the question of whether they belong the class I is still open.

We shall describe now an attempt to classify representations belonging to class I made in Ismagilov (1971). We shall show that any such representation ρ can be realized using the scheme A.1.3 (the formula (1.6)), where the space Ω is the space of all closed subsets of S^1 with Borel measure μ_ρ (the Borel structure on Ω is generated by subsets $\Omega_F \subset \Omega$, where Ω_F is the set of all closed subsets of $F(F \in \Omega)$).

Let us choose a representation $\rho \in \mathrm{I}$ acting on a space H. Let F be a closed subset of S^1, let $D(F)$ be the group of all diffeomorphisms of S^1 which are the identity on a neighbourhood of F, let H_F be the set of all $D(F)$-invariant vectors in H and let P_F be the projection on H_F.

Lemma 1.1.

A) *If $F_1 \supset F_2$, then $P_{F_1} P_{F_2} = P_{F_2}$.*
B) *If $F_1 \supset F_2 \supset \ldots$ and $F = \cap F_j$, then $P_F = P_{F_1} P_{F_2} \ldots$ (the proof is trivial).*

Lemma 1.2. *Let D be the group of diffeomorphisms of the interval I which are the identity near the end points, endowed with the topology of inductive limit. Let D_0 be its subgroup consisting of all diffeomorphisms which are identities in a neighbourhood of a point $c \in I$. Then for any $q \in D$, there exist $a_k, b_k \in D_0$ and $w_k \in D$ such that we have $\lim_{k \to \infty} a_k w_k b_k = q$ and $\lim_{k \to \infty} w_k = 1$.*

Corollary 1.1. *If a continuous function f on D is constant on all two-sided equivalence classes with respect to D_0, then $f = $ const.*

Lemma 1.3. *Let c be a point; then $P_{F \cup c} = P_F$.*

Proof. Let $v \in H_{F \cup c}$. Then $\langle \rho(g)v, v \rangle = f(g)$ is a function on $D(F)$ which is constant on two-sided equivalence classes with respect to $D(F \cup c)$.

As a consequence, we get

Lemma 1.4. *If F is closed, if K is a union of a finite number of closed intervals and if K' is the closure of $S^1 \backslash K$, then $P_{F \cup K} P_{F \cup K'} = P_F$.*

Lemma 1.5. $P_{F_1} P_{F_2} = P_{F_1 \cap F_2}$.

Proof. Lemma 1.1.B implies that the proof reduces to the case where F_2 is union of a finite number of closed intervals (any closed set is an intersection of such subsets). Suppose that F_2' is the closure of $S^1 \backslash F_2$. Lemma 1.1.A and Lemma 1.4 then imply

$$P_{F_2} P_{F_1} = P_{F_2}(P_{F_1 \cup F_2} P_{F_1 \cup F_2'}) = P_{F_2} P_{F_1 \cup F_2'},$$
$$P_{F_1 \cap F_2} = P_{(F_1 \cap F_2) \cup F_2} P_{(F_1 \cap F_2) \cup F_2'} = P_{F_2} P_{F_1 \cup F_2'}.$$

Now let R be the closure (in the norm topology) of the commutative algebra of operators in H generated by all projections H_F. Using the Gel'fand transform, the algebra R is realized as an algebra of continuous functions on a compact space $\mathcal{M}$ with measure ν and the space H is the space $L^2(\mathcal{M}, \mu)$ of vector-valued functions. The representation ρ has the form (1.6).

Let χ_F be the function on $\mathcal{M}$ corresponding to the operator P_F. We have clearly $\chi_F^2 = \chi_F$, so χ_F is the characteristic function of an open and closed set F^*. We shall now construct a map $\Phi : \mathcal{M} \to \Omega$. If $\omega \in \mathcal{M}$, then $\Phi(\omega)$ is the intersection of all closed subsets of F such that $\omega \in F^*$. It induces a probability measure μ_ρ on Ω, so we have constructed a realization of ρ on the space of vector-valued functions on Ω.

Suppose now that the measure μ on Ω is concentrated on n-point subsets of S^1. This implies that if the representation ρ has a dense Gårding subspace, then ρ belongs to the class K. This can be considered as a positive answer, in the case of representations of class I, to the problem given in Conjecture 1 in A.1.2. An intermediate result is of independent interest: any irreducible representation of the group $D(\varphi_1, \ldots, \varphi_k)$ (see A.1.2) is trivial on a subgroup of the form $A(\varphi_1, l_1, \ldots, \varphi_k, l_k)$.

The above considerations can be transferred to arbitrary compact manifolds (the main additional complications appear in Lemma 1.2). Also, our construction is not far from an equivalence between the class I and a certain family of dynamical systems on Ω (see Ismagilov (1971) and further papers by him).

§2. Verma Modules over the Virasoro Algebra

A substantial part of this Section (2.2–2.4) is used only in 5.2.

2.0. Definitions. We shall consider the following subalgebras in the Virasoro algebra $\mathcal{L}$ (they are analogues of upper and lower triangular subalgebras of semisimple Lie algebras):

$$N_+ = \{L_n, \; n > 0\} \quad B_+ = \{z, \; L_n, \; n \geq 0\},$$
$$N_- = \{L_n, \; n < 0\} \quad B_- = \{z, \; L_n, n \leq 0\},$$

where the brackets $\{\;\}$ denote the span of the corresponding elements. Let $\chi_{h,c}$ be the character $B_- \to \mathbb{C}$ given by the conditions $\chi_{h,c}(L_n) = 0$ for $n < 0$, $\chi_{h,c}(z) = c$, $\chi_{h,c}(L_0) = h$. The symbol $U(\mathfrak{g})$ denotes, as usual, the universal enveloping algebra of the Lie algebra $\mathfrak{g}$.

Definition.

1. Let M be a graded module over $\mathcal{L}$. A vector v is called a *singular vector* of weight (h, c) if $N_- v = 0$, $L_0 v = hv$, $zv = cv$.
2. A singular vector v of weight (h, c) in a module M is called a *highest weight vector* if v is a cyclic vector. i.e. $U(N_+)v = M$. The module M is called a *module with highest weight* (h, c). The number c is called the *level of the module M*.
3. The *Verma module* $M(h, c)$ is the $\mathcal{L}$-module induced by the character $\chi_{h,c}$ of the subalgebra B_-. In other words, $M(h, c) = U(\mathcal{L}) \otimes_{U(B_-)} \mathbb{C}$, where $\mathbb{C}$ is the $U(B_-)$-module defined by the character $\chi_{h,c}$ (then $1 \otimes 1$ is the highest weight vector of weight (h, c) in $M(h, c)$.)
4. Let $L(h, c)$ be the unique irreducible module with highest weight (h, c). If $L(h, c) = M(h, c)$, then the module is called *nonsingular*.

2.1. Properties of Verma Modules. The following properties of Verma modules are evident, see also Feigin and Fuks (1982).

α) Let v be the highest weight vector of $M(h,c)$. Then vectors of the form

$$L_1^{k_1} L_2^{k_2} \ldots L_n^{k_n} v, \quad \text{where } k_j \geq 0, \tag{2.1}$$

form a basis of $M(h,c)$. The grading of $M(h,c)$ is given by the relation $\deg (L_1^k \ldots L_n^{k_n} v) = k_1 + 2k_2 + \ldots + n k_n$. The dimension of the m-th homogeneous component is equal to $p(m)$ (i.e to the number of corresponding partitions of the number m). If $q \in U(N_+)$, $w \in M(h,c)$, $q \neq 0$, $w \neq 0$, then $qw \neq 0$. The module $M(h,c)$ is isomorphic, as a vector space, to $U(N_+)$. The element z acts on $M(h,c)$ as a scalar operator c.

β) Any module with highest weight (h,c) is a quotient module of $M(h,c)$. In any highest weight module M, there exists a unique maximal submodule M_0 and then $M/M_0 = L(h,c)$.

γ) Any submodule of a highest weight module M contains a singular vector. Conversely, let w be a singular vector of M, then $U(N_+)w$ is a submodule of M. If $M = M(h,c)$; then $U(N_+)w \simeq M(h+l,c)$, where $l = \deg (w)$. In particular, any morphism of Verma modules is either trivial or injective.

δ) Any two submodules of $M(h,c)$ have nontrivial intersection (this follows from growth estimates for $p(n)$).

ε) For any highest weight module M, there exists a unique (up to a scalar multiple) nonzero invariant bilinear form $\langle \cdot, \cdot \rangle$ (*the Shapovalov bilinear form*). The invariance properties of the form imply that the scalar product $\langle L_1^{k_1} \ldots L_n^{k_n} v, L_1^{l_1} \ldots L_n^{l_n} v \rangle$ is equal to the coefficient λ in the expression $L_{-n}^{k_n} \ldots L_{-1}^{k_1} L_1^{l_1} \ldots L_n^{l_n} v = \lambda v$ (see Feigin and Fuks (1984)). The kernel of the form is the maximal submodule of M. The homogeneous components of M are pairwise orthogonal. The Shapovalov form is nonsingular on $L(h,c)$.

ζ) If $h \in \mathbb{R}$, $c \in \mathbb{R}$, then there exists a unique (up to a scalar multiple) nonzero invariant hermitian form on $M(h,c)$ (*the Shapovalov hermitian form*). In the basis (2.1), the matrices of the Shapovalov hermitian and bilinear forms coincide.

η) There exists a unitary invariant structure on $L(h,c)$ if and only if the Shapovalov hermitian form is positive definite.

2.2. Determinants of the Shapovalov Form. Let $D_n(h,c)$ denote the determinant of the Shapovalov form restricted to the n-th homogeneous component of $M(h,c)$ in the basis (2.1).

Theorem 2.1. (V.G.Kac, B.L.Feigin, D.B.Fuks, see Feigin and Fuks (1982))
We have

$$D_n(h,c) = A \prod_{\alpha\beta \leq n, \, 0 < \alpha \leq \beta} \Phi_{\alpha,\beta}^{p(n-\alpha\beta)}, \tag{2.2}$$

where $A > 0$ and

$$\Phi_{\alpha,\beta}(h,c) = \begin{cases} \left(h + \frac{c-13}{24}(\beta^2 - 1) + \frac{1}{2}(\alpha\beta - 1)\right) \times \\ \times \left(h + \frac{c-13}{24}(\alpha^2 - 1) + \frac{1}{2}(\alpha\beta - 1)\right) + \\ + \frac{1}{16}(\alpha^2 - \beta^2)^2, & \text{if } \alpha \neq \beta; \\ h + \frac{c-1}{24}(\alpha^2 - 1), & \text{if } \alpha = \beta. \end{cases} \qquad (2.3)$$

Corollary.

1. *There exists a singular vector of degree less or equal to k in $M(h,c)$ if and only if there exist positive integers α, β such that $\alpha\beta \leq k$, $\Phi_{\alpha,\beta}(h,c) = 0$.*
2. *There exists a submodule of $M(h,c)$ isomorphic to $M(h+l,c)$ if and only if there exist nonnegative integers*

$$0 = l_0 < l_1 < \ldots < l_m = l, \ \alpha_1, \ldots, \alpha_m, \ \beta_1, \ldots, \beta_m$$

such that

$$\alpha_j \beta_j = l_j - l_{j-1} \quad \text{and} \quad \Phi_{\alpha_j,\beta_j}(h + l_{j-1}, c) = 0.$$

More precise information can be found in 2.4.

2.3. Properties of Quadrics. The net of quadrics $\Phi_{\alpha,\beta}(h,c) = 0$ in the plane is itself a beautiful geometrical object. We shall only discuss some of its properties. We shall extend the definition of $\Phi_{\alpha,\beta}$, by the same formula (2.3) to all integral values of α, β. We clearly have

$$\Phi_{\alpha,\beta} = \Phi_{\beta,\alpha} = \Phi_{-\alpha,-\beta} = \Phi_{-\beta,-\alpha}.$$

α) If $\Phi_{-\alpha,\beta}(h,c) = 0$, $\alpha > 0$, $\beta > 0$, then $\Phi_{\alpha,\beta}(h - \alpha\beta, c) = 0$. So $\Phi_{-\alpha,\beta}(h,c) = 0$ implies that $M(h,c) \subset M(h - \alpha\beta, c)$.

β) (Duality) $\Phi_{\alpha,\beta}(h,c) = \Phi_{-\alpha,\beta}(1 - h, 26 - c)$.

γ) The line $\Phi_{\alpha,\alpha} = 0$ passes through the point $(0,1)$.

δ) The quadric $\Phi_{\alpha,\beta}, \alpha \neq \pm\beta$ is a hyperbola. One branch of it lies below the line $c = 1$ and touches it in the point $h = (\alpha - \beta)^2/4, c = 1$; the other branch lies above its tangent $c = 26$. Its asymptotes are parallel to the lines $\Phi_{\alpha,\alpha} = 0$ and $\Phi_{\beta,\beta} = 0$ (see Fig. 1).

ε) The line $\Phi_{0,0} = 0$ is a common tangent to all hyperbolas $\Phi_{\alpha,\beta} = 0$.

ζ) All intersection points of the quadrics $\Phi_{\alpha,\beta} = 0$ and $\Phi_{\gamma,\delta} = 0$ lie in the real plane ($h \in \mathbb{R}$, $c \in \mathbb{R}$) and are given by the formulas

$$c_{1,2} = 1 - \frac{6((\alpha \pm \gamma) - (\beta \pm \delta))^2}{(\alpha \pm \gamma)(\beta \pm \delta)},$$

$$h_{1,2} = \frac{(\gamma\beta - \alpha\delta)^2 - ((\alpha \pm \gamma) - (\beta \pm \delta))^2}{4(\alpha \pm \gamma)(\beta \pm \delta)}.$$

Another pair of solutions is obtained by permuting γ and δ.

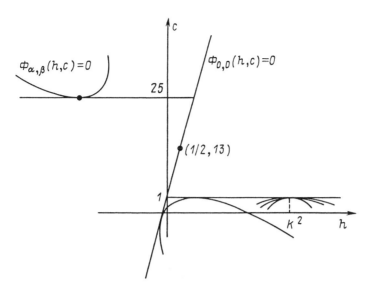

Fig. 1

η) If two different quadrics $\Phi_{\alpha,\beta} = 0$ pass through the point (h_0, c_0), then there are an infinite number of quadrics going through this point. All pairs of integers (α, β) satisfying the equation $\Phi_{\alpha,\beta}(h_0, c_0) = 0$ lie on four lines in $\mathbb{R}^2$ (these lines are permuted by maps of the following form: $(\alpha, \beta) \to (-\beta, -\alpha)$ and $(\alpha, \beta) \to (\beta, \alpha)$).

θ) Let $\alpha > 0$, $\beta > 0$. Then the family $\Phi_{\alpha,\beta} = 0$ is dense in the domain $\{c \le 1, h \ge \frac{c-1}{24}\}$ and locally finite in $\{c > 25, h < \frac{c-1}{24}\}$. In the strip $1 < c < 25$, our family consists of a locally finite set of segments of the lines $\Phi_{\alpha,\alpha} = 0$. No part of the quadrics $\Phi_{\alpha,\beta} = 0$, $\alpha > 0$, $\beta > 0$ lies in $\{c > 1, h > 0\}$ and $\{c < 1, h < \frac{c-1}{24}\}$. For $\alpha < 0$, $\beta > 0$ we have the dual property.

Corollary.

1.   *The Verma modules are irreducible for the following values of (h, c) :*

$$\{h > 0,\ c > 1\}, \quad \left\{c < 1,\ h < \frac{c-1}{24}\right\}.$$

2.   *If (h, c) belongs to one of the sets*

$$\{c < 25,\ h < 1\}, \quad \left\{c > 25, h > \frac{c-1}{24}\right\},$$

then the Verma module $M(h, c)$ is not a proper submodule of any other Verma module.

2.4. The Structure of Verma Modules. Let us consider the Verma module $M(h, c)$.

If $\Phi_{\alpha,\beta}(h, c) \neq 0$ for all α, β, then the module $M(h, c)$ is irreducible and is not a submodule of any other Verma module.

If there exists exactly one pair α, β such that $\Phi_{\alpha,\beta}(h, c) = 0$, then there are three possibilities:

1) If $\alpha > 0$ and $\beta > 0$, then the module $M(h, c)$ contains exactly one submodule $M(h + \alpha\beta, c)$.
2) If $\alpha < 0$ and $\beta > 0$, then $M(h, c)$ is a submodule of exactly one module $M(h + \alpha\beta, c)$.
3) If $\alpha = 0$ or $\beta = 0$, then $M(h, c)$ is irreducible and it is not a submodule of any other Verma module.

A more interesting case is the next one in which an infinite number of quadrics $\Phi_{\alpha,\beta} = 0$ pass through a point (h, c) (see 2.3. η).

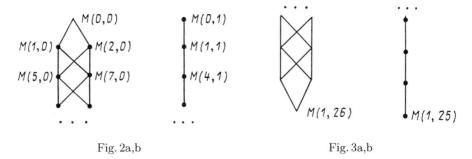

Fig. 2a,b   Fig. 3a,b

Example 2.1. The module $M(0, 0)$ contains all modules which have the form $M\left(\frac{3k^2 \pm k}{2}, 0\right)$. The structure of submodules is shown in Fig. 2a).

Example 2.2. The module $M(0, 1)$ contains all modules which have the form $M(k^2, 1)$, $k \in \mathbb{Z}$. The structure of submodules is shown in Fig. 2b).

Example 2.3. The module $M(1, 26)$ is contained in any module of the form $M\left(1 - \frac{3k^2 \pm k}{2}, 26\right)$. The structure of modules containing $M(1, 26)$ is shown in Fig. 3a).

Example 2.4. The module $M(1, 25)$ is contained in all modules of the form $M(1 - k^2, 25)$. The structure of modules containing $M(1, 25)$ is shown in Fig. 3b).

Theorem 2.2. (Feigin, Fuks)

1) *Let a point (h, c) belong to an infinite number of quadrics $\Phi_{\alpha,\beta} = 0$. Then*
   a) *If $c \leq 1$, then $M(h, c)$ belongs to one of the maximal chains of type 2a) or 2b) (see Fig. 2).*
   b) *If $c \geq 26$, then $M(h, c)$ belongs to one of the maximal chains of type 3a) or 3b).*
   If (h, c) belongs to at least one quadric of type $\Phi_{\alpha,0} = 0$, then one of the cases 2b) or 3b) occurs. In the opposite case, one of the cases 2a) or 3b) occurs. There are no nontrivial morphisms among modules belonging to different chains.
2) $\dim \operatorname{Hom}(M(h_1, c), M(h_2, c)) \leq 1.$
3) *Any submodule of $M(h, c)$ is generated by its singular vectors.*

Remarks.

a) Let a module $M(h, c)$ belong to a straight chain (of type 2b) or 3b)); then for any singular vector of order n lying in it, there exist α, β such that $\alpha\beta = n$, $\Phi_{\alpha,\beta}(h, c) = 0$.
b) Let $M(h, c)$ belong to a branched chain and suppose, for definitness, that $c < 1$. Then the same remark as above is true "over one floor", i.e. it is true for those points of Fig. 2a) having the property that it is necessary to go through an odd number of edges on the way to the top point.

Using these remarks and Theorem 2.2, it is not difficult to find the chain of submodules (or the chain of modules containing it) for any module $M(h, c)$.

On the proof of Theorem 2.2. The computation of the degrees of singular vectors is a mechanical but tiresome task. The main difficulty here is to estimate on the order of zeros of $D_n(h, c)$.

Corollary. (Duality.) If $M(h, c) \supset M(r, c)$, then

$$M(1 - r, 26 - c) \supset M(1 - h, 26 - c).$$

§3. Methods of Second Quantization

In 3.1–3.6, the main technical tools used in the paper are discussed.

3.1. Infinite-Dimensional (G, K)-Pairs.

3.1.1. It is not as easy to find a suitable infinite-dimensional analogue of the groups $U(n)$, $O(n)$, $Sp(n)$ etc. as it might appear. We shall present the approach developed in Ol'shanskij (1983), see also Ol'shanskij (1990), where the reader can also find a discussion of a connection of this theory and other approaches to infinite dimensional groups.

Let U(∞), O(∞, $\mathbb{R}$), GL(∞) etc. denote respectively the full unitary, full orthogonal, full general linear, etc. groups on a Hilbert space.

Definition. Let $G \supset K$ be two groups of the mentioned types. Then the symbol (G, K) denotes the subgroup of G consisting of all operators having the form $A(1 + T)$ where $A \in K$ and T is a Hilbert-Schmidt operator.

3.1.2. Example. Let H be a Hilbert space with scalar product $\langle \cdot, \cdot \rangle$. Then U($\infty$) (Sp($2\infty$, $\mathbb{R}$), resp. O(2∞, $\mathbb{R}$)) is the group of all operators on H preserving $\langle \cdot, \cdot \rangle$ (Im $\langle \cdot, \cdot \rangle$, resp. Re $\langle \cdot, \cdot \rangle$). This leads to the definition of the groups (Sp(2∞, $\mathbb{R}$), U(∞)) and (O(2∞, $\mathbb{R}$), U(∞)).

3.2. The Bosonic Fock Space.

3.2.1. Definition. Let H be a complex Hilbert space with scalar product $\langle \cdot, \cdot \rangle$. The *bosonic Fock space* $F(H)$ or simply F is the space of all holomorphic functions on H with scalar product

$$\{f, g\} = \int f(z)\overline{g(z)}\exp\left(-\langle z, z \rangle\right) \prod \frac{dz_i d\overline{z_i}}{\pi}. \tag{3.1}$$

The function $f(z) = 1$ is called the *vacuum vector*. The space H will often be identified with l_C^2 and an element $z = (z_1, z_2, \ldots)$ will then be considered as a row vector.

Remark. Integration with respect to the Gauss measure in (3.1) is well defined (see Berezin (1965)). Nevertheless, it is more suitable to treat (3.1) as a formal expression: if $H = \mathbb{C}^n$, everything is well defined and it is then sufficient to consider the completion of $\lim_{n\to\infty} F(\mathbb{C}^n) = \cup F(\mathbb{C}^n)$.

3.2.2. All bounded and many unbounded operators on F can be represented in the form

$$Af(z) = \int K(z, \overline{u})f(u)\exp\left(-\langle u, u \rangle\right) \prod \frac{du_i d\overline{u_i}}{\pi}, \tag{3.2}$$

where the *kernel* $K(z, \overline{u})$ is a function on $H \times H$ which is holomorphic in z and antiholomorphic in u (see Berezin (1965)), we then have

$$K(v, \overline{u}) = \{\exp\left(\langle z, u \rangle\right), A\exp\left(\langle z, v \rangle\right)\}. \tag{3.3}$$

3.2.3. The *creation and annihilation operators* in F are given by the formulas

$$\widehat{a}_k f(z) = z_k f(z), \quad \widehat{a}_k^* f(z) = \frac{\partial}{\partial z_k} f(z). \tag{3.4}$$

It is easy to see that the operator $\widehat{a}_k^*$ is really the adjoint of $\widehat{a}_k$. Given an element $\gamma = (\alpha, \beta) \in l_C^2 \oplus l_C^2$, we introduce the operator

$$\widehat{a}(\gamma) = \sum \alpha_j \widehat{a}_j + \sum \beta_j \widehat{a}_j^*. \tag{3.5}$$

It is not difficult to verify that

$$[\widehat{a}_k, \widehat{a}_l^*] = [\widehat{a}_k, \widehat{a}_l] = [\widehat{a}_k^*, \widehat{a}_l^*] = 0, \ k \neq l$$
$$[\widehat{a}_k^*, \widehat{a}_k] = 1.$$

(3.6)

The real Lie algebra with the basis $r, p_i, q_i \, (1 \leq i < n + 1)$ and relations $[p_i, q_i] = r$ (all other commutators being trivial) is called the *Heisenberg algebra* $\mathrm{Heis}_n \, (1 \leq n \leq \infty)$. There is an equivalent definition: if H is a complex n-dimensional Hilbert space, then Heis_n is defined as the vector space $H \oplus \mathbb{R}$ with bracket given by

$$[(u, \alpha), (v, \beta)] = (0, \mathrm{Im} \, \langle u, v \rangle).$$

(3.7)

The corresponding Lie group (*Heisenberg group*) can be interpreted as the space $H \oplus \mathbb{R}$ with the operation given by

$$(a_1, c_1)(a_2, c_2) = (a_1 + a_2, c_1 + c_2 + \mathrm{Im} \, \langle a_1, a_2 \rangle).$$

The *unitary Fock representation* of the algebra Heis is given by

$$p_k \to \frac{1}{2}(a_k + a_k^*), \quad q_k \to \frac{1}{2}(a_k - a_k^*), \quad r \to -\mathrm{i},$$

and on the Lie group level by

$$G(a, c)f(z) = f(z + a) \exp \left(-\langle z, a \rangle - \frac{1}{2}\langle a, a \rangle + \mathrm{i}c \right).$$

(3.8)

Remarks.

1. See A.1.3.
2. If $n < \infty$, then the Fock representation π is the unique representation of Heis_n satisfying the condition that r is represented by the identity operator (see Kashiwara and Vergne (1978)). On the other hand, for Heis_∞, there are a pathological number of such representations.

3.2.4. All bounded and many unbounded operators on F can be represented in the *Wick normal form*, i.e. in the form of operator series

$$A = L(\widehat{a}, \widehat{a}^*) = \sum p_{i_1 \ldots i_k, j_1 \ldots j_l} \prod_{\mu=1}^{k} \widehat{a}_{i_\mu} \prod_{\nu=1}^{l} \widehat{a}_{j_\nu}^*,$$

where $i_1 \leq i_2 \leq \ldots \leq i_k$, $j_1 \leq j_2 \leq \ldots \leq j_l$, $p_{i_1 \ldots j_l} \in \mathbb{C}$. The symbol L and the kernel K of the operator A are connected by the relation

$$K(z, \overline{u}) = L(z, \overline{u}) \exp \left(\langle z, u \rangle \right).$$

(3.9)

3.2.5. Vertex operators. Let $\alpha, \beta \in l_\mathbb{C}^2$. Let us introduce the following map on $F(l_\mathbb{C}^2)$:

$$\Theta'(\alpha, \beta)f(z) = f(z - \alpha) \exp \left(\langle z, \beta \rangle \right).$$

(3.10)

If $\alpha = \bar{\beta}$, then this reduces to the unitary operators $G(a,c)$ (see (3.8)). If $\alpha \neq \bar{\beta}$, the operators $\Theta'(\alpha,\beta)$ are unbounded but well-defined. Their common dense domain is the set of all $f \in F$ for which there exists ε such that $|f(z)| < C\exp((1-\varepsilon)\langle z,z\rangle/2)$. It is easy to check that the operators $\Theta'(\alpha,\beta)$ generate a holomorphic representation of the complexified Heisenberg group.

If we forget the conditions $\alpha,\beta \in l_C^2$, then $\Theta'(\alpha,\beta)$ is still a well-defined operator mapping the space of all holomorphic functions in a finite number of variables to the space of formal power series. We shall encounter distributions with values in such operators in Sect.8.

Using the Wick normal form (directly or using (3.3) and (3.9)), we obtain the expressions (up to a scalar multiple)

$$\Theta(\alpha,\beta) = \exp\left(\sum \beta_j \hat{a}_j\right) \exp\left(-\sum \alpha_j \hat{a}_j^*\right). \tag{3.11}$$

It is easy to see that

$$\begin{aligned}
[\hat{a}_j, \Theta(\alpha,\beta)] &= \alpha_j \Theta(\alpha,\beta), \\
[\hat{a}_j^*, \Theta(\alpha,\beta)] &= \beta_j \Theta(\alpha,\beta),
\end{aligned} \tag{3.12}$$

$$\Theta(\alpha,\beta)\Theta(\gamma,\delta) = \exp\left(-\sum \alpha_j \delta_j\right)\Theta(\alpha+\gamma, \beta+\delta). \tag{3.13}$$

It is important to express the singular operators $\Theta(\alpha,\beta)$ as a limit of ordinary ones. This can conveniently be done using *G.Segal's regularization*

$$\Theta(\alpha,\beta) = \lim_{\lambda \to 1} \exp\left(\sum_{j=1}^{\infty} \beta_j \lambda^{-j} \hat{a}_j\right) \exp\left(-\sum_{j=1}^{\infty} \alpha_j \lambda^{-j} \hat{a}_j^*\right). \tag{3.14}$$

Operators of the form (3.10) and (3.11) will be called *vertex operators*. If the algebra Heis acts on the space of all operators by commutation, then the vertex operators are exactly its eigenvectors. In fact, the conditions (3.12) define a system of differential equations for the Wick symbol of the operator.

3.3. The Weil Representation of $(\mathrm{Sp}(2\infty, \mathbb{R}), \mathrm{U}(\infty))$.

3.3.1. The Weil representation of $\mathrm{Sp}(2n, \mathbb{R})$. Formula (3.7) implies that the group $\mathrm{Sp}(2n, \mathbb{R})$ acts on Heis_n by automorphisms of the form $B(a,c) = (Ba, c)$, where $B \in \mathrm{Sp}(2n, \mathbb{R})$, $(a,c) \in \mathrm{Heis}_n$. The point 3.2.3, Rem.2 shows that the representations π and $\pi \circ B$ (where $B \in \mathrm{Sp}(2n, \mathbb{R})$) are equivalent. Let $W(B)$ be an intertwining operator (a unitary operator defined up to a scalar multiple of the form $e^{i\varphi}$). Then $B \to W(B)$ is a projective unitary representation of $\mathrm{Sp}(2n, \mathbb{R})$ (see Kashiwara and Vergne (1978)). We shall realize below the same procedure in the case of $\mathrm{Sp}(2\infty, \mathbb{R})$, the only difference being that $\pi \simeq \pi \circ B$ only if $B \in (\mathrm{Sp}(2\infty, \mathbb{R}), \mathrm{U}(\infty))$.

3.3.2. We shall realize $\mathrm{Sp}(2\infty, \mathbb{R})$ as the group of matrices in $l_C^2 \oplus l_C^2$ preserving the skew-form $\begin{pmatrix} 0 & \mathbb{1} \\ -\mathbb{1} & 0 \end{pmatrix}$ and the real subspace l_R of all vectors of the form $(a, \bar{a})$. These matrices have the form:

$$A = \begin{pmatrix} \Phi & \Psi \\ \overline{\Psi} & \overline{\Phi} \end{pmatrix}, \text{where } A \begin{pmatrix} 0 & \mathbb{1} \\ -\mathbb{1} & 0 \end{pmatrix} A^t = \begin{pmatrix} 0 & \mathbb{1} \\ -\mathbb{1} & 0 \end{pmatrix}. \tag{3.15}$$

The subgroup $U(\infty) \subset Sp(2\infty, \mathbb{R})$ consists of all matrices with $\Psi = 0$.

Remark. To see a connection between the described realization of the group $(Sp(2\infty, \mathbb{R}), U(\infty))$ and the realization discussed in 3.1.2, let us consider the space H of 3.1.2 as a real space equipped with the operator of a complex structure $I : Iv = iv$. Choose a real basis in H having the form $e_1, e_2, \ldots, Ie_1, Ie_2, \ldots$. Then the set

$$\frac{ie_1 + Ie_1}{\sqrt{2}}, \frac{ie_2 + Ie_2}{\sqrt{2}}, \ldots, \frac{ie_1 - Ie_1}{\sqrt{2}}, \frac{ie_2 - Ie_2}{\sqrt{2}}, \ldots, \tag{3.16}$$

is a basis of the complexification H_C of H and in this basis, the symplectic operator has the form (3.15), where I is given by the matrix

$$\begin{pmatrix} i & 0 \\ 0 & -i \end{pmatrix}$$

and the space l_R is H.

Theorem 3.1. (Berezin (1965)) *Let $A \in (Sp(2\infty, \mathbb{R}), U(\infty))$ be given by a matrix of the form (3.15). Let $W(A)$ be the operator in $F(l_C^2)$ with kernel*

$$\det (\Phi \Phi^*)^{-\frac{1}{4}} \exp \left\{ \frac{1}{2}(z\overline{u}) \begin{pmatrix} \overline{\Psi}\Phi^{-1} & (\Phi^t)^{-1} \\ \Phi^{-1} & -\Phi^{-1}\Psi \end{pmatrix} \begin{pmatrix} z^t \\ \overline{u}^t \end{pmatrix} \right\}.$$

Then the map $A \to W(A)$ is a projective unitary representation of the group $(Sp(2\infty, \mathbb{R}), U(\infty))$ on F such that for any $\gamma \in l_C^2 \oplus l_C^2$, we have the relation

$$W(A)\widehat{a}(\gamma)W(A)^{-1} = \widehat{a}(\gamma A^t).$$

3.3.3. The operators of the form (3.8) can be combined together with the operators $W(A)$; then we get a projective unitary representation of the group $(Sp(2\infty, \mathbb{R}), U(\infty)) \ltimes l_{\mathbb{R}}^2$ or, in the realization 3.1.2, of the group

$$(Sp(2\infty, \mathbb{R}), U(\infty)) \ltimes H.$$

3.3.4. It is natural to consider the set sp of all matrices of the form

$$\begin{pmatrix} L & M \\ \overline{M} & \overline{L} \end{pmatrix},$$

where $L = -L^*$, $M = M^t$, M is a Hilbert-Schmidt operator and L is, in general, unbounded, as the Lie algebra of the group $(Sp(2\infty, \mathbb{R}), U(\infty))$. The Lie algebra sp_C then consists of all matrices

$$\begin{pmatrix} L & M \\ N & -L^t \end{pmatrix}, \quad M = M^t, \quad N = N^t, \tag{3.17}$$

where M and N are Hilbert-Schmidt operators. The Weil representation of the Lie algebra sp_C is given by

$$dW\left[\begin{pmatrix} L & M \\ N & -L^t \end{pmatrix}\right] = \sum l_{ij} z_i \frac{\partial}{\partial z_j} + \frac{1}{2}\sum m_{ij} z_i z_j + \frac{1}{2}\sum n_{ij}\frac{\partial}{\partial z_i}\frac{\partial}{\partial z_j}. \quad (3.18)$$

A discussion of the correctness and the connection between Lie groups and Lie algebras can be found in Berezin (1965), Sect.6. If we denote

$$A_1 = \begin{pmatrix} L_1 & M_1 \\ N_1 & -L_1^t \end{pmatrix}, \quad A_2 = \begin{pmatrix} L_2 & M_2 \\ N_2 & -L_2^t \end{pmatrix},$$

then

$$[dW(A_1),\, dW(A_2)] = dW([A_1, A_2]) + \frac{1}{2}\mathrm{tr}\,(N_2 M_1 - N_1 M_2). \quad (3.19)$$

3.4. The Fermionic Fock Space.

3.4.1. Let $\xi_1, \xi_2, \ldots$ be holomorphic anticommuting variables, i.e.

$$\xi_i \xi_j = -\xi_j \xi_i, \ \xi_i \overline{\xi_j} = -\overline{\xi_j}\xi_i, \ \overline{\xi_i}\,\overline{\xi_j} = -\overline{\xi_j}\,\overline{\xi_i}, \ \overline{\xi_i \xi_j} = -\overline{\xi_j}\,\overline{\xi_i}$$

(in particular $\xi_i^2 = 0$). Formal power series in the variables $\xi_1, \xi_2, \ldots$ are called functions of ξ_i. The *left derivation* is defined by

$$\frac{\partial}{\partial \xi_i}\xi_i f(\xi) = f(\xi), \quad \frac{\partial}{\partial \xi_i} f(\xi) = 0,$$

where $f(\xi)$ does not depend on ξ_i. The integral $\int \prod_{i=1}^{j} \xi_{k_i}\overline{\xi_{k_i}}d\xi$ is equal to $(-1)^{j+1}$ by definition. If the factors are permuted, the sign is changed accordingly. The integrals of other monomials are, by definition, zero.

The space of linear forms in the variables ξ will be denoted by H.

Definition. The *fermionic Fock space* $\Lambda(H)$ or Λ is the space of all functions of the variables $\xi_1, \xi_2, \ldots$ with the scalar product

$$\langle f, g \rangle = \int f(\xi)\overline{g(\xi)}d\xi.$$

Remark. The length of every monomial $\xi_i \ldots \xi_{i_k}$, where $i_1 < \ldots < i_k$, is equal to 1. Different monomials are pairwise orthogonal.

3.4.2. All bounded and many unbounded operators in Λ can be written in the form

$$Af(\xi) = \int K(\xi, \overline{\eta})f(\eta)d\eta.$$

The expression $K(\xi, \overline{\eta})$ is called the *kernel* of the operator A. The *creation and annihilation operators* $\hat{a}_k$, $\hat{a}_k^*$, $\hat{a}(\gamma)$ and the *Wick normal form* of an operator are defined in the same way as was done in the bosonic case (see (3.4) and (3.5)). Finally, let us note that

$$F(\oplus H_i) = \otimes F(H_i) \quad \Lambda(\oplus H_i) = \otimes \Lambda(H_i). \tag{3.20}$$

3.4.3. Suppose that $f = \sum f_k(\xi) \in \Lambda$, where the $f_k(\xi)$ are homogeneous forms of degree k in the variables ξ. Let $\Lambda_0(H)$ denote the dense subset of $\Lambda(H)$ consisting of all functions $f \in \Lambda(H)$ such that the norm $||f_k||$ approaches zero more rapidly than any sequence of the form $\exp(-Ck)$. Let us introduce the family of seminorms $P_C(f) = \max_k(||f_k|| \exp(Ck))$ on $\Lambda_0(H)$. Then the space Λ_0 is a Fréchet space (i.e. a complete locally convex space).

3.5. Spinor Representations of $(O(2\infty, \mathbb{R}), U(\infty))$, $(O(2\infty + 1, \mathbb{R}), U(\infty))$ and $(O(2\infty, \mathbb{C}), GL(\infty, \mathbb{C}))$.

3.5.1. We shall realize the group $O(2\infty, \mathbb{C})$ as the group of all matrices in $l_C^2 \oplus l_C^2$ preserving the symmetric bilinear form with matrix $\begin{pmatrix} 0 & \mathbb{1} \\ \mathbb{1} & 0 \end{pmatrix}$. Its subgroup $O(2\infty, \mathbb{R})$ consists of all operators preserving the real subspace of all vectors of the form $(a, \bar{a})$. So

$$Q = \begin{pmatrix} A & B \\ C & D \end{pmatrix} \in O(2\infty, \mathbb{C}) \quad \Longleftrightarrow \quad Q\begin{pmatrix} 0 & \mathbb{1} \\ \mathbb{1} & 0 \end{pmatrix} Q^t = \begin{pmatrix} 0 & \mathbb{1} \\ \mathbb{1} & 0 \end{pmatrix}.$$

The matrix Q belongs to $O(2\infty, \mathbb{R})$ if and only if $D = \bar{A}$, $\bar{C} = B$. Finally, the subgroups $U(\infty) \subset O(2\infty, \mathbb{R})$ and $GL(\infty, \mathbb{C}) \subset O(2\infty, \mathbb{C})$ are characterized by the conditions $B = C = 0$.

Theorem 3.2. *Let $Q \in (O(2\infty, \mathbb{C}), GL(\infty, \mathbb{C}))$. Let $\mathrm{Spin}(Q)$ be the operator with kernel*

$$\det(AA^*)^{-\frac{1}{4}} \exp\left\{ \frac{1}{2} (\xi \bar{\eta}) \begin{pmatrix} CA^{-1} & (A^t)^{-1} \\ A^{-1} & A^{-1}B \end{pmatrix} \begin{pmatrix} \xi^t \\ \bar{\eta}^t \end{pmatrix} \right\}, \tag{3.21}$$

where $\xi = (\xi_1, \xi_2, \ldots)$, $\eta = (\eta_1, \eta_2, \ldots)$. Then:

a) *The map $Q \to \mathrm{Spin}(Q)$ defines a unitary projective representation of the pair $(O(2\infty, \mathbb{R}), U(\infty))$ in $\Lambda(l_C^2)$ (see Berezin (1965)).*
b) *The map $Q \to \mathrm{Spin}(Q)$ defines a representation of the group*

$$(O(2\infty, \mathbb{C}), GL(\infty, \mathbb{C}))$$

by bounded operators in $\Lambda_0(l_C^2)$ (see Neretin (1986)).

In both cases, we have

$$\mathrm{Spin}(Q)\hat{a}(\gamma)\mathrm{Spin}(Q)^{-1} = \hat{a}(\gamma Q^t).$$

Remark. Formula (3.21) is only defined on an open dense subset of the pair $(O(2\infty, \mathbb{R}), U(\infty))$ and $(O(2\infty, \mathbb{C}), GL(\infty, \mathbb{C}))$.

3.5.2. The "Lie algebra" of the group $(O(2\infty, \mathbb{C}), GL(\infty, \mathbb{C}))$ consists of all matrices of the form $\begin{pmatrix} L & M \\ N & -L^t \end{pmatrix}$, where M and N are Hilbert-Schmidt operators, $M = -M^t$, $N = -N^t$ and L is, in general, unbounded. The subalgebra corresponding to the subgroup $(O(2\infty, \mathbb{R}), U(\infty))$ is characterized by the conditions $M = \overline{N}$, $L = -L^*$. On the level of Lie algebras, the spinor representation is given by the formula

$$
d \operatorname{Spin} \left[\begin{pmatrix} L & M \\ N & -L^t \end{pmatrix} \right] = \sum l_{ij} \xi_i \frac{\partial}{\partial \xi_j} + \frac{1}{2} \sum m_{ij} \xi_i \xi_j +
$$
$$
+ \frac{1}{2} \sum n_{ij} \frac{\partial}{\partial \xi_i} \frac{\partial}{\partial \xi_j} \qquad (3.22)
$$

3.5.3. The representations Spin and W can clearly be decomposed into the sum of two irreducible representations $\operatorname{Spin} = \operatorname{Spin}_+ \oplus \operatorname{Spin}_-$, resp. $W = W_+ \oplus W_-$; the irreducible components are realized in the space of even, resp. odd functions.

3.5.4. The representations $\operatorname{Spin}_\pm$ are analogues of the half-spin representations of $O(2n)$ (see Zhelobenko and Shtern (1983)). We shall now construct an analogue of the spin representation of $O(2n + 1)$. Let the group $O(2\infty + 1)$ act in $l_C^2 \oplus \mathbb{R}$; its subgroup $U(\infty)$ is realized as the group of operators which are unitary on l_C^2 and act as the identity on $\mathbb{R}$. Then the *spin representation* Spin_0 of the group $(O(2\infty + 1), U(\infty))$ is defined as the restriction of either of the representations $\operatorname{Spin}_\pm$ of the group $(O(2\infty + 2, \mathbb{R}), U(\infty + 1))$.

3.6. Representations of (G, K)-Pairs.

3.6.1. It is clear that far from all (G, K)-pairs covered in the Def. 3.1.1 are interesting. An (incomplete) list of *"good" (G, K)-pairs* can be found in Ol'shanskij (1983) (where the spaces G/K are "symmetric spaces of infinite rank"). We shall not make the notion of a "good" (G, K)-pair more precise; let us note only that the pairs from Sect.4

$$
(G(\infty), K(\infty)) = (O(2\infty, \mathbb{R}), U(\infty)), \; (U(2\infty), U(\infty) \times U(\infty)),
$$
$$
(U(\infty) \times U(\infty), U(\infty)), \; (O(2\infty + 1, \mathbb{R}), U(\infty)),
$$
$$
(U(\infty), O(\infty, \mathbb{R})), \; (U(2\infty), UH(\infty))
$$

(these are all pairs of compact type, i.e. $G(n)$ is compact) and

$$
(GL(\infty, \mathbb{R}), O(\infty, \mathbb{R})), \; (GL(\infty, \mathbb{C}), U(\infty)), \; (Sp(2\infty, \mathbb{R}), U(\infty))
$$

(pairs of noncompact type) are all good in the sense of G.I.Ol'shanskij. Let us note that the group $K(n)$ should be compact for all good pairs.

3.6.2. Let ρ be the defining representation of $K(\infty) = O(\infty, \mathbb{R}), U(\infty)$, resp. $UH(\infty)$, where $UH(\infty)$ is the unitary group with respect to the quaternionic scalar product. Any subrepresentation of a representation of the form $\rho^{\otimes l} \otimes (\rho^*)^{\otimes m}$ will be called a tensor representation of $K(\infty)$. If moreover $K(\infty) = U(\infty) \times U(\infty)$, then its tensor representations are defined as products of the tensor representations of $U(\infty)$.

Let us call by *Harish-Chandra module* of (G, K) a unitary representation μ of the group G such that the restriction $\mu|_K$ is a sum of tensor representations. For every good (G, K)-pair of compact type, three series of Harish-Chandra modules were constructed in Ol'shanskij (1983): bosonic, fermionic and an "intermediate" one (using the imbedding into $(\mathrm{Sp}(2\infty, \mathbb{R}), U(\infty))$, resp. into $(O(2\infty, \mathbb{R}), U(\infty))$, resp. into the group $U(H) \ltimes S^2 H$, where $U(H)$ is the unitary group of the Hilbert space H and the symbol $S^2 H$ denotes the symmetric square of H). For pairs of noncompact type, there exists only the bosonic series. We restrict ourselves only to the simplest examples.

3.6.3. Example. Let P be a quaternionic Hilbert space. i.e. a real vector space with orthogonal operators I, J, K such that $I^2 = J^2 = K^2 = -1$, $IJ = -JI = K$. Let $U(2\infty)$ consist of all orthogonal operators commuting with I. Then its subgroup $UH(\infty)$ is the group of all operators commuting with I, J, K. The space P will be considered as a complex Hilbert space with the complex structure given by the operator $I' = \alpha + \beta I + \gamma J + \delta K$, where $\alpha^2 + \beta^2 + \gamma^2 + \delta^2 = 1$. It is easy to see that we obtain an imbedding of $(U(2\infty), UH(\infty))$ into $(O(4\infty, \mathbb{R}), U(2\infty))$. Restricting the representation $\mathrm{Spin}_\pm$ to $(U(2\infty), UH(\infty))$, we get an irreducible representation (for $I' \neq I$).

3.6.4. Example. Let us consider the group $\mathrm{Sp}(2\infty, \mathbb{R})$ realized as the group of matrices in $l_R^2 \oplus l_R^2$ preserving the skew-symmetric form $\begin{pmatrix} 0 & \mathbb{1} \\ -\mathbb{1} & 0 \end{pmatrix}$. Suppose that $I(a, b) = (-b, a)$. The group $U(\infty) \subset \mathrm{Sp}(2\infty, \mathbb{R})$ consists of all matrices commuting with I (this is the model described in 3.1.2). The following formula determines a series of imbeddings of $(GL(\infty, \mathbb{R}), O(\infty, \mathbb{R}))$ into $(\mathrm{Sp}(2\infty, \mathbb{R}), U(\infty))$:

$$\tau_\omega(g) = \begin{pmatrix} \cosh \omega & \sinh \omega \\ \sinh \omega & \cosh \omega \end{pmatrix} \begin{pmatrix} g & 0 \\ 0 & g^{t-1} \end{pmatrix} \begin{pmatrix} \cosh \omega & \sinh \omega \\ \sinh \omega & \cosh \omega \end{pmatrix}^{-1}.$$

The restriction of the modules $W_\pm$ to the pair $(GL(\infty, \mathbb{R}), O(\infty, \mathbb{R}))$ defines the series $\mathrm{Exp}_\omega^\pm$ of pairwise inequivalent representations of the pair $(GL(\infty, \mathbb{R}), O(\infty, \mathbb{R}))$, see also A.3.2.

Remark. The restriction of $\mathrm{Exp}_\omega = \mathrm{Exp}_\omega^+ \oplus \mathrm{Exp}_\omega^-$ to $O(\infty, \mathbb{R})$ is clearly isomorphic to $\oplus S^k \rho$, where ρ is the standard representation of $O(\infty, \mathbb{R})$ and $S^k \rho$ is its symmetric power.

3.6.5. (Fundamental module over $\mathrm{gl}(\infty)$) Let H be a Hilbert space and let $\overline{H}$ denote the same space with the conjugate complex structure. Then the identity map $H \oplus \overline{H} \to H \oplus H$ induces an imbedding of the pair $(U(2\infty), U(\infty) \times U(\infty))$ to $(O(4\infty, \mathbb{R}), U(2\infty))$. If we restrict the representations $\mathrm{Spin}_\pm$ to $(U(2\infty), U(\infty) \times U(\infty))$, then we get a pair of irreducible representations of $(U(2\infty), U(\infty) \times U(\infty))$.

3.6.6. The following fact (due to N.I.Nessonov and G.I.Olshanskii) partially explains the definition of (G, K)-pair. Let GL_0 (resp. U_0) be the group of all (resp. all unitary) operators in l_C^2 preserving all basis vectors of the form $(\dots, 0, 1, 0, 0, \dots)$ except for a finite number of them. A representation of GL_0 is called spherical if its restriction to U_0 contains an invariant vector. Any spherical representation of GL_0 can be extended to a Harish-Chandra module over $(\mathrm{GL}(\infty, \mathbb{C}), U(\infty))$. All such representations (and generally all Harish-Chandra modules over good (G, K)-pairs) admit a full classification.

Remark. The cocycle construction, described in A.3.1, Ex. 3, together with the imbedding into $(\mathrm{GL}(\infty, \mathbb{C}), U(\infty)) \ltimes l_C^2$ offer examples of nonspherical irreducible representations of GL_0.

Appendices

A.3.1. Araki's Scheme and Multiplicative Integral. Let $U(H)$ be the unitary group of a Hilbert space H. Let $\mathrm{Isom}(H) = U(H) \ltimes H$ be the group of transformations of H generated by $U(H)$ and by translations. The projective unitary representation Ex of the group $\mathrm{Isom}(H)$ on $F(H)$ is constructed in the following way: the translations act by formula (3.8) and elements of $U(H)$ by $T(A)f(z) = f(Az)$, $A \in U(H)$.

Remark. It is clear that $U(H) \ltimes H \subset (\mathrm{Sp}(2\infty, \mathbb{R}), U(\infty)) \ltimes H$. It is not difficult to check that the representation Ex can be obtained by the restriction of the Weil representation, see 3.3.3.

Affine actions of groups. Let the group G be imbedded into $\mathrm{Isom}(H)$. This means that G acts on H by isometries of the form

$$Z(g)v = \rho(g)v + \gamma(g),$$

where $v \in H$, $g \in G$, $\gamma(g) \in H$ and $\rho(g)$ is a unitary representation of G. It is easy to verify that nontrivial affine actions of G on H with unitary part ρ are in one-to-one correspondence with elements of the group $H^1(G, H) = Z^1(G, H)/B^1(G, H)$, where $Z^1(G, H)$ consists of maps (*affine cocycles*) $\gamma : G \to H$ satisfying the condition

$$\gamma(g_1 g_2) = \rho(g_1)\gamma(g_2) + \gamma(g_1)$$

and its subgroup $B^1(G, H)$ consists of maps of the form

$$\gamma(g) = \rho(g)v - v, \tag{3.23}$$

where $v \in H$. These maps $\gamma(g)$ correspond to translations of the origin.

Example 1. Suppose that the group $\mathrm{PSL}_2(\mathbb{R})$ acts on the space H_0 of all real functions on S^1 with zero mean (see 4.6) by the formula

$$T(q)f(\varphi) = f(q(\varphi))q'(\varphi).$$

It is a unitary representation with respect to the scalar product (4.9). Then $\gamma(q) = q'(\varphi) - 1$ is an affine cocycle (see also 5.1.1). Finally, $\gamma(q)$ has the form (3.23), but $f(\varphi) \equiv 1 \notin H$.

Example 2. Let us consider a nonunitary representation T_0 of $\mathrm{SL}_2(\mathbb{R})$ having highest weight 0 (see 5.5). It is a reducible representation and the unitary representation T_1 with highest weight 1 acts in a subspace H_0 of codimension 1. Let v be a highest weight vector of the representation T_0. Then $\gamma(g) = T_0(g)v - v \in H_0$, hence $\gamma(g)$ is an affine cocycle for T_1.

Remark. Example 1 can be obtained from Example 2 if the Hilbert space H_0 is considered as a real space.

For semisimple Lie groups, affine cocycles exist for the series $\mathrm{SO}(n,1)$ and $\mathrm{SU}(n,1)$; they are constructed in a similar way.

Example 3. Let us consider a sequence of numbers $(a_1, a_2, a_3, \ldots)$. Then the defining representation of the group U_0 (see 3.6.6.) in l^2 has an affine cocycle $\gamma(g) = ga - a$ (I.M.Gel'fand, M.I.Graev).

Araki's scheme. For every imbedding σ of an infinite-dimensional group G into the group $\mathrm{Isom}\,(H)$, there is the corresponding unitary projective representation $Ex \circ \sigma$ of the group G in $F(H)$. Let us discuss here the two main examples.

A. Let P be a Lie group acting on a space M by affine isometries. Let X be a space endowed with a measure. Then the group $G = \mathcal{F}(X;P)$ of all measurable P - valued functions on X clearly acts on $H = L^2(X, M)$ by affine isometries (see Vershik et al. (1973)).

B. Let K be a compact matrix Lie group, let V^n be a Riemannian manifold and let $G = C^\infty(V^n, K)$. The Hilbert space H consists of all 1-forms on V^n with values in the Lie algebra of the group P. Then the affine cocycle $\gamma \in H^1(G, H)$ is given by the formula $\gamma(u) = u^{-1}du$, where du is the differential of u (R.S.Ismagilov, see Albeverio et al. (1983)).

Remark. The choice of various riemannian metrics in Ex. B (or different choice of measures in Ex. A) leads to inequivalent representations.

Multiplicative integral. It is possible to look at Ex. A and Ex. B from another point of view. It is well-known that irreducible representations of a direct sum of groups $\oplus G_i$, where $1 \leq i \leq k$, have the form $\otimes \pi_i$, where π_i is an irreducible representation of G_i. The groups of type $\mathcal{F}(X, P)$ can be considered as a continuous direct sum of groups isomorphic to P. On the other hand, formula (3.20) allows one to define a continuous tensor product or multiplicative integral of Hilbert spaces and representations

$$\int^{\otimes} F(H_x)d\mu_x \overset{\text{def}}{=} F\left(\int H_x d\mu_x\right).$$

This is just the idea used in Ex. A and Ex. B.

A.3.2. The Space $L^2(\mathbb{R}^\infty)$ as a Model of the Fock Space. Let $\mathbb{R}^\infty$ denote the product of a countable number of copies of $\mathbb{R}$. We shall consider the Gaussian measure μ on $\mathbb{R}^\infty$; it is a countable product of Gaussian measures with density $\frac{1}{\sqrt{2\pi}}\exp(-x^2/2)$. Suppose that A is the matrix of a bounded operator in l^2. Then the Kolmogorov-Khinchin theorem on series of stochastic variables implies that the formal product of the matrix A and a column-vector $x \in \mathbb{R}^\infty$ is well-defined for almost all $x \in \mathbb{R}^\infty$. As a consequence of the Feldman-Hájek theorem on equivalence of Gaussian measures (see Kuo (1975), Guichardet (1972)), we obtain that the measure μ is quasi-invariant with respect to operators $A \in (\mathrm{GL}(\infty, \mathbb{R}), \mathrm{O}(\infty, \mathbb{R}))$. Using the scheme of A.1.3, we get a series of unitary representations of $(\mathrm{GL}(\infty, \mathbb{R}), \mathrm{O}(\infty, \mathbb{R}))$ on $L^2(\mathbb{R}^\infty)$. These representations are equivalent to the representations Exp_ω (see 3.6.4) (for the proof, it is sufficient to compute the spherical functions corresponding to the vaccum).

The described realization of the Fock space is less convenient than the model 3.2.1 (due to Bargmann). Another construction can be found in Guichardet (1972), it does not differ significantly from our model. A very nice construction is treated in Vershik et al. (1983).

§4. Almost Invariant Structures[1]

The main aim of this section is to construct imbeddings of the group Diff into certain (G, K)-pairs. Using information contained in 3.6, any such imbedding leads to the construction of a series of unitary representations of the group Diff.

4.1. The Almost Invariant Scalar Product (an Imbedding of Diff into $(\mathrm{GL}(\infty, \mathbb{R}), \mathrm{O}(\infty, \mathbb{R})))$. Let Diff act on the space C of all real smooth functions on S^1 by the formula

$$T_{0,s}(q)f(\varphi) = f(q(\varphi))q'(\varphi)^{\frac{1+s}{2}}, \tag{4.1}$$

where $0 < s < 1$, $q \in$ Diff. Let us define the space H_s as the completion of C in the norm given by the scalar product

$$\langle f_1, f_2 \rangle_s = \int_0^{2\pi} \int_0^{2\pi} \left| \sin\left(\frac{\varphi_1 - \varphi_2}{2}\right) \right|^{s-1} f_1(\varphi_1)\overline{f_2(\varphi_2)} d\varphi_1 d\varphi_2. \tag{4.2}$$

Lemma. *Suppose that* $\mathrm{Re}\,\nu = 0$, $\alpha \in \mathbb{R}$ *and*

---

[1] For new results on almost invariant structures see Neretin (1993)

$$F_{\nu,\alpha}(\psi) = \exp\left(2\pi i \alpha(1 + [\psi/2\pi])\right)|\sin\psi/2|^{\nu-1}, \qquad (4.3)$$

where [ ] denotes the integer part of a number. Then

$$F_{\nu,\alpha}(\psi) = \sum_{n=-\infty}^{\infty} \frac{\pi 2^{-\nu}}{B\left(\frac{\nu+1}{2} - n - \alpha, \frac{\nu+1}{2} + n + \alpha\right)} e^{i(n+\alpha)\psi}. \qquad (4.4)$$

Proof. Both sides of (4.4) satisfy the condition $f(\psi+2\pi) = \exp\left(2\pi i \alpha\right)f(\psi)$. So it is sufficient to decompose $F_{\nu,\alpha}$ into a Fourier series on $[0, 2\pi]$ (see any tables of integrals).

Substituting $\nu = s$, $\alpha = 0$, $\psi = \varphi_1 - \varphi_2$ in Lemma 4.1, we get that vectors of the form $\exp\left(in\varphi\right)$ are pairwise orthogonal and that their lengths are equal to $\sqrt{c_n(s)}$, where

$$c_n(s) = \frac{\Gamma\left(n + \frac{1-s}{2}\right)}{\Gamma\left(n + \frac{1+s}{2}\right)}. \qquad (4.5)$$

Remark. The operator $T_{0,s}(q)$ preserves the form (4.2) if and only if $q \in \mathrm{PSL}_2(\mathbb{R})$. The restriction of $T_{0,s}$ to $\mathrm{PSL}_2(\mathbb{R})$ is a representation of the complementary series of representations of $\mathrm{PSL}_2(\mathbb{R})$. It turns out that the other operators $T_{0,s}(q)$ "almost preserve " the scalar product (4.2).

Theorem 4.1. *The operators $T_{0,s}(q)$ belong to $(\mathrm{GL}(\infty, \mathbb{R}), O(\infty, \mathbb{R}))$.*

Proof. The element $T_{0,s}(q)$ belongs to $(\mathrm{GL}(\infty, \mathbb{R}), O(\infty, \mathbb{R}))$ if and only if $L(q) = T_{0,s}^*(q)T_{0,s}(q) - \mathbb{1}$ is a Hilbert-Schmidt operator in H_s. A direct computation gives

$$\langle L(q)f_1, f_2 \rangle_s = \langle T_{0,s}(q)f_1, T_{0,s}(q)f_2 \rangle_s - \langle f_1, f_2 \rangle_s =$$

$$= \int_0^{2\pi} \int_0^{2\pi} \left[\left(\frac{p'(\psi_1)p'(\psi_2)}{\sin^2\left(\frac{p(\psi_1)-p(\psi_2)}{2}\right)} \right)^{\frac{1-s}{2}} - \frac{1}{|\sin\frac{\psi_1-\psi_2}{2}|^{1-s}} \right] \times$$

$$\times f_1(\psi_1)\overline{f_2(\psi_2)}d\psi_1 d\psi_2,$$

where p is the inverse map to q. Let us denote the expression inside the brackets by $Z(\psi_1, \psi_2)$, let l_{mn} be the matrix coefficients of $L(q)$ with respect to the orthonormal basis $e^{in\varphi}/\sqrt{c_n}$ (see (4.5)) and let z_{mn} be the Fourier coefficients of $Z(\psi_1, \psi_2)$. We have now to prove that $\sum |l_{mn}|^2 = \sum |z_{mn}|^2/c_n c_m < \infty$. A direct computation shows that $Z(\psi_1, \psi_2)$ has the form

$$\left|\sin\frac{\psi_1 - \psi_2}{2}\right|^{1+s} \lambda(\psi_1, \psi_2), \quad \text{where} \quad \lambda(\psi_1, \psi_2) \in C^\infty(S^1 \times S^1).$$

Making the substitution $\mu_1 = \psi_1 - \psi_2$, $\mu_2 = \psi_2$ (which is well-defined on the torus), we can represent $Z(\mu_1, \mu_2)$ in the form $|\sin\mu_1/2|^{1+s}P(\mu_2) + Q(\mu_1, \mu_2)$, where $P(\mu_2) \in C^\infty(S^1)$ and $Q(\mu_1, \mu_2) \in C^\infty(S^1 \times S^1)$. Let y_{lk} be the Fourier coefficients of $Z(\mu_1, \mu_2)$; then clearly $z_{mn} = y_{m,m-n}$. Noting that $c_s \sim n^{-s}$ for $n \to \infty$ and estimating the Fourier coefficients of $P(\mu_2)$ and $Q(\mu_1, \mu_2)$ in a standard way, we obtain estimates for the numbers y_{kl}, z_{mn}, l_{mn}.

4.2. The Almost Invariant Quaternionic Structures. A function on S^1 is called odd if the condition $f(\varphi + \pi) = -f(\varphi)$ is satisifed. The subspace of all odd functions in $L^2(S^1)$ will be denoted by L^2_-. A diffeomorphism $q : S^1 \to S^1$ is called even if $q(\pi + \varphi) = q(\varphi) + \pi$. The group of all even diffeomorphisms of S^1 is isomorphic to $\mathrm{Diff}^{(2)}$ (see 1.2).

Suppose that the group $\mathrm{Diff}^{(2)}$ acts on L^2_- by the formula

$$P_{1/2,s}(q)f(\varphi) = f(q(\varphi))q'(\varphi)^{\frac{1+is}{2}}, \tag{4.6}$$

where $q \in \mathrm{Diff}^{(2)}$. It is clear that $P_{1/2,s}(q)$ is a unitary operator in L^2_-. Let us consider the transformation

$$J_s f(\varphi) = \frac{1}{\Gamma(1+is)} \int_0^{2\pi} \frac{\overline{f(\psi)}\mathrm{sgn}(\sin(\varphi - \psi))d\psi}{|\sin(\varphi - \psi)|^{1+is}}. \tag{4.7}$$

The integral in (4.7) is divergent and it is necessary to define it either as the principal value of the integral or by analytic continuation in s from the domain $\mathrm{Im}\,(is) < 0$.

Using Lemma 4.1 with $\nu = is$, $\alpha = 1/2$, it is not difficult to verify that J_s is an antilinear orthogonal operator and that $J_s^2 = -1$. In this way, the structure of a quaternionic Hilbert space is defined on L^2_- (the quaternionic imaginary units being $i = \mathrm{i}, j = J_s$ and $k = \mathrm{i}J_s$)).

Remark. The operator J_s (hence all quaternionic imaginary units) commutes with the action of the subgroup $\mathrm{SL}_2(\mathbb{R})$ of $\mathrm{Diff}^{(2)}$. In other words, the described unitary representation of $\mathrm{SL}_2(\mathbb{R})$ (a representation belonging to the odd principal series, see Zhelobenko and Shtern (1983)) is of quaternionic type (see Sect. 6.2 of the first paper of this volume).

Theorem 4.2. *The formula (4.6) gives an imbedding of the group $\mathrm{Diff}^{(2)}$ into $(\mathrm{U}(2\infty), \mathrm{UH}(\infty))$.*

Proof. It is sufficient to compute the kernel of the operator $[P_{1/2,s}(q), J_s]$ and to check that the kernel is bounded.

4.3. The Imbedding of Diff into $(\mathrm{U}(2\infty), \mathrm{U}(\infty) \times \mathrm{U}(\infty))$. Let the group Diff act on $L^2(S^1)$ by the formula

$$P_{0,s}(q)f(\varphi) = f(q(\varphi))q'(\varphi)^{\frac{1+is}{2}}. \tag{4.8}$$

The space $L^2(S^1)$ can be decomposed into a direct sum $H_+ \oplus H_-$, where H_+ is generated by $\exp(\mathrm{i}n\varphi), n \geq 0$, and H_- is generated by $\exp(\mathrm{i}n\varphi), n < 0$. It turns out that the decomposition is almost invariant.

Theorem 4.3. *Let* $U(\infty) \times U(\infty)$ *be the group of all unitary operators in* $L^2(S^1)$ *preserving the subspaces* $H_\pm$. *Then the formula (4.8) gives an imbedding of* Diff *into* $(U(2\infty), U(\infty) \times U(\infty))$.

For the proof, it is sufficient to verify that $[P_{0,s}(q), I]$, where I is given by formula (4.10) below, is a Hilbert-Schmidt operator.

4.4. The Almost Invariant Indefinite Form. Let us consider the representation $T_{0,s}$ of the group Diff (see 4.1) in the space with the hermitian form (4.2), where s is an arbitrary real number, $s \neq 2k+1$, $s \neq 0$ and the integral (4.2) is defined for $s < 0$ by analytic continuation in s. Then expression (4.2) is an indefinite scalar product with index of inertia equal to (p, ∞), where $p = \left| \left[\frac{s+1}{2} \right] \right|$.

Lemma 4.2. *Formula (4.1) gives an imbedding of* Diff *into the group* $(GL(\infty, \mathbb{R}), O(p, \infty))$.

Remark. The group $O(p, \infty)$ is not of a compact type, which means that the group $(GL(\infty, \mathbb{R}), O(p, \infty))$ is not a good (G, K)-pair. Let us look at this example from another point of view. The index of inertia is finite, so there exists $K(\varphi_1, \varphi_2) \in C^\infty(S^1, S^1)$ such that the scalar product

$$\langle f_1, f_2 \rangle =$$
$$= \int_0^{2\pi} \int_0^{2\pi} \left(\left| \sin \left(\frac{\varphi_1 - \varphi_2}{2} \right) \right|^{s-1} + K(\varphi_1, \varphi_2) \right) f_1(\varphi_1) \overline{f_2(\varphi_2)} d\varphi_1 d\varphi_2$$

is positive definite. It is not difficult to prove that the scalar product is almost invariant and that it does not depend, up to natural equivalence, on the choice of $K(\varphi_1, \varphi_2)$ ($\langle \cdot, \cdot \rangle \sim (\cdot, \cdot)$ if there exists a Hilbert-Schmidt operator T such that $\langle (\mathbb{1} + T)f, g \rangle = (f, g)$). Then the case $-1 < s < 1$ is characterized by the fact that there exists a canonical $PSL_2(\mathbb{R})$-representative of the equivalence class of almost invariant structures.

4.5. The List of Known Almost Invariant Structures.

A. The "principal series". Let $P^0_{\alpha,s}$ denote the restriction of the module $P_{\alpha,s}$ of the trivial series (see 1.4) to $PSL_2^{(\infty)}(\mathbb{R})$. It is easy to see that $P^0_{\alpha,s}$ is a representation of $PSL_2^{(\infty)}(\mathbb{R})$ belonging to the *principal* unitary *series* (see Zhelobenko and Shtern (1983)). The representations $P^0_{\alpha,s}$ and $P^0_{\alpha,-s}$ are equivalent and an intertwining operator $A_{\alpha,s} : P^0_{\alpha,s} \to P^0_{\alpha,-s}$ is given by the formula

$$A_{\alpha,s} f(\varphi) = \frac{1}{\Gamma(1 + is)} \int_{-\pi}^{\pi} F_{is,\alpha}(\varphi - \psi) f(\psi) d\psi,$$

where F is defined by the expression (4.3) and the integral is interpreted in the sense of analytic continuation in s from the domain $\text{Re}(is) > 0$.

Theorem 4.4. *The operator*

$$A_{\alpha,s} P_{\alpha,s}(q) - P_{\alpha,-s}(q) A_{\alpha,s}$$

is a Hilbert-Schmidt operator.

It determines imbeddings of $\text{Diff}^{(\infty)}$ into the following groups:

α) $(U(\infty) \times U(\infty), U(\infty))$ with (α, s) arbitrary. The group $\text{Diff}^{(\infty)}$ acts on $H_\alpha \oplus H_\alpha$ by the formula $P_{\alpha,s} \oplus P_{\alpha,-s}$. Two copies of the space H_α are identified using the operator $A_{\alpha,s}$. The group $U(\infty) \times U(\infty)$ preserves each summand H_α and $U(\infty)$ is imbedded diagonally.

β) $(U(2\infty), UH(\infty))$, $\alpha = 1/2$ (see 4.2), $J_s f = A_{1/2,s} \overline{f}$.

γ) $(U(\infty), O(\infty, \mathbb{R}))$, $\alpha = 0$. The antilinear operator $Bf = A_{0,s}(\overline{f})$ satisfies the relation $B^2 = 1$. The subgroup $O(\infty, \mathbb{R})$ of the unitary group of the space $H_0 = L^2(S^1)$ consists of operators preserving the real subspaces $\text{Ker}\,(B \pm 1\!\!1)$.

Remark. The representation $P_{0,s}^0$ of the group $SL_2(\mathbb{R})$ is of real type; the subspaces $\text{Ker}\,(B \pm 1\!\!1)$ are $SL_2(\mathbb{R})$-invariant.

δ) $(O(2\infty, \mathbb{R}), U(\infty))$, $\alpha = 1/2$, $s = 0$ (see 4.7.1)

B. The complementary series. Suppose that $T_{\alpha,s}^0$ is the restriction of $T_{\alpha,s}$ to $\text{PSL}_2^{(\infty)}(\mathbb{R})$ (see 1.4). Then for $2\alpha \pm s \neq 2k + 1$, $k \in \mathbb{Z}$, the module $P_{\alpha,s}^0$ admits the invariant nonsingular hermitian form

$$\langle f, g \rangle = - \int_{-\pi}^{\pi} \int_{-\pi}^{\pi} F_{s,\alpha}(\varphi - \psi) f(\varphi) \overline{g(\psi)} d\varphi d\psi,$$

where $F_{s,\alpha}$ is given by the formula (4.3). If $0 < s < 1 - 2|\alpha|$, then the form is positive definite and almost invariant (see (4.4)). This implies a homomorphism of $\text{Diff}^{(\infty)}$ into the following groups:

α) $(GL(\infty, \mathbb{C}), U(\infty))$, $0 < s < 1 - 2|\alpha|$;
β) $(GL(\infty, \mathbb{R}), O(\infty, \mathbb{R}))$, $\alpha = 0$, $0 < s < 1$ (see 4.1);
γ) $(Sp(2\infty, \mathbb{R}), U(\infty))$, $\alpha = 0$, $s = 1$ (see 4.6);
δ) $(Sp(2\infty, \mathbb{R}), U(\infty))$, $\alpha = 0$, $s = -1$ (see 8.4.6).

C. The indefinite series. Let us consider, for definiteness, the case

$$-\frac{1}{2} \leq \alpha \leq \frac{1}{2}, \; s > 0, \; 2\alpha \pm s \neq 2k + 1.$$

Then for every (s, α) that does not satisfy $0 < s < 1 - 2|\alpha|$, we get a homomorphism of $\text{Diff}^{(\infty)}$ to the following groups:

α) $(GL(\infty, \mathbb{C}), U(p, \infty))$ for $p = \left[\frac{s+1}{2} + \alpha\right] = \left[\frac{s+1}{2} - \alpha\right]$;
β) $(GL(2\infty, \mathbb{C}), U(\infty, \infty))$ for $\left[\frac{s+1}{2} + \alpha\right] \neq \left[\frac{s+1}{2} - \alpha\right]$;
γ) $(GL(\infty, \mathbb{R}); O(p, \infty))$ for $\alpha = 0$, $p = \left[\frac{s+1}{2}\right]$ (see 4.4);
δ) $(GL(2\infty, \mathbb{R}), O(\infty, \infty))$ for $\alpha = 1/2$.

D. The construction described in 4.3 can be automatically transferred to an arbitrary module $P_{\alpha,s}$. It gives a homomorphism of $\mathrm{Diff}^{(\infty)}$ into the following groups:

α) $(\mathrm{U}(2\infty), \mathrm{U}(\infty) \times \mathrm{U}(\infty))$, where (α, s) is an arbitrary pair;
β) $(\mathrm{O}(2\infty, \mathbb{R}), \mathrm{U}(\infty))$, $\alpha = 1/2$, $s = 0$ (see 4.7.1);
γ) $(\mathrm{O}(2\infty + 1, \mathbb{R}), \mathrm{U}(\infty))$, $\alpha = 0$, $s = 1$ (see 4.7.2).

4.6. The Imbedding of Diff into $(\mathrm{Sp}(2\infty, \mathbb{R}), \mathrm{U}(\infty))$. We shall consider the space H_0 of real functions on S^1 with zero mean $\left(\int_0^{2\pi} f(\varphi)d\varphi = 0 \right)$ and with scalar product

$$\langle f, g \rangle = -\frac{\sqrt{2}}{\pi} \int_0^{2\pi} \int_0^{2\pi} \ln \left| \sin \frac{\varphi - \psi}{2} \right| f(\varphi)\overline{g(\psi)}d\varphi d\psi. \tag{4.9}$$

It is easy to see that $\langle e^{in\varphi}, e^{in\varphi} \rangle = 1/|n|$ (the vectors $e^{in\varphi}$ belong to the complexification of H_0). The *Hilbert transform* in H_0 is defined by

$$If(\varphi) = \frac{1}{\pi}\mathrm{v.p.} \int_0^{2\pi} \mathrm{ctg}\left(\frac{\varphi - \psi}{2} \right) f(\psi)d\psi. \tag{4.10}$$

The integral above is, generally speaking, divergent and it is necessary to interpret it as the principal value. It is not difficult to verify that the operator I is orthogonal, $I^2 = -1$, $I\exp(in\varphi) = i\,\mathrm{sgn}(n)\exp(in\,\varphi)$. Hence the operator I defines a complex structure on the space H_0. The complex scalar product in H_0 is defined by $\{f, g\} = \langle f, g \rangle + i\langle f, Ig \rangle$. It is easy to compute the symplectic form $\langle f, Ig \rangle$:

$$\langle f, Ig \rangle = \frac{\sqrt{2}}{\pi} \int_0^{2\pi} \int_0^{2\pi} f(\varphi)g(\psi)d\varphi d\psi. \tag{4.11}$$

Let the group Diff act on H_0 by real linear maps of the form

$$Q(p)f(\varphi) = f(p(\varphi))p'(\varphi),$$

where $p \in \mathrm{Diff}$. The group $(\mathrm{Sp}(2\infty, \mathbb{R}), \mathrm{U}(\infty))$ will be realized as in 3.1.2.

Theorem 4.5. (Segal (1981), Neretin (1983)) *The operator $Q(p)$ belongs to the group*

$$(\mathrm{Sp}(2\infty, \mathbb{R}), \mathrm{U}(\infty)).$$

Proof. It is possible to verify directly that the symplectic form (4.11) is invariant. It is also necessary to check that $L(q) = Q^t(q)Q(q) - \mathbb{1}$ is a Hilbert-Schmidt operator. We have

$$\langle L(q)f_1, f_2 \rangle =$$
$$= -\frac{\sqrt{2}}{\pi} \int_0^{2\pi} \int_0^{2\pi} \left[\ln \left| \sin \frac{q(\varphi_1) - q(\varphi_2)}{2} \right| - \ln \left| \sin \frac{\varphi_1 - \varphi_2}{2} \right| \right] \times$$
$$\times f_1(\varphi_1)f_2(\varphi_2)d\varphi_1 d\varphi_2,$$

where q is the inverse map to p. The expression inside the brackets is smooth, so its Fourier coefficients are rapidly decreasing.

Remark. Suppose that the parameter s (see 4.1) approaches 1. The operator $T_{0,1}(q)$ preserves H_0 and $T_{0,1|H_0} = Q$. Let us choose a constant $C(s)$ in such a way that $C(s)\langle e^{i\varphi}, e^{i\varphi}\rangle = 1$. Then the scalar product (4.9) on H_0 coincides with $\lim_{s\to 1} C(s)\langle \cdot, \cdot\rangle_s$. The Hilbert transform almost commutes with $T_{0,s}(q)$ for any s (in analogy with 4.3), but for $s = 1$, the symplectic form becomes invariant. Note also that I commutes with the action $Q(p)$ of the group $\mathrm{PSL}_2(\mathbb{R})$.

4.7. The Imbedding of $\mathrm{Diff}^{(2)}$ into $(\mathrm{O}(2\infty, \mathbb{R}), \mathrm{U}(\infty))$.

4.7.1. Let L^2_- denote the subspace of real functions in $L^2(S^1)$ satisfying the condition $f(\varphi + \pi) = -f(\varphi)$. Let the group $\mathrm{Diff}^{(2)}$ act on L^2_- by the formula

$$P_{1/2,0}(q)f(\varphi) = f(q(\varphi))q'(\varphi)^{1/2}. \tag{4.12}$$

The Hilbert transform (4.10) gives a complex structure on L^2_-. Suppose that the group $(\mathrm{O}(2\infty, \mathbb{R}), \mathrm{U}(\infty))$ is realized as in 3.1.2.

Theorem 4.6. (Neretin (1983)). *The formula (4.12) gives an imbedding of the group $\mathrm{Diff}^{(2)}$ into $(\mathrm{O}(2\infty, \mathbb{R}), \mathrm{U}(\infty))$.*

Remark. The described construction is almost identical to the construction in 4.2 (the case $s = 0$). But for $s = 0$, the image of $\mathrm{Diff}^{(2)}$ clearly belongs to the smaller group $(\mathrm{O}(2\infty, \mathbb{R}), \mathrm{U}(\infty))$.

4.7.2. The imbedding of the group Diff into $(\mathrm{O}(2\infty+1, \mathbb{R}), \mathrm{U}(\infty))$. Let the group Diff act on the space $L^2(S^1)$ by the formula

$$P_{0,0}(q)f(\varphi) = f(q(\varphi))q'(\varphi)^{1/2}. \tag{4.13}$$

The Hilbert transform (4.10) annihilates the one-dimensional subspace of constants and gives a complex structure on the space of functions with zero mean. The group $(\mathrm{O}(2\infty+1, \mathbb{R}), \mathrm{U}(\infty))$ is defined in the same way as in 3.5.6.

Theorem 4.7. (R.S.Ismagilov) *The operators $P_{0,0}(q)$, $q \in \mathrm{Diff}$ belong to the group $(\mathrm{O}(2\infty + 1, \mathbb{R}), \mathrm{U}(\infty))$.*

§5. Unitary Representations of Diff

In 5.1–5.3, highest weight representations are studied, while in 5.4, we discuss an example of a representation without a highest weight. The subsection 5.6 contains the description of general representations corresponding to almost invariant structures.

5.1. Constructions of Representations with Highest Weights.

5.1.1. The bosonic construction.

The notation of 4.6 will be used here. The formula

$$Z_{\alpha,\beta}(q)f(\varphi) = f(q(\varphi))q'(\varphi) + \alpha(q'(\varphi) - 1) + \beta q''(\varphi)/q'(\varphi), \qquad (5.1)$$

where $q \in \mathrm{Diff}$, $\alpha, \beta \in \mathbb{R}$, defines an affine transformation of the space H_0. It is easy to check that the map $q \to Z_{\alpha,\beta}(q)$ is a homomorphism of the group Diff into the affine group of the space H_0. Theorem 4.5 then implies that $Z_{\alpha,\beta}(q) \in M = (\mathrm{Sp}(2\infty, \mathbb{R}), \mathrm{U}(\infty)) \ltimes H_0$. Recall that H_0 is a complex vector space, the multiplication by imaginary unit i being defined by the Hilbert transform. Using 3.3.3, we get a projective unitary representation of M and hence also of Diff on the Fock space $F(H_0)$. The representation of Diff constructed in this way will be denoted by $N_{\alpha,\beta}$.

Theorem 5.1. (Neretin (1983)). *The vacuum vector of $F(H_0)$ is a singular vector of the representation $N_{\alpha,\beta}$ with weight*

$$(h, c) = (\tfrac{1}{2}(\alpha^2 + \beta^2),\ 1 + 12\beta^2).$$

If $(h, c) \neq (k^2/4, 1)$, where $k \in \mathbb{Z}$, then the representation $N_{\alpha,\beta}$ is irreducible. Otherwise, $N_{\alpha,\beta} = \oplus L((k + 2j)^2/4, 1)$, where $j = 0, 1, 2, \ldots$.

Remark. The representations $L(h, c)$ constructed above fill in the domain $c \geq 1$, $h \geq \frac{c-1}{24}$.

Let us first write down the final formulas for the generators of the Virasoro algebra in $F(l^2)$:

$$L_n = \sum_{k=1}^{\infty} \sqrt{k(n+k)}\, z_{n+k} \frac{\partial}{\partial z_k} + \qquad (5.2)$$
$$+ \frac{1}{2}\sum_{j=1}^{n-1} \sqrt{j(n-j)}\, z_i z_j + \sqrt{n}(\alpha + i\beta n) z_n,$$

$$L_0 = \sum_{k=1}^{\infty} k z_k \frac{\partial}{\partial z_k} + \frac{1}{2}(\alpha^2 + \beta^2). \qquad (5.3)$$

For $n > 0$, $L_{-n} = L_n^*$, (see 3.2.3).

We shall make the computation in the case $\alpha = \beta = 0$. Firstly, we have to show how the group $(\mathrm{Sp}(2\infty, \mathbb{R}), \mathrm{U}(\infty))$ can be represented in the form (3.15). Repeating the procedure described in the Remark in 3.3.2 for the basis $e_n = \sqrt{2/n}\cos n\varphi$, $n > 0$, in H_0, we get the following analogue of the basis (3.16): $f_n = e^{in\varphi}/\sqrt{n}$ corresponds to $(ie_n + Ie_n)/\sqrt{2}$ and $f_{-n} = e^{-in\varphi}/\sqrt{n}$ corresponds to $(ie_n - Ie_n)/\sqrt{2}$. On the level of the Lie algebra Vect_C, the action (5.1) looks like (see 1.0)

$$\pi(\alpha(\varphi))f(\varphi) = \alpha(\varphi)f'(\varphi) + \alpha'(\varphi)f(\varphi).$$

So the relations $Y_k e^{in\varphi} = (n+k)e^{i(n+k)\varphi}$ or $Y_k f_n = \sqrt{|(n+k)k|} f_{n+k}$ hold for the generators $Y_k \in \mathrm{Vect}_C$. Then the matrices (3.17) and the corresponding operators (3.18) can be written down. The relation (3.19) implies the commutation relations $[Y_n, Y_m] = (m-n)Y_{m+n} + c_{m,n}$. It is easy to compute the numbers $c_{m,n}$ from (3.19); we get as the result, as it turns out, exactly the commutation relations of the Virasoro algebra (1.2). If α or β is not equal to 0 (as well as in 5.1.2 and 5.1.3), a constant should be added to Y_0 to obtain (1.2). It is easy to compute its value.

Remark. If we introduce the operators $a_k f = \sqrt{k} z_k f$, $\widehat{a}_k f = \sqrt{k} \frac{\partial}{\partial z_k} f$, then the expressions (5.2) and (5.3) simplify and we get the *Virasoro formula*, found in 1970, which was the beginning of the theory considered in the paper.

5.1.2. The fermionic construction. We shall use the notation of 4.7. The expressions for $\mathrm{Spin}_\pm(P_{1/2,0}(q))$ and $\mathrm{Spin}_0(P_{0,0}(q))$ (see (4.12),(4.13), 3.5.3, 3.5.4) evidently determine a unitary projective representation of Diff .

Theorem 5.2. $\mathrm{Spin}_+(P_{1/2,0}(q)) \simeq L(0, 1/2)$,

$$\mathrm{Spin}_0(P_{0,0}(q)) \simeq L\left(\frac{1}{16}, \frac{1}{2}\right), \quad \mathrm{Spin}_-(P_{1/2,0}(q)) \simeq L(1/2, 1/2).$$

To write the action of the Virasoro algebra, it is necessary to linearize (4.12) and (4.13), to write them in matrix form and to substitute them into (3.22). For $L(0, 1/2)$ and $L(1/2, 1/2)$, we get the following expressions for the generators of the Virasoro algebra, $n > 0$:

$$L_n = \sum_{k \geq 0} \left(k + \frac{n}{2}\right) \xi_{n+k} \frac{\partial}{\partial \xi_k} + \frac{1}{4} \sum_{\alpha+\beta=n} (\alpha - \beta)\xi_\alpha \xi_\beta, \qquad (5.4)$$

where the index of ξ takes values in the set $1/2, 3/2, 5/2, \ldots$. Next, we have $L_0 = \sum k \xi_k \frac{\partial}{\partial \xi_k}$, $L_{-n} = L_n^*$ (see 3.4.2). The module $L(0, 1/2)$ is realized in the space of even functions, the highest weight vector being $f(\xi) = 1$. The module $L(1/2, 1/2)$ is realized in the space of odd functions, the highest weight vector being $f(\xi) = \xi_{1/2}$. The irreducibility follows from Theorem 5.3.

Finally, let us note that the formulas for $L(1/16, 1/2)$ are basically the same, only that the indices at ξ take the values $0, 1, 2, \ldots$, at the same time ξ_0 should be replaced by $\xi_0/\sqrt{2}$ and $\frac{\partial}{\partial \xi_0}$ by $\frac{1}{\sqrt{2}} \frac{\partial}{\partial \xi_0}$.

5.1.3. The two-fermionic construction. As was explained in 4.3 or in more general situation in 4.5 Dα), the group $\mathrm{Diff}^{(\infty)}$ can be imbedded into the group $(\mathrm{U}(2\infty), \mathrm{U}(\infty) \times \mathrm{U}(\infty))$, the unitary representation of which was constructed in 3.6.5. We shall describe now the unitary representation of the Virasoro algebra corresponding to the representation $P_{\alpha,s}$.

Let p take values in the set $\{n + \alpha \mid n \in \mathbb{Z}\}$. We shall introduce the operators A_p in the Fock space Λ : $A_p f = \xi_p f$ if $p \geq 0$ and $A_p f = \frac{\partial}{\partial \xi_p} f$ if $p < 0$. Then

$$L_k = \sum_p \left(p + \frac{1+is}{2} k \right) A_{p+k} A_p^*$$

for $k \neq 0$, $L_0 = \sum |p| \xi_p \frac{\partial}{\partial \xi_p} + c'$, where c' depends on α and s. These representations are highly reducible.

Remark. The above construction gives a realization of all the representations 5.1.1 and, evidently, only of them. See also 9.3. Another variant of the construction can be found in Feigin and Fuks (1982).

5.2. The Conditions for Unitarizability of Modules $L(h, c)$. Let us first note that in $\Omega = \{(h, c) : h > 0, c > 1\}$, all the modules $L(h, c)$ are nonsingular (see 2.3) and some of them are unitarizable (see 5.1.1). Hence, by continuity, all the modules in $\overline{\Omega} = \{(h, c) : h \geq 0, c \geq 1\}$ are unitarizable. As well, we know that the modules $L\left(\frac{1}{16}, \frac{1}{2}\right), L\left(0, \frac{1}{2}\right), L\left(\frac{1}{2}, \frac{1}{2}\right)$ and the one-dimensional module $L(0, 0)$ are also unitarizable (see 5.1.2).

Theorem 5.3. *The modules $L(h, c)$ are unitarizable if and only if one of the following conditions is satisfied:*

$\alpha)$ $h \geq 0$, $c \geq 1$;
$\beta)$ $c = 1 - \frac{6}{p(p+1)}$, $h = \frac{(\alpha p - \beta(p+1))^2 - 1}{4p(p+1)}$, $\qquad$ (5.5)
$\qquad$ *where $\alpha, \beta, p \in \mathbb{Z}$, $p \geq 2$, $1 \leq \alpha \leq p$, $1 \leq \beta \leq p - 1$.*

The proof of the necessary conditions was done by the author in 1983 and by D.Friedan, Z.Qiu and S.Shenker (see Lepowsky et al. (1985)). The sufficient conditions were proved (see 7.3 and 7.4) by P.Goddard, A.Kent and D.Olive (see Goddard et al. (1986)).

The proof of the necessary conditions. We shall use here the notation and facts from Sections 2.2–2.4. Let L be a highest weight module. Let $\sigma_n(L)$ denote the sign of the determinant of the Shapovalov form in the n-th homogeneous component of the module $L, \sigma_n(L) = 0, \pm 1$. To prove that L is not unitarizable, it is sufficient to find a positive integer n with the property $\sigma_n(L) = -1$.

The number $\alpha\beta$ will be called the level of the quadric $\Phi_{\alpha,\beta} = 0$.

Lemma 5.1. *Let us consider all points (h, c) of a quadric $\Phi_{\alpha,\beta}(h, c) = 0$.*

A) *The singular vector of order $\alpha\beta$ in $M(h, c)$ depends analytically on (h, c).*
B) *If the hyperbola Γ given by the equation $\Phi_{\alpha,\beta} = 0$ does not intersect any quadrics $\Phi_{\gamma,\delta} = 0$ of the level $\leq n + \alpha\beta$, then for all $(h, c) \in \Gamma$ we have*

$$\sigma_{n+\alpha\beta}(M(h, c)/M(h + \alpha\beta, c)) =$$
$$= \operatorname{sgn} \frac{D_{n+\alpha\beta}(h, c)}{D_n(h + \alpha\beta, c)(\Phi_{\alpha,\beta}(h, c)\lambda)^{p(n)}}, \qquad (5.6)$$

where λ does not depend on n.

Proof. Part A) follows from Theorem 2.2, part 2) and part B) follows from estimates of the orders of zeros of $D_k(h, c)$.

If we remove all curves $\Phi_{\alpha,\beta}(h, c) = 0$ of levels $\leq n$, the plane splits into connected open sets. Let Σ_n be the domain containing Ω, let $\overline{\Sigma}_n$ be its closure and $\partial \Sigma_n$ its boundary.

Lemma 5.2.

A) *If $(h, c) \in \overline{\Sigma}_n$, $k \leq n$, then $\sigma_k(L(h, c)) = +1$.*
B) *Any point in $\Sigma_n \backslash \overline{\Sigma}_{n+1}$ is separated from Σ_{n+1} by exactly one quadric of level $n + 1$.*

Corollary 5.1.   *If $(h, c) \in \Sigma_n \backslash \overline{\Sigma}_{n+1}$, then $\sigma_{n+1}(M(h, c)) = -1$, so $L(h, c)$ is not unitarizable.*

The curve $\Phi_{\alpha,\beta} = 0$ is cut by all quadrics of level $\leq \alpha\beta$ into pieces. Let $I_{\alpha,\beta}$ denote the piece containing the point $B_{\alpha-\beta} = (\frac{1}{4}(\alpha - \beta)^2, 1)$. We know that $\cap \overline{\Sigma}_n = \Omega$, hence the only "suspicious" points where unitarizable modules can lie are points in $\cup \partial \Sigma_n = \cup I_{\alpha,\beta}$.

The property of the first place of degeneracy. We shall consider all intersection points of the curve $\Phi_{\alpha,\beta} = 0$ with all quadrics of level less than or equal to $k + \alpha\beta$. Let $A^+_{\alpha,\beta,k}$ denote the nearest intersection point "from the right" to $B_{\alpha-\beta}$ ("from the right" means that we go along the curve $\Phi_{\alpha,\beta} = 0$ in the direction indicated by the increasing parameter h). Similarly, let $A^-_{\alpha,\beta,k}$ denote the nearest intersection point from the left (in a small neighbourhood of the point $B_{\alpha-\beta}$ on the curve $\Phi_{\alpha,\beta} = 0$, the modules corresponding to $A^{\pm}_{\alpha,\beta,k}$ are the "first" degenerate modules).

Lemma 5.3.   *The set of all points of the form $A^+_{\alpha,\beta,k}$ coincides with the set of all points of the form $A^-_{\alpha,\beta,k}$ and coincides with the set of singular points described in Theorem 5.3.*

Suppose that n is increasing. Then the sign in formula (5.6) can become negative if the sign of the numerator or the denominator changes. However, before the arc $(B_{\alpha-\beta}, A^+_{\alpha,\beta,0})$ of the curve intersects the first quadric $\Phi_{\gamma,\delta} = 0$, the signs of the numerator and the denominator change simultaneously (the factor $\Phi_{\alpha+\omega,\beta+\omega}(h, c) < 0$ in the numerator appears at the same time as the factor $\Phi_{\omega,\alpha+\beta+\omega}(h, c) < 0$ in the denominator).

Let $\Phi_{\gamma,\delta} = 0$ be a quadric of lowest level intersecting $(B_{\alpha-\beta}, A^+_{\alpha,\beta,0})$, and let $\gamma\delta - \alpha\beta = l$. Then the intersection point is $A^+_{\alpha,\beta,l}$. Thus, on the arc $(B_{\alpha,\beta}, A^+_{\alpha,\beta,l})$, the relation $\sigma_{\alpha\beta+l}(M(h, c)/M(h + \alpha\beta, c)) = +1$ holds. By analyticity in $A^+_{\alpha,\beta,0}$, the sign of the expression $\sigma_{\alpha\beta+l}(M(h, c)/M(h + \alpha\beta, c))$ changes. Repeating the same process, we get the following lemma.

Lemma 5.4.   *On the arc of the quadric $\Phi_{\alpha,\beta} = 0$ bounded by the neighbouring points $A^+_{\alpha,\beta,m-1}$ and $A^+_{\alpha,\beta,m}$, the relation $\sigma_m(L(h, c)) = -1$ holds.*

The same is true also for $A_{\alpha,\beta,m}^{-}$, hence the proof of Theorem 5.3. is finished.

5.3. The Integrability.

Theorem 5.4. (Goodman and Wallach (1985)). *Any unitarizable module $L(h,c)$ integrates to a projective unitary representation of the group Diff.*

Sketch of the Proof Let $\mathcal{L}_{\mathbb{R}} = \text{Vect} \ltimes \mathbb{R} \subset \mathcal{L}$ (see 1.3). Operators in $i\mathcal{L}_{\mathbb{R}}$ are essentially selfadjoint on $L(h,c)$ and have a common dense Gårding subspace (see Goodman and Wallach (1985)). In any other case, integrability would follow (see the Remark after Theorem 7.2), but the group Diff is not generated by its one-parameter subgroups, see A.1.1.

It is sufficient to prove that for any highest weight representation π and any $q \in$ Diff, the representations π and $\pi \circ \text{Ad}_q$ are equivalent (if $U(q)$ is an intertwining operator, then the desired representation of Diff is given by $q \to U(q)$). This is, in turn, equivalent to the following property: the minimal eigenvalues of the operators $\pi(L_0)$ and $\pi(\text{Ad}_q(L_0))$ coincide. If the latter property holds for the representations π_1 and $\pi_1 \otimes \pi$, then it also holds for π. So everything now follows from Theorem 5.1.

Remark. In the proof, we are basically showing that the notion of "highest weight module" does not depend on the choice of a basis in $\mathcal{L}$.

5.4. An Example of a Series of Unitary Representations of Diff Without a Highest Weight.

A series of imbeddings $T_{0,s}$ of the group Diff into $(\text{GL}(\infty,\mathbb{R}), \text{O}(\infty,\mathbb{R}))$ was constructed in 4.1, a series of unitary representations $\text{Exp}_{\omega}^{\pm}$ of the group $(\text{GL}(\infty,\mathbb{R}), \text{O}(\infty,\mathbb{R}))$ was constructed in 3.6.4. In this way, the composition $\text{Exp}_{\omega}^{\pm} \circ T_{0,s}$ is a unitary representation of the group Diff.

We shall write down explicit formulas for the Lie algebra Vect_C in the case $\omega = 0$. Let us choose the orthonormal basis $f_n = e^{in\varphi}\sqrt{c_n}$ in $(H_s)_C$, where c_n is given by formula (4.5). The generators Y_p are (in the basis f_n) given by the matrices (see 1.0)

$$A_p f_n = \sqrt{\frac{c_{n+p}}{c_n}}\left(n + p\frac{1+s}{2}\right)f_{n+p}.$$

The dual operators to A_{-p} act as

$$A_{-p}^{*}f_n = \sqrt{\frac{c_n}{c_{n+p}}}\left(n + p\frac{1-s}{2}\right)f_{n+p}.$$

Going through the construction of the representations Exp_{ω}, the generators Y_p of the algebra Vect_C will be written in the form (3.17)

$$Y_p \sim \begin{pmatrix} A_p + A_{-p}^{*} & A_p - A_{-p}^{*} \\ A_{-p}^{*} - A_p & A_p + A_{-p}^{*} \end{pmatrix}.$$

So the formula (3.18) can be used to get a representation of Vect_C by differential operators in a countable number of variables. We prefer, however, a slightly different notation. Let us introduce the operators $\hat{a}_k$ and $\hat{a}_k^*$, $k \in \mathbb{Z}$, $\hat{a}_k f = \sqrt{c_k^{-1}} z_k f$, $\hat{a}_k^* f = \sqrt{c_k^{-1}} \frac{\partial}{\partial z_k} f$. Then

$$Y_p = \sum_n (u_{p,n} + v_{p,n}) a_{p+n} a_n^* + \frac{1}{2} \sum_n (u_{p,n} - v_{p,n}) a_n a_{p-n} +$$

$$+ \frac{1}{2} \sum (v_{p,n} - u_{p,n}) a_n^* a_{-p-n}^*, \tag{5.7}$$

where

$$u_{p,n} = c_n^{-1} \left(n + p \frac{1+s}{2} \right), \quad v_{p,n} = c_{p+n}^{-1} \left(n + p \frac{1-s}{2} \right).$$

It is clear that all coefficients are rational expressions in n. If $p = 0$, ± 1, all sums but the first are equal to zero.

Conjecture. The representations $\pi_{\omega,s}^{\pm} = \mathrm{Exp}_{\omega}^{\pm} \circ T_{0,s}$ are irreducible.

The restriction of $\pi_{\omega,s} = \pi_{\omega,s}^+ \oplus \pi_{\omega,s}^-$ to $\mathrm{PSL}_2(\mathbb{R})$ has quite an interesting structure. It decomposes into a direct integral (with infinite multiplicity) over the principal series plus an infinite sum (with infinite multiplicity) over the discrete series plus a finite discrete sum

$$I \oplus \left(\oplus_k T_{0,1-k(1-s)}^0 \right),$$

where I is the one-dimensional representation, $T_{0,\alpha}^0$ was introduced in 4.5. B and k satisfies the condition $k(1 - s) < 1$. Using 3.6.4, the problem of the decomposition reduces to the decomposition of $\oplus S^k T_{0,s}^0$ (see 5.5).

Lemma 5.5. *The cyclic hull of any of the subspaces I and $T_{0,1-k(1-s)}^0$ with respect to the group Diff is irreducible.*

Proof. Let M denote the cyclic hull of I and suppose that $M = M_1 \oplus M_2$. The projections of I onto M_1 and M_2 have to be $\mathrm{PSL}_2(\mathbb{R})$-invariant one-dimensional subspaces. But here is only one such subspace. So $I \subset M_1$ or $I \subset M_2$, hence $M = M_1$ or $M = M_2$.

Remark. It follows from A.3.2 that our series can be realized in the space L^2 with respect to the Gaussian measure on an infinite-dimensional space.

5.5. Tensor Products of Unitary Representations of $\mathrm{PSL}_2^{(\infty)}(\mathbb{R})$. The principal series $P_{\alpha,s}^0$ and the complementary series $T_{\lambda,s}^0$ of representations of $\mathrm{PSL}_2^{(\infty)}(\mathbb{R})$ were introduced in 4.5. Also (see Zhelobenko and Shtern (1983)), the group $\mathrm{PSL}_2^{(\infty)}(\mathbb{R})$ has the series $T_\lambda, 0 < \lambda < \infty$, of *unitary highest weight representations* (if $Y_0, Y_{\pm 1} \in \mathrm{Vect}_C$ is a basis of sl_2 – see 1.2 – then there exists a vector v in the space of the representation T_λ such that $Y_0 v = \lambda v$, $Y_1 v = 0$))

and the series $T'_\lambda, 0 < \lambda < \infty$, of *unitary lowest weight representations* (i.e. correspondingly $Y_0 v = -\lambda v$, $Y_{-1} v = 0$). Suppose now that $0 < \lambda < 1/2$. We shall change the notation; the representation T_λ will be denoted by $T^0_{\lambda,1-2\lambda}$ and T'_λ by $T^0_{-\lambda,1-2\lambda}$. The set of all representations $T^0_{\alpha,s}$, where $s > 0$, $s \pm 2|\alpha| \leq 1$, will be called the *generalized complementary series*.

Theorem 5.5. (Pukanszky et al.) *Let A_1, A_2 be irreducible unitary representations of $\mathrm{PSL}_2^{(\infty)}(\mathbb{R})$ such that $\dim A_i > 1$. Then*

A) $P^0_{\alpha,s}$ *cannot be part of the decomposition of $A_1 \otimes A_2$ as an irreducible subrepresentation.*

B) *The representations of the generalized complementary series can belong only to the discrete part of the decomposition of $A_1 \otimes A_2$ and can have at most multiplicity one. All such imbeddings are given by*

$$T^0_{\alpha,s} \otimes T^0_{\beta,t} \supset T^0_{\alpha+\beta,s+t-1},$$

where $s + t > 1$.

5.6. Representations of Diff$^{(\infty)}$ Corresponding to Almost Invariant Structures. Let μ be an imbedding of Diff$^{(\infty)}$ into a good (G,K)-pair (see 4.5, series A,B,D) and let R be a representation of the group (G,K). Then $R \circ \mu$ is a unitary representation of the group Diff.

In the case of the series C of almost invariant structures, it is necessary to note for the construction of the representations that

Cα) $(\mathrm{GL}(\infty,\mathbb{C}),\mathrm{U}(p,\infty)) = (\mathrm{GL}(\infty,\mathbb{C}),\mathrm{U}(\infty))$,

Cβ) $\mathrm{Im}\ \mathrm{Diff}^{(\infty)} \subset (\mathrm{GL}(2\infty,\mathbb{C}),\mathrm{U}(\infty) \times \mathrm{U}(\infty)) \subset$
$\subset (\mathrm{GL}(2\infty,\mathbb{C}),\mathrm{U}(2\infty))$,

Cγ) $(\mathrm{GL}(\infty,\mathbb{R}),\mathrm{O}(p,\infty)) = (\mathrm{GL}(\infty,\mathbb{R}),\mathrm{O}(\infty,\mathbb{R}))$,

Cδ) $\mathrm{Im}\ \mathrm{Diff}^{(2)} \subset (\mathrm{GL}(2\infty,\mathbb{R}),\mathrm{O}(\infty,\mathbb{R}) \times \mathrm{O}(\infty,\mathbb{R})) \subset$
$\subset (\mathrm{GL}(2\infty,\mathbb{R}),\mathrm{O}(2\infty,\mathbb{R}))$.

In the case of the series A and B, it is easy to compute the spectrum of the restriction of $R \circ \mu$ to $\mathrm{PSL}_2(\mathbb{R})$, because the image of $\mathrm{PSL}_2(\mathbb{R}) \subset$ Diff belongs to the subgroup $K \subset (G,K)$ in these cases (see 3.6.2 and Theorem 5.5). Analogues of Lemma 5.5 are obtained in this way.

The problem of the restriction of $R \circ \mu$ to subgroups of the form $\mathrm{PSL}_2^{(n)}(\mathbb{R})$ in Diff (see 1.1) is solvable for the series A, and sometimes also for the series B and C. The scheme for calculations is based on the replacement of a Diff- invariant structure by an equivalent $\mathrm{PSL}_2^{(n)}(\mathbb{R})$-invariant structure.

Analogues of formulas (5.7) can be written down using the corresponding substitution into formulas (3.18) and (3.22).

Appendices

A.5.1. On p-adic Analogues. Let $\mathbb{Q}_p\mathbb{P}^1$ denote the p-adic projective line. Let An_p be the group of all analytic 1-1 maps of $\mathbb{Q}_p\mathbb{P}^1$ into itself such that their inverses are also analytic. Let us consider the space H_s of all real-valued functions on $\mathbb{Q}_p\mathbb{P}^1$ with the scalar product

$$\langle f, g \rangle = \int_{Q_p} \int_{Q_p} |z_1 - z_2|^{s-1} f(z_1) g(z_2) dz_1 dz_2,$$

where $0 < s < 1$. Let An_p act on H_s by the formula

$$T_s(q)f(z) = f(q(z))|q'(z)|^{\frac{1+s}{2}}.$$

It is not difficult to verify that the operators $T_s^*(q)T_s(q) - \mathbb{1}$ are finite-dimensional, hence $T_s(q) \in (\mathrm{GL}(\infty, \mathbb{R}), \mathrm{O}(\infty, \mathbb{R}))$. Lemma 5.5 remains valid in this case.

It is possible to weaken the condition of analyticity.

Let J_p be the *Bruhat-Tits tree*, i.e. a tree (connected graph without cycles) such that there are $(p + 1)$ edges entering every vertex of the graph. A sequence $(b_1, b_2, \ldots)$ of points will be called a curve in J_p if b_i and b_{i+1} are connected by an edge and $b_i \neq b_{i+2}$. Two curves are equivalent if they coincide starting from an index. An absolute A_p of the tree J_p is the set of all equivalence classes of curves. It is easy to establish a natural 1-1 correspondence between A_p and $\mathbb{Q}_p\mathbb{P}^1$. (Choose a point $a \in A_p$. Then the set of all paths from a to another point of the absolute identifies in an evident, but not canonical, way with the set of all p-adic numbers.)

There are deep analogies between the described construction and the Lobachevskij plane on the level of geometry as well as on the level of representation theory (P.Cartier, G.I.Ol'shanskij). An analogue of $\mathrm{SL}_2(\mathbb{R})$ is the group of all automorphisms of a graph of J_p. The group $\mathrm{Diff}\,(A_p)$ is defined as the group of all 1-1 transformations of A_p which can be extended to a neighbourhood of the absolute (the precise definition is: $q \in \mathrm{Diff}\,(A_p)$ if there exist finite subgraphs C, D in J_p and an isomorphism $Q : (J_p \backslash C) \to (J_p \backslash D)$ inducing a map q on A_p.)

The imbedding of An_p into $(\mathrm{GL}(\infty, \mathbb{R}), \mathrm{O}(\infty, \mathbb{R}))$ can be easily extended to the group $\mathrm{Diff}\,(A_p)$.

§6. Affine Lie Algebras

6.0. The Notation for Sections 6–8. Let G be a complex simple Lie group of rank l and $\mathfrak{g}$ its Lie algebra. Let us choose a fixed Cartan subalgebra $\mathfrak{h} \subset \mathfrak{g}$, let Δ denote the root system and Δ_+ the set of all positive roots. Let $\alpha_1, \ldots, \alpha_l$ be the set of simple roots and α_+ the maximal root, let X_α denote a root vector corresponding to the root α and $\mathfrak{g}_\alpha$ the corresponding root space, let W be the Weyl group. The Killing form $B(\cdot,\cdot)$ is normalized in such a way that the length squared of the long root is equal to 2 (the normalization is important). Let K denote the maximal compact subgroup of the group G, $\mathfrak{k}$ its Lie algebra and $\mathbf{T}$ a maximal torus in K.

6.1. Affine Lie Algebras. Let $\tilde{\mathfrak{g}} = C^{(\infty)}(S^1, \mathfrak{g})$ be the space of all smooth functions on S^1 with values in $\mathfrak{g}$. The pointwise commutation defines the structure of a Lie algebra on $\tilde{\mathfrak{g}}$. If $Y_1, \ldots, Y_p$ is a basis of $\mathfrak{g}$, then the functions of the form

$$Y_j e^{in\varphi}, \quad n \in \mathbb{Z}, \quad j = 1, \ldots, p,$$

form a basis of $\tilde{\mathfrak{g}}$. We shall define the map $c : \tilde{\mathfrak{g}} \times \tilde{\mathfrak{g}} \to \mathbb{C}$ by

$$c(f, g) = \frac{1}{2\pi i} \int_0^{2\pi} B\left(\frac{d\,f(\varphi)}{d\,\varphi}, g(\varphi)\right) d\varphi,$$

it is not difficult to verify that $c \in H^2(\tilde{\mathfrak{g}}, \mathbb{C})$ or, in other words, that the formula

$$[(f_1, d_1), (f_2, d_2)] = ([f_1, f_2], c(f_1, f_2))$$

defines the structure of a Lie algebra on the space $\mathfrak{g} \oplus \mathbb{C}$. The Lie algebra constructed in this way is called an *affine Lie algebra* and is denoted by $\widehat{\mathfrak{g}}$. The central element $(0, 1)$ of the algebra $\widehat{\mathfrak{g}}$ is denoted by z.

Another definition of the affine Lie algebra is realized on the space $\breve{\mathfrak{g}}$ of all formal Laurent polynomials with coefficients in $\mathfrak{g}$, i.e. the set of all expressions of the form

$$p(t) = \sum_{j=-\infty}^{\infty} X_j t^j,$$

where $X_j \in \mathfrak{g}$ and the sum has only a finite number of nontrival terms. The bracket is given by $[X_1 t^k, X_2 t^l] = [X_1, X_2] t^{k+l}$. The cocycle c is defined by the formula

$$c(p(t), q(t)) = \operatorname{res} B(p'(t), q(t)),$$

where res denotes the residue. The two described realizations of affine Lie algebras are connected by the substitution $t = e^{i\varphi}$.

The original algebra $\mathfrak{g}$ is often identified with the subalgebra of $\widehat{\mathfrak{g}}$ consisting of all vectors of the form Xt^0. Finally, let us introduce the subalgebra $\widehat{\mathfrak{k}}$ of $\widehat{\mathfrak{g}}$ generated by the vectors $P\cos n\varphi, P\sin n\varphi$ and iz, where $P \in \mathfrak{k}$.

Notation. Let $X \in \mathfrak{g}$ and $\theta(\varphi) \in C^\infty(S^1)$. Then $X\theta(\varphi)$ will be denoted by $X(\theta)$ and $Xe^{ik\varphi}$ by $X^{(k)}$.

6.2. Lie Groups Corresponding to Affine Lie Algebras. It is natural to consider $C^\infty(S^1, G)$ as the Lie group corresponding to the Lie algebra $C^\infty(S^1, \mathfrak{g})$. Its "compact form" is the group $C^\infty(S^1, K)$. A countable family of central extensions of $C^\infty(S^1, K)$ by the circle $\mathbb{R}/2\pi\mathbb{Z}$ corresponds to the Lie algebra $\widehat{\mathfrak{k}}$, the universal central extension $\widehat{K}$ being among them. This extension is given by a topologically nontrivial fibre bundle $\widehat{K} \to C^\infty(S^1, K)$ and cannot be lifted to a fibration with fiber $\mathbb{R}$ (its Chern class is discussed in Segal (1981)). Analogously, there exists a universal extension $\widehat{G}$ of the group $C^\infty(S^1, G)$ by the multiplicative group of $\mathbb{C}$.

No good formulas for multiplication in $\widehat{K}$ and $\widehat{G}$ are known (the constructions of 8.1 or 9.4 make it possible to write an explicit formula for the cocycle using Fredholm determinants but the desire to find something better remains. However, the existence of unsatisfying formulas does not imply that good ones do not exist). As a small consolation, note that it is easy to write formulas for the adjoint and coadjoint actions of $\widehat{K}$ and $\widehat{G}$.

6.3. Root Systems. Many properties of affine Lie algebras are reminiscent of the ordinary semisimple Lie algebras. Let us introduce the corresponding definitions. The abelian subalgebra $\widehat{\mathfrak{h}} = \mathfrak{h} \oplus \mathbb{C}z$ is called the *Cartan subalgebra* of $\widehat{\mathfrak{g}}$.

The *root system* $\widetilde{\Delta}$ of the algebra $\widehat{\mathfrak{g}}$ is defined by

$$\widetilde{\Delta} = \{m\delta + \gamma, m \in \mathbb{Z}, \gamma \in \Delta \cup 0\} \setminus \{0\},$$

where δ is a formal variable. The *root subspace* $\widehat{\mathfrak{g}}_{\gamma + m\delta}$ in $\widehat{\mathfrak{g}}$ corresponding to the root $\gamma + m\delta$ is the space $\mathfrak{g}_\gamma t^m$ for $\gamma \neq 0$, resp. $\mathfrak{h}t^m$ for $\gamma = 0$ (in the latter case, the root space is not one-dimensional, with the exception of the case $\mathfrak{g} = \mathrm{sl}_2$). The *simple roots of* $\widehat{\mathfrak{g}}$ are the simple roots $\alpha_1, \ldots, \alpha_l$ of the algebra $\mathfrak{g}$ and $\delta - \alpha_+$, where α_+ is the maximal root. A root $\gamma + m\delta$ is called *positive* if $m > 0$ or $m = 0$ and $\gamma \in \Delta_+$.

We shall consider all modules over $\widehat{\mathfrak{g}}$ as graded modules in the following sense: the module M is assumed to be decomposable into a sum of *weight subspaces* $M = \oplus M_{p,l,k}$, where $l, k \in \mathbb{Z}$, and p belongs to the lattice of all weights of $\mathfrak{g}$ such that for every $x \in M_{p,l,k}$ we have:

1. $hx = p(h)\,x$, if $h \in \mathfrak{h}$, and $zx = kx$;
2. $X_\alpha^{(s)} M_{p,l,k} \subset M_{p+\alpha, l+s, k}$.

The number k, which is constant for every irreducible module, will be called the *level* of the module.

Let $\mu = \gamma + k\delta$ be a positive root, $\gamma \neq 0$. Let us consider the algebra $A^{(\mu)}$ generated by $X_\gamma^{(k)}$, $X_{-\gamma}^{(-k)}$, $H_\mu = [X_\gamma^{(k)}, X_{-\gamma}^{(-k)}]$. It is clear that $A^{(\mu)}$ is isomorphic to sl_2. Let us choose a suitable multiple h_μ of H_μ in such a way that $[h_\mu, X_\gamma^{(k)}] = 2X_\gamma^{(k)}$. Let $\mathrm{SU}_2^{(\mu)}$ denote the subgroup of $\widehat{K}$ corresponding to the subalgebra $A^{(\mu)}$ and let $\omega^{(\mu)}$ denote the generator of the Weyl group of $\mathrm{SU}_2^{(\mu)}$.

If the restriction of a $\widehat{\mathfrak{g}}$-module V to any subalgebra $A^{(\mu)}$ is a direct sum of finite-dimensional $A^{(\mu)}$-modules, then V is called *locally nilpotent*.

Appendices

A.6.1. Dynkin Diagrams. Dynkin diagrams for the series $\widehat{A}_n, \widehat{B}_n, \widehat{C}_n, \widehat{D}_n$ are shown in Fig. 4.

The points correspond to simple roots, the one on the left corresponds to the root $\delta - \alpha_+$. Let us consider projections of roots, which are elements of $\mathfrak{h}^* \oplus \mathbb{C}\delta$, to $\mathfrak{h}^*$. The number of lines connecting a pair of points is equal to $4\cos^2\varphi$, where φ is the angle between the projections of roots onto $\mathfrak{h}^*$ and the arrow points towards the short projection. The numbers marking points are the coefficients of linear dependence of projections (see also 7.1, Remark 1).

A.6.2. Outer Automorphisms. For any automorphism of the Dynkin diagram, there exists the corresponding outer automorphism of $\widehat{\mathfrak{g}}$. The example bellow illustrates clearly what they look like.

Example. $(\widehat{A}_{n-1} = \widehat{\mathfrak{sl}}_n)$ Let us consider the function $r : [0, 2\pi] \to \mathrm{SU}_n$ given by

$$
r(\varphi) = \begin{pmatrix} e^{i\frac{n-1}{n}\varphi} & & & \\ & e^{\frac{-i\varphi}{n}} & & \\ & & \ddots & \\ & & & e^{\frac{-i\varphi}{n}} \end{pmatrix}.
$$

The formula $r(2\pi) = \exp\left(-\frac{2\pi i}{n}\right) r(0)$ shows that the mapping σ given by $\sigma : A(\varphi) \to r(\varphi)^{-1} A(\varphi) r(\varphi)$ gives a well-defined outer automorphism of $\widehat{\mathfrak{sl}}_n$.

A.6.3. The Affine Weyl Group. (See Frenkel and Kac (1980), Kac (1983)) In the group $\widehat{K}$, there exist elements r having the following property (ψ): if V is any $\mathfrak{g}$-module integrable to a representation of the group $\widehat{K}$ and if $V = \oplus V_\lambda$ is its weight decomposition, then r permutes the weight spaces, i.e. it maps V_λ onto $V_{r(\lambda)}$. Thus such an element r determines a permutation $\lambda \to r(\lambda)$ of the weight lattice. The set of all such permutations is called the *affine Weyl group* $\widehat{W}$ of the algebra $\widehat{\mathfrak{g}}$.

Note now that the elements $w^{(\mu)}$, introduced in 6.3, have the property (ψ). The corresponding maps of the weight lattice of level k are given by the formula

$$
w^{(\alpha+l\delta)}(a + s\delta) = a + s\delta - 2(\langle a, \alpha \rangle + ks)\frac{\alpha + l\delta}{\langle \alpha, \alpha \rangle}.
$$

The elements $w^{(\mu)}$ generate the group $\widehat{W}$ (as can be shown using the adjoint representation of $\widehat{K}$).

Remark. If Q denotes the group of all maps of S^1 into $\mathbf{T}^l$ that are one-parameter subgroups, then $\widehat{W} = W \ltimes Q$. The elements $y \in W$ act by $y(a + s\delta) = ya + s\delta$ and the set $Q \simeq \mathbb{Z}^l$ consists of all transformations of the form

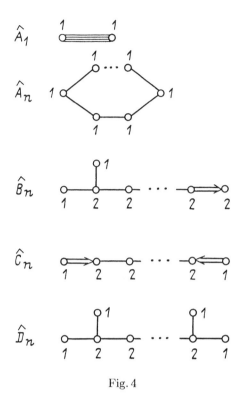

Fig. 4

$$T_\alpha(a + s\delta) = a + s\delta - \left(\langle a, \alpha \rangle + \frac{ks}{2}\langle \alpha, \alpha \rangle\right)\delta,$$

where α belongs to the lattice in $\mathfrak{h}^*$ generated by all elements of the form $2\gamma/\langle\gamma, \gamma\rangle$, where $\gamma \in \Delta$.

Remark. The condition of $\hat{K}$-integrability of the module is superfluous. It is sufficient to assume that the module is locally nilpotent.

§7. Representations of the Group $\mathrm{Diff} \ltimes C^\infty(S^1, K)$

7.1. Highest Weight Modules of $\hat{\mathfrak{g}}$. Let $\alpha_0 = \delta - \alpha_+, \alpha_1, \ldots, \alpha_l$ be the simple roots of $\hat{\mathfrak{g}}$. The weight $\Lambda \in \mathfrak{h}^*$ is called *dominant* if the $n_i = \Lambda(h_{\alpha_i})$ are nonnegative integers (the vectors h_{α_i} were introduced in 6.3).

Definition. Let Λ be a dominant weight. We say that an irreducible module $L(\Lambda) = L(n_0, \ldots, n_l)$ is a *highest weight module* if there exists a vector v (the *highest weight vector*) such that

1. $hv = \Lambda(h)v$ for all $h \in \hat{\mathfrak{h}}$;
2. $\mathfrak{g}_{\gamma+m\delta}v = 0$ for all positive roots $\gamma + m\delta \in \tilde{\Delta}$.

Modules of the form $L(0, \ldots, 0, 1, 0, \ldots)$ are called *fundamental*. One of them, namely $L(1, 0, \ldots)$, is called *basic*.

Remark 1. The modules $L(\Lambda)$ are locally nilpotent. This is just a consequence of the condition defining the fundamental weights Λ.

Remark 2. The level c of a module $L(\Lambda)$ is a nonnegative integer and $c = 0$ only for the one-dimensional module $L(0, \ldots, 0)$. The level of a fundamental module is equal to the number labelling the corresponding point in the Dynkin diagram.

Remark 3. The basic module is the simplest of all the modules $L(\Lambda)$. The highest weight condition is satisfied here in a stronger sense:

$$(\mathfrak{g}t^m)v = 0 \quad \text{for all } m \geq 0, \text{ and } zv = v.$$

For any module $L(\Lambda)$, there exists an invariant hermitian form called the *Shapovalov form*.

Theorem 7.1. (Garland (1980)). *The Shapovalov form on $L(\Lambda)$ is positive definite.*

Theorem 7.2. (Goodman and Wallach (1984)). *The representation $L(\Lambda)$ can be integrated to a projective unitary representation ρ_Λ of $C^\infty(S^1, K)$.*

Remark. (Segal (1981)). To prove Theorem 7.2, it is sufficient to show that the operators iX, $X \in \mathfrak{k}$, are essentially selfadjoint. In fact, the group $C^\infty(S^1, K)$ is generated by its one-parameter subgroups. Let us suppose that $Q \in C^\infty(S^1, K)$ has the form

$$Q = r_1(t) \ldots r_k(t), \tag{7.1}$$

where $r_i(t)$ is the one-parameter subgroup with generator X_i. Then we shall define $\rho_\Lambda(Q)$ by $\exp(t_1 X_1) \ldots \exp(t X_k)$ and it is sufficient to verify the compatibility of the definition. If Q is represented by (7.1) in two different ways and if U_1 and U_2 are the corresponding unitary operators, then $U_1 U_2^{-1}$ commutes with the action of $\widehat{\mathfrak{g}}$, hence $U_1 = \lambda U_2$.

7.2. The Action of Diff on the Modules $L(\Lambda)$. (The construction of Sugawara-G.Segal.) Let us first recall that the group Diff acts on the group $C^\infty(S^1, K)$ by automorphisms: if $q \in$ Diff and if $f(\varphi) \in C^\infty(S^1, K)$, then the automorphism $A(q)$ is given by the formula $A(q)f(\varphi) = f(q(\varphi))$.

It turns out that the representation $\rho_\Lambda \circ A(q)$ is equivalent to ρ_Λ. Let $U(q)$ be the corresponding intertwining operator. Then the map $q \to U(q)$ is clearly a unitary projective representation of the group Diff.

Theorem 7.3. (Goodman and Wallach (1984), Segal (1981)) *The construction described above gives a projective unitary representation of the group Diff $\ltimes C^\infty(S^1, K)$.*

Remark 1. There is an evident analogy with the Weil representation and the spinor representation. Moreover, all our constructions of representations of the group Diff were, in fact, reduced to the automorphisms of commutation or anticommutation relations. We encounter here "generalized commutation relations", namely the commutation relations of an affine Lie algebra.

Remark 2. The proof of Theorem 7.3 is a consequence of the following algebraic construction together with Theorem 7.2 and Theorem 5.4.

We would now like to clarify how the generators L_m of the Virasoro algebra $\mathcal{L}$ act on $L(\Lambda)$. The definition of the operators $U(q)$ implies that the relations

$$[L_m, Y^{(n)}] = -nY^{(m+n)} \tag{7.2}$$

should be satisfied, where Y is an arbitrary element of $\mathfrak{g}$ and $Y^{(n)}$ is interpreted as an operator on $L(\Lambda)$. The relations (7.2) force us to search for the operators L_m as quadratic expressions in the operators $Y_i^{(n)}$. Let Y_i be a basis of $\mathfrak{g}$, and let X_i be the dual basis, i.e. $B(Y_i, X_i) = \delta_{ij}$. Then

$$L_m = -\frac{1}{\kappa(\mathfrak{g}) + k}\left[\sum_i\left(\sum_{s > [-m/2]} Y_i^{(-s)}X_i^{(m+s)} + \frac{\theta}{2}Y_i^{(\frac{m}{2})}X_i^{(\frac{m}{2})}\right)\right], \tag{7.3}$$

where k is the level of $L(\Lambda)$, $\theta = 1$ for m even and $\theta = 0$ for m odd.

The constant $\kappa(\mathfrak{g})$ is the *dual Coxeter number* which is defined in one of the following ways.

1. $\kappa(\mathfrak{sl}_n) = n$, $\kappa(\mathfrak{so}_n) = n - 2$, $\kappa(\mathfrak{sp}_{2n}) = n + 1$, $\kappa(E_6) = 12$, $\kappa(E_7) = 18$, $\kappa(E_8) = 30$, $\kappa(F_4) = 9$, $\kappa(G_2) = 4$.
2. The number $\kappa(\mathfrak{g})$ is the eigenvalue of the Casimir operator $\frac{1}{2}\sum Y_i X^i$ in the adjoint representation of $\mathfrak{g}$.

The operators L_m give a unitary representation of the Virasoro algebra $\mathcal{L}$ on $L(\Lambda)$ of the level

$$c = \frac{k \dim \mathfrak{g}}{k + \kappa(\mathfrak{g})}.$$

It is also clear that the described representation is a direct sum of highest weight representations.

7.3. The Subtraction of Virasoro Algebras (Goddard-Kent-Olive construction)

7.3.1. Consider the module $L(\Lambda)$ of $\widehat{\mathfrak{g}}$ and the corresponding algebra of operators L_m (see 7.2).

Let $\mathfrak{a}$ be a simple subalgebra of $\mathfrak{g}$ and let us apply the construction of 7.2 to the restriction of $L(\Lambda)$ to $\widehat{\mathfrak{a}}$. We shall get another (the second one!) set of operators L'_m. Let us note now that for any $Y \in \mathfrak{a}$, the relations (7.2) imply that

$$[L_m, Y^{(n)}] = -nY^{(m+n)}, \quad [L'_m, Y^{(n)}] = -nY^{(m+n)}.$$

This means that $K_m = L_m - L'_m$ commutes with $\hat{\mathfrak{a}}$ and it implies, in turn, that K_m commutes with L'_j (because the L'_j are expressed using operators from $\mathfrak{a}$). So

$$[L_m, L_n] = [L'_m, L'_n] + [K_m, K_n]$$

and the operators K_m generate a representation of the Virasoro algebra, which will be denoted by P.

The relations $L_m^* = -L_{-m}$ and $(L'_m)^* = -L'_{-m}$ then imply the relation $K_m^* = -K_{-m}$, i.e. the representation P is unitary. We know that if we take s sufficiently large, then $(L_m - L'_m)^s v = 0$ for any homogeneous vector v and for every $m < 0$, hence our representation P is a sum of highest weight representations.

The level of P is clearly given by

$$c_P = \frac{k \dim \mathfrak{g}}{k + \kappa(\mathfrak{g})} - \frac{l \dim \mathfrak{a}}{l + \kappa(\mathfrak{a})}, \qquad (7.4)$$

where l is the level of the $\hat{\mathfrak{a}}$-module $L(\Lambda)$. To compute l, it is necessary to know the length of the long root of $\mathfrak{a}$ in $\mathfrak{g}$ (see 6.0); note that $l/k \in \mathbb{Z}$.

7.3.2. Let now $\mathfrak{a}$ be a reductive algebra, $\mathfrak{a} = \mathfrak{a}_1 \oplus \mathfrak{a}_2 \oplus \ldots$; then we construct the operators $L_m^{(i)}$ for every $\mathfrak{a}_i$ and take $L'_m = \sum L_m^{(i)}$. (If there is an abelian subalgebra among the $\mathfrak{a}_i$, then we set $\kappa(\mathfrak{a}_i) = 0$; this is just the Virasoro construction, see 5.1.1). Then we can repeat the procedure described in 7.3.1.

7.4. The Existence of the Discrete Series of Modules $L(h, c)$. Consider now the algebra $\mathfrak{g} = \mathrm{sp}\,(2m+2)$ and its subalgebra $\mathrm{sp}(2m) \times \mathrm{sp}\,(2)$, take the basic module over $\mathfrak{g}$ and apply the above construction. Using the formula (7.4) (with $k = l = 1$), we get

$$c_P = 1 - \frac{6}{(m+3)(m+2)}.$$

In this way we obtain the full set of numbers c from Theorem 5.4. It was shown in Goddard et al. (1986) that if we apply the construction to the algebra $\widehat{\mathfrak{sl}}_2 \oplus \widehat{\mathfrak{sl}}_2$ and its diagonal subalgebra $\widehat{\mathfrak{sl}}_2$, we obtain the complete discrete series of unitarizable modules.

Appendices

A.7.1. The Weyl Character Formula. (Kac (1983)) Let $\rho \in \hat{\mathfrak{h}}^*$ be defined by the condition $\rho(h_{\alpha_i}) = 1$, where $\alpha_0, \ldots, \alpha_l$ are the simple roots. Let $K(\lambda)$ be the Kostant partition function, i.e. the number of partitions of the weight $\lambda \in \hat{\mathfrak{h}}^* \oplus \mathbb{C}\delta$ as a sum of positive roots. Then the multiplicity m_λ of the weight λ in $L(\Lambda)$ (i.e. the dimension of the weight space $L(\Lambda)_\lambda$) is equal to

$$m_\lambda(\Lambda) = \sum_{w \in \widetilde{W}} (\det \omega) K(\omega(\Lambda + \rho) - (\lambda + \rho))$$

(the function det on the affine Weyl group $\widetilde{W}$ is defined by the conditions $\det \omega^{(\gamma)} = -1$, $\det (\omega_1 \cdot \omega_2) = \det \omega_1 \cdot \det \omega_2$).

§8. The Constructions of Basic Modules

8.1. The Fermionic Construction. (I.B.Frenkel, R.S.Ismagilov) Let H denote the space of functions on S^1 with values in $\mathbb{R}^n$ satisfying the conditions $f(\varphi + \pi) = -f(\varphi)$. The scalar product in H is defined by

$$\{f, g\} = \int_0^{2\pi} \langle f(\varphi), g(\varphi) \rangle d\varphi.$$

Let us consider the group G of all transformations of the space H having the form $f(\varphi) \to A(\varphi)f(\varphi)$, where $A(\varphi)$ is a smooth function on S^1 with values in $SO(n)$ satisfying the condition $A(\varphi + \pi) = A(\varphi)$. We clearly have $G \simeq C^\infty(S^1, SO(n))$. Let the group $\text{Diff}^{(2)}$ of all diffeomorphisms of S^1 satisfying the condition $q(\varphi + \pi) = q(\varphi) + \pi$ act on H by the formula

$$T(q)f(\varphi) = f(q(\varphi))q'(\varphi)^{1/2}.$$

It is not difficult to verify that the group $\text{Diff}^{(2)}$ normalizes G, so the group of transformations of the space H generated by G and $\text{Diff}^{(2)}$ has the form $\text{Diff}^{(2)} \ltimes G$. It is also clear that the group $\text{Diff} \ltimes C^\infty(C^1, SO(n))$ of 7.2 is isomorphic to $(\text{Diff}^{(2)} \ltimes G)/\mathbb{Z}_2$, where the group $\mathbb{Z}_2$ is generated by the diffeomorphism $q(\varphi) = \varphi + \pi$.

The complex structure of H is defined using the Hilbert transform I:

$$If(\varphi) = \frac{1}{\pi} \int_0^{2\pi} \cot \left(\frac{\varphi - \psi}{2} \right) f(\psi) d\psi \tag{8.1}$$

and $I^2 = -1$. The group $O(2\infty, \mathbb{R})$ will be realized as the orthogonal group of the space H and its subgroup $U(\infty)$ as the set of all transformations commuting with I.

Theorem 8.1. *The group $\text{Diff}^{(2)} \ltimes G$ is contained in $(O(2\infty, \mathbb{R}), U(\infty))$.*

Proof. A direct computation shows that

$$[A(\varphi), I]f(\varphi) = \int_0^{2\pi} [A(\varphi) - A(\psi)] \cot \left(\frac{\varphi - \psi}{2} \right) f(\psi) d\psi.$$

The kernel of this operator is bounded, so it is a Hilbert-Schmidt operator. This implies that $G \subset (O(2\infty, \mathbb{R}), U(\infty))$. As for $\text{Diff}^{(2)}$, see Sect.4.

Restricting now the spinor representation $(O(2\infty, \mathbb{R}), U(\infty))$ to the subgroup $\mathrm{Diff}^{(2)} \ltimes G$, we get a unitary projective representation of $\mathrm{Diff}^{(2)} \ltimes G$ on $\Lambda(H)$. We shall write down the explicit formulas for it on the Lie algebra level.

Let $f_1, \ldots, f_n$ be a basis of $\mathbb{R}^n$; then $f_j e^{ik\varphi}$, $k = \pm 1, \pm 3, \ldots$ is a basis in $H_C \subset L^2(S^1, \mathbb{C}^n)$. Let e_{ij} be the $n \times n$ matrix with only nontrivial element 1 in the i-th row and the j-th column. Set $E_{ij} = e_{ij} - e_{ji}$; then $E_{\alpha\beta}^{(l)} = E_{\alpha\beta} e^{2il\varphi}$ is a basis of $C^\infty(S^1, \mathfrak{so}_n)$. Let Y_n be the generators of Vect_C (see 1.0).

$$E_{\alpha\beta}^{(l)}(f_j e^{ik\varphi}) = (f_\alpha \delta_{j\beta} - f_\beta \delta_{j\alpha})e^{i(k+2l)\varphi},$$

$$Y_n(f_j e^{ik\varphi}) = \frac{1}{2}(k+n)f_j e^{i(k+2n)\varphi}.$$

The block form of a matrix in formula (3.22) corresponds to the splitting of the basis $f_j e^{ik\varphi}$ into positive and negative k. We can now apply formulas (3.22). Let $\xi_k^{(\nu)}$ be anticommuting variables, $k = 1, 3, 5, 7, \ldots$ and $\nu = 1, 2, \ldots, n$. Then the affine Lie algebra $\widehat{\mathfrak{so}}_n$ acts on $\Lambda(H)$ by the operators

$$E_{\alpha\beta}^{(l)} = \sum_{k>0}\left(\xi_{k+2l}^{(\alpha)}\frac{\partial}{\partial\xi_k^{(\beta)}} - \xi_{k+2l}^{(\beta)}\frac{\partial}{\partial\xi_k^{(\alpha)}}\right) + \frac{1}{2}\sum_{\mu+\nu=2l}(\xi_\mu^{(\alpha)}\xi_\nu^{(\beta)} - \xi_\mu^{(\beta)}\xi_\nu^{(\alpha)}),$$

where $l \geq 0$, $E_{\alpha\beta}^{(-l)} = -(E_{\alpha\beta}^{(l)})^*$. The generators of the Virasoro algebra act on $\Lambda(H)$ by the formula

$$L_n = \sum_\nu \left[\frac{1}{2}\sum_{k>0}(k+n)\xi_{k+2n}^{(\nu)}\frac{\partial}{\partial\xi_k^{(\nu)}} + \frac{1}{4}\sum_{k+l=2n}(k-l)\xi_k^{(\nu)}\xi_l^{(\nu)}\right],$$

where $n > 0$, $L_{-n} = L_n^*$.

The constructed representation of $\widehat{\mathfrak{so}}_n$ is reducible. If n is even, then its irreducible subrepresentations are realized in even and odd functions of $\xi_k^{(\nu)}$ (see Frenkel (1981)) and they can both be transformed into the basic module using an outer automorphism of $\widehat{\mathfrak{so}}_n$.

Remark. The construction admits a variation in the spirit of 4.7.2.

8.2. Right Tori. Let P be the space of all maximal tori in K. Let $p(\varphi)$ be a smooth map from S^1 into P. The *right torus* $\mathbf{T}_p$ in $C^\infty(S^1, K)$ is the set of all $f \in C^\infty(S^1, K)$ such that $f(\varphi) \in p(\varphi)$. A right torus is a maximal abelian subalgebra of $C^\infty(S^1, K)$. The converse is not true. It is not difficult to see that conjugacy classes of right tori are indexed by conjugacy classes of the Weyl group of algebra $\mathfrak{g}$.

Example. There exist two right tori $\mathbf{T}_1$ and $\mathbf{T}_2$ in $C^\infty(S^1, SU_2)$ that are not conjugate. They are defined as the centralizers of the two paths

$$M_1(\varphi) = \begin{pmatrix} i & 0 \\ 0 & -i \end{pmatrix}, \quad M_2(\varphi) = \begin{pmatrix} 0 & e^{\frac{i\varphi}{2}} \\ e^{-\frac{i\varphi}{2}} & 0 \end{pmatrix}.$$

It is important to note that the curve $M_2(\varphi) : [0, 2\pi] \to SU_2$ is not closed but that $M_2(0)$ and $M_2(2\pi)$ are regular elements of the same torus in SU_2.

If $\mathbf{T}_p$ is a right torus, then the corresponding subalgebra $\mathfrak{t}_p \subset C^\infty(S^1, \mathfrak{g})$ is, of course, abelian, but its extension $\widehat{\mathfrak{t}}_p \subset \widehat{\mathfrak{g}}$ is already isomorphic to an algebra of the form $\text{Heis}_\infty \oplus \mathbb{C}^n$, where $\mathbb{C}^n$ is an abelian algebra (see 3.2.3). The restriction of the module $L(\Lambda)$ to $\widehat{\mathfrak{t}}_p$ is a direct sum of Fock modules.

The principal idea of Sections 8.3.–8.5 is to extend representations from $\widehat{\mathfrak{t}}_p$ to the whole algebra $\widehat{\mathfrak{g}}$. For the algebra sl_2, the extension is constructed for the torus $\mathbf{T}_2$ in 8.3 and for the torus $\mathbf{T}_1$ in 8.4.

8.3. The Twisted Vertex Construction for $\widehat{sl}_2$. Let us consider the elements

$$E_0(\theta) = \begin{pmatrix} \theta & 0 \\ 0 & -\theta \end{pmatrix}, \quad E_-(\theta) = \begin{pmatrix} 0 & 0 \\ \theta & 0 \end{pmatrix}, \quad E_+(\theta) = \begin{pmatrix} 0 & \theta \\ 0 & 0 \end{pmatrix}$$

where $\theta(\varphi) \in C^\infty(S^1)$; they belong to $C^\infty(S^1, sl_2)$. Suppose that, as usual, $E_\mu^{(k)} = E_\mu(e^{ik\varphi})$. The basic module Λ_0 of the algebra $\widehat{sl}_2$ is the irreducible module containing a highest weight vector w satisfying the conditions

$$E_0^{(k)} w = E_+^{(k)} w = E_-^{(k)} w, \quad zw = w, \quad k \geq 0.$$

8.3.1. The Algebraic Variant. (Lepowsky-Wilson construction, see Lepowsky and Wilson (1978)) Let us change the grading of $\widehat{sl}_2$ in the following way: $\deg E_+^{(j)} = 2j + 1$, $\deg E_0^{(j)} = 2j$, $\deg E_-^{(j)} = 2j - 1$, $\deg z = 0$ and choose a new basis

$$B_{2j+1} = E_+^{(j)} + E_-^{(j+1)}, \quad C_{2j+1} = -E_+^{(j)} + E_-^{(j+1)},$$

$$H_{2j} = E_0^{(j)} - \frac{1}{2} z \delta_{j,0}.$$

The elements B_l, C_l, H_k have the following commutation relations

$$\begin{array}{ll}
[B_l, B_k] = l\delta_{l,-k} z, & [B_l, H_k] = 2C_{l+k}, \\
[C_l, C_k] = -l\delta_{l,-k} z, & [C_l, H_k] = 2B_{l+k}, \\
[H_l, H_k] = l\delta_{l,-k} z, & [B_l, C_k] = 2H_{l+k}.
\end{array} \tag{8.2}$$

The algebra $\widehat{S}$ generated by the elements B_k is clearly isomorphic to Heis_∞. It turns out that the restriction $\Lambda_0|_{\widehat{S}}$ is an irreducible Fock representation of $\widehat{S}$ (such properties are usually proved using formal characters). Let us consider the bosonic Fock space F consisting of all holomorphic functions in the variables $z_1, z_3, z_5, \ldots$. We introduce the operators

$$B_j f = \sqrt{j} \frac{\partial}{\partial z_j} f, \quad B_{-j} f = \sqrt{j} z_j f,$$

where $j = 1, 3, 5, \ldots$. Using the relations (8.2), we can show that the operator-valued generating function

$$X(\lambda) = \sum H_{2k}\lambda^{2k} + \sum C_{2k+1}\lambda^{2k+1} \tag{8.3}$$

satisfies the equation

$$[B_j, X(\lambda)] = 2X(\lambda)\lambda^{-j}$$

and using (3.12), we can write down the explicit expression for the Wick normal form of $X(\lambda)$:

$$X(\lambda) = C\exp\left(2\sum \frac{\lambda^k}{\sqrt{k}}z_k\right)\exp\left(-2\sum \frac{\lambda^{-k}}{\sqrt{k}}\frac{\partial}{\partial z_k}\right), \tag{8.4}$$

where $k = 1, 3, 5, \ldots$. The homogeneous components of $X(\lambda)$ (with respect to λ) are the operators C_{2j+1} and H_{2j}. The homogeneity of C_{2j+1}, H_{2j} implies that C does not depend on λ. Finally, applying $X(\lambda)$ to w, we get $C = -1/2$.

8.3.2. The Analytic ("Physical") Variant.

Let us consider the abelian subalgebra S of $C^\infty(S^1, \mathrm{sl}_2)$ consisting of all functions of the form $B(\theta) = E_+(\theta) + e^{i\varphi}E_-(\theta)$ and its adjoint representation on $C^\infty(S^1, \mathrm{sl}_2)$. The corresponding eigenfunctions are

$$x_\nu(\varphi) = \begin{pmatrix} \delta_\nu(\varphi) & -e^{\frac{-i\nu}{2}}\delta_\nu(\varphi) \\ e^{\frac{i\nu}{2}}\delta_\nu(\varphi) & -\delta_\nu(\varphi) \end{pmatrix},$$

where δ_ν is the delta function concentrated at the point ν :

$$[B(\theta), x_\nu(\varphi)] = 2\theta(\nu)\exp\left(\frac{i\nu}{2}\right)x_\nu(\varphi). \tag{8.5}$$

Remark. The well-known relation $\delta_0(\varphi) = \sum e^{in\varphi}$ implies that the distribution $x_\nu(\varphi)$ is just $X\left(e^{\frac{i\varphi}{2}}\right)$, see (8.3).

Let us consider the Fock representation of the central extension $\widehat{S}$ of the algebra S, introduced in 8.3.1. The operators F corresponding to $x_\nu(\varphi)$ can be computed using (8.5) and (3.12). As a result, we get expression (8.4), where $\lambda = \exp\left(i\frac{\varphi-\nu}{2}\right)$. So we are left with the task of constructing the operators of the representations corresponding to "genuine" elements of $C^\infty(S^1, \mathrm{sl}_2)$. These are just the expressions

$$H_{2k} = \frac{1}{2}\int_0^{4\pi} e^{ik\nu}x_\nu(\varphi)d\nu, \quad C_{2k+1} = \frac{1}{2}\int_0^{4\pi} e^{i(k+\frac{1}{2})\nu}x_\nu(\varphi)d\nu.$$

The commutation relations for H_j and C_l can be verified using Segal's regularization (3.10) for $x_\nu(\varphi)$ and (3.13); the computation is similar to that in 8.4.

8.4. Segal's Construction for $\widehat{\mathrm{sl}}_2$.

Let us consider the disconnected group $M = C^\infty(S^1, \mathbf{T})$, where $\mathbf{T}$ is the set of diagonal matrices in SU_2 and

$\mathbf{T} \simeq \mathbb{R}/2\pi\mathbb{Z}$. The important property of the basic representation Λ_0 of the group $C^\infty(S^1, \mathrm{SU}_2)$ (which will be constructed bellow) is that its restriction to M is irreducible (this will be clear from the proof). Our problem is to guess what this restriction looks like and then to reconstruct the representation Λ_0 itself. We shall use the notation of the beginning of 8.3.

8.4.1. The Group $\widehat{M}$. Let us consider the subset Γ of $C^\infty(\mathbb{R}^1)$ consisting of all functions f satisfying the condition

$$\Delta_f = f(\varphi + 2\pi) - f(\varphi) \in \mathbb{Z}.$$

Then any function of the form $\exp(if)$, where $f \in \Gamma$, can be identified with an element of the group M. We define a skew-symmetric Diff-invariant form on Γ by

$$S(f,g) = \frac{1}{4\pi} \int_0^{2\pi} (f'(\theta)g(\theta) - f(\theta)g'(\theta))d\theta +$$

$$+ \frac{1}{4\pi}(f(2\pi)g(0) - f(0)g(2\pi)).$$

Let us define the group $\widehat{M}$ as the space $M \times \mathbf{T}$ with the group operation

$$(e^{if}, \lambda)(e^{ig}, \mu) = (e^{i(f+g)}, \lambda\mu e^{-iS(f,g)}).$$

Remark. The group $\widehat{M}$ is the universal Diff-invariant extension of M.

It is natural to identify the space Γ_0 consisting of all functions f such that $\Delta_f = 0$ with the Lie algebra of the group M. Let us introduce a degenerate nonnegative scalar product on Γ_0 by the formula

$$\langle f,g \rangle = \frac{1}{\pi} \int_0^{2\pi} \int_0^{2\pi} \cot\left(\frac{\varphi - \psi}{2}\right) f'(\varphi)g(\psi)d\varphi d\psi.$$

Then $S(f,g) = \frac{1}{2\pi}\int f'g d\varphi = i\langle f, Ig \rangle$, where I is the Hilbert transform defined by (4.10). We have $S(e^{in\varphi}, e^{im\varphi}) = n\delta_{n,-m}$. Let $\Gamma_0/\mathbb{R}$ be the quotient space of Γ_0 by the kernel of the hermitian form $\langle \cdot, \cdot \rangle$ (which is equal to the space of constants). Then $\Gamma_0/\mathbb{R}$ is a complex Hilbert space with complex structure given by the operator I.

The space $\Gamma_0 \oplus \mathbb{R}$ with the operation

$$[(f_1, c_2), (f_2, c_2)] = (0, S(f_1, f_2))$$

can be naturally considered as the Lie algebra $\mathfrak{m}$ of the group $\widehat{M}$.

The elements (λ, c), where λ is a constant, form the center of the algebra $\mathfrak{m}$; hence $\mathfrak{m} \simeq \mathrm{Heis}_\infty \oplus \mathbb{R}$ (see (3.7)).

8.4.2. The Fock Representation $\widehat{M}$. Let H_k, $k = 0, \pm1, \pm2, \ldots$, be a countable number of copies of the Fock space $F(\Gamma_0/\mathbb{R})$ (see 8.4.1). Let the space $H = \oplus H_k$ be realized as the space of all functions of the variables $u, z_1, z_2, \ldots$, that are holomorphic in z_l and which are Laurent series in u. The space H_j consists of functions of the form $u^j f(z)$. The representation T of the group $\widehat{M}$ on H is completely defined by the following conditions:

$$E_0^{(k)} f = \sqrt{2k} z_k f, \quad E_0^{(-k)} f = \sqrt{2k} \frac{\partial}{\partial z_k} f,$$

$$E_0^{(0)} f = 2u \frac{d}{du} f, \quad T(e^{i\varphi}, 0) f = uf,$$

(8.6)

where $k > 0$ and $E_0^{(k)}$ were introduced in 8.3.

8.4.3. The Extension of the Representation from $\widehat{M}$ to $\widehat{sl}_2$. Let us suppose that $v \in H_j, \theta \in C^\infty(S^1)$, then

$$E_0^{(0)} E_+(\theta)v = \{2E_+(\theta) + E_+(\theta)E_0\}v = (2j+2)E_+(\theta)v.$$

So $E_+(\theta)$ maps H_j into H_{j+1} and $E_-(\theta)$ maps H_j into H_{j-1}. Moreover,

$$[E_0(\theta), E_+(\delta_\nu(\varphi))] = 2\theta(\nu)E_+(\delta_\nu(\varphi)),$$

where δ_ν is the delta function concentrated at the point ν. Using now (8.6) and (3.12), we find an explicit expression (up to a constant) for the operator on H corresponding to $E_+(\delta_\nu(\varphi))$:

$$E_\nu^+ = u \exp\left(2i\nu u \frac{\partial}{\partial u}\right) \exp\left(\sqrt{2}\sum \frac{e^{i\nu k}}{\sqrt{k}} z_k\right) \times$$

$$\times \exp\left(-\sqrt{2}\sum \frac{e^{-i\nu k}}{\sqrt{k}} \frac{\partial}{\partial z_k}\right).$$

(8.7)

Let us now write down formal definitions. Let E_ν^+ be given by the formula (8.7). Let $\theta(\nu)$ be a trigonometric polynomial.

Define the operators

$$E_+(\theta) = \int_0^{2\pi} \theta(\nu) E_\nu^+ d\nu.$$

Even if the operator δ-function E_ν^+ is highly singular, the operators $E_+(\theta)$ are ordinary unbounded operators on H mapping the polynomial algebra in the variables $u, u^{-1}, z_1, z_2, z_3, \ldots$ into itself. Set $E_\nu^- = (E_\nu^+)^*$, $E_-(\theta) = E_+^*(\theta)$.

Theorem 8.2. *Let θ be a trigonometric polynomial. Then the operators $E_0(\theta), E_+(\theta), E_-(\theta)$ satisfy the commutation relations of the algebra $\widehat{sl}_2$.*

8.4.4. We shall now prove the most complicated part of the theorem (details can be found in the original paper Segal (1981)):

$$[E_+(\varphi), E_-(\psi)] = E_0(\varphi\psi) + iS(\varphi, \psi).$$

Applying Segal's regularization to E_+^ν and E_-^ν (see 3.2.5), we introduce the operators

$$E_{\nu,t}^+ = u \exp\left(2i\nu u \frac{\partial}{\partial u}\right) \exp\left(-\sqrt{2}\sum \frac{(te^{i\nu})^k}{\sqrt{k}} z_k\right) \times$$

$$\times \exp\left(\sqrt{2}\sum \frac{(te^{-i\nu})^k}{\sqrt{k}} \frac{\partial}{\partial z_k}\right),$$

$$E_{\nu,t}^- = (E_{\nu,t}^+)^*,$$

where $t < 1$. Then

$$E_\nu^+ = \lim_{t\to 1} E_{\nu,t}^+, \quad E_+(\theta) = \lim_{t\to 1} \int_0^{2\pi} \theta(\nu) E_{\nu,t}^+ d\nu.$$

Applying (3.13), we get

$$[E_{\nu,t}^+, E_{\mu,t}^-] = c_{\mu,\nu,t} Q_{\mu,\nu,t}\left(z, \frac{\partial}{\partial z}\right),$$

where

$$c_{\mu,\nu,t} = -\frac{i}{t^2} \frac{\partial}{\partial \nu} \frac{1-t^4}{1-2t^2 \cos(\mu-\nu)+t^4},$$

$$Q_{\mu,\nu,t} = u \exp\left(2i(\mu-\nu)u\frac{\partial}{\partial u}\right) \exp\left(\sqrt{2}\sum \frac{t^k(e^{ik\nu}-e^{ik\mu})}{\sqrt{k}} z_k\right) \times$$

$$\times \exp\left(-\sqrt{2}\sum \frac{t^k(e^{-ik\nu}-e^{-ik\mu})}{\sqrt{k}} \frac{\partial}{\partial z_k}\right).$$

Using $\lim_{t\to 1} c_{\mu,\nu,t} = 2\pi i \delta'(\mu-\nu)$ and $\delta'(x)f(x) = \delta'(x)f(0) - \delta(x)f'(0)$, we get

$$\lim_{t\to 1}[E_{\nu,t}^+, E_{\mu,t}^-] = 2\pi i \delta'(\mu-\nu) -$$

$$-2\pi i \delta(\mu-\nu)\frac{\partial}{\partial \nu}\left[\sqrt{2}\sum \frac{e^{i\nu k}}{\sqrt{k}} z_k - \sqrt{2}\sum \frac{e^{-i\nu k}}{\sqrt{k}} \frac{\partial}{\partial z_k} + 2i\nu u \frac{\partial}{\partial u}\right] =$$

$$= 2\pi i \delta'(\mu-\nu) + 4\pi^2 \delta(\mu-\nu) E_0(\delta_\mu),$$

which is the desired formula.

Remark. Let

$$\gamma_{\mu,t}(\varphi) = \frac{t - e^{i(\varphi-\nu)}}{1 - te^{i(\varphi-\nu)}}$$

and $h_{\nu,t} = (\gamma_{\nu,t}, 1) \in \widehat{M}$. Then

$$E_{\nu,t}^+ = (1-t^2)^{-1} T(h_{\nu,t}).$$

8.4.5. To prove that the operators

$$iE_0(\theta), \quad \frac{1}{2}(E_+(\theta) + E_-(\theta)), \quad \frac{1}{2i}(E_+(\theta) - E_-(\theta)),$$

where $\theta \in C^\infty(S^1)$, are essentially selfadjoint needs only standard techniques. The remark after Theorem 7.2 shows that it implies the integrability of our representation to a unitary projective representation of $C^\infty(S^1, K)$.

8.4.6. The Action of Diff on Λ_0. Let Diff act on $\mathfrak{m}$ by automorphisms of the form $T(q)(f(\varphi), c)) = (f(q(\varphi)), c)$. It is not difficult to show that $T(q)$ belongs to the group $(\mathrm{Sp}(2\infty, \mathbb{R}), \mathrm{U}(\infty))$ of the space Γ_0, so we have an almost invariant structure of 4.5 Bδ) (Kazhdan, Segal). Together with the construction of 4.1 (due to the author), this is one of the first examples of almost invariant structures.

We shall now construct the action of Diff on the subspaces $H_k \subset H$. In fact, it is not $\mathfrak{m}$ but the quotient algebra of $\mathfrak{m}$ by the subalgebra A_k consisting of all elements of the form $(c, -2kc)$, where c is a constant, that acts on every H_k. In this way, the group Diff is imbedded into the full group of automorphisms of the algebra $\mathfrak{m}/A_k \simeq \mathrm{Heis}_\infty$. This means that we have constructed an imbedding of Diff into $(\mathrm{Sp}(2\infty, \mathbb{R}), \mathrm{U}(\infty)) \ltimes (\Gamma_0/\mathbb{R})$, and hence, because of 3.3.3 (see also 5.1), a unitary representation of Diff on $H_k \simeq F(\Gamma_0/\mathbb{R})$. The representation of the Virasoro algebra on H_k is given by the formulas (5.2) – (5.3) with $\beta = 0, \alpha = \sqrt{2k}$, hence is isomorphic to $\oplus_{j \geq k} L(j^2, 1)$.

So we are left only with the task of verifying the formulas (7.2).

Remark. Let us consider the restriction of the constructed projective representation D of the group Diff $\ltimes C^\infty(S^1, \mathrm{SU}_2)$ to the subgroup Diff $\oplus \mathrm{SU}_2$. The operators $E_\pm^{(0)}$ map H_j into $H_{j\pm 1}$ and preserve the weight subspaces of Diff, so we get

$$D|_{\mathrm{Diff} \otimes \mathrm{SU}_2} = \otimes_{q=0}^\infty (L(q^2, 1) \otimes V_{2q+1}),$$

where V_{2q+1} is the representation of SU_2 of dimension $2q + 1$.

8.5. Closing Remarks. It is relatively easy to transfer the vertex constructions of basic modules, described in 8.3 and 8.4, to the algebras belonging to the series $\widehat{A}_n, \widehat{D}_n, \widehat{E}_k$ (they are distinguished by the fact that all their roots have the same length) and it is possible to extend it to the other affine Lie algebras (see Kac et al. (1981), Segal (1981), Frenkel and Kac (1980), Kac (1983)). The vertex constructions are labelled by conjugacy classes of the Weyl group (see 8.2 and Lepowsky (1985)).

Applying the outer automorphisms (see A.6.2), we can also get explicit constructions for some fundamental modules (all of them for $\widehat{A}_n$). Explicit realizations of other fundamental modules by differential operators of the type of Sections 8.1 – 8.3 are not known to the author (even if there are also other approaches possible, see Lepowsky and Wilson (1978) or the paper by Misra in Lepowsky et al. (1985)).

The construction of all fundamental modules would make possible the construction of all other modules $L(\Lambda)$ as subrepresentations of tensor products (note that suitable explicit constructions are also missing for representations of compact Lie groups). Complete clarity is still not achieved even in the case of the algebra $\hat{A}_1$ (see Lepowsky and Wilson (1978)).

The abundance of realizations of basic modules leads to the interesting problem of the construction of intertwining operators among different realizations (see e.g. Frenkel (1981)).

Appendices

A.8.1. Almost Invariant Structures and Multiplicative Integrals. The constructions of representations of the group $C^\infty(S^1, K)$ in A.3.1 are not Diff-invariant in the sense of 7.2, i.e. they cannot be extended to the group Diff $\ltimes C^\infty(S^1, K)$. Most of the almost invariant structures of Sect.4 can be used for constructions of unitary representations of Diff $\ltimes C^\infty(S^1, K)$, where K is compact.

Example. Let $0 < s < 1$. Consider the space $H_s^{(n)}$ of all functions on S^1 with values in $\mathbb{R}^n$ with the scalar product

$$\{f, g\}_s = \int_0^{2\pi} \int_0^{2\pi} \sin^{s-1}\left(\frac{\varphi - \psi}{2}\right) \langle f(\varphi), g(\psi)\rangle d\varphi d\psi.$$

Let $C^\infty(S^1, \mathrm{SO}(n, \mathbb{R}))$ act on $H_s^{(n)}$ in the trivial way and let the action of Diff be given by the formula

$$T(q)f(\varphi) = f(q(\varphi))q'(\varphi)^{\frac{1+s}{2}}.$$

It implies that the group Diff $\ltimes C^\infty(S^1, \mathrm{SO}(n, \mathbb{R}))$ is contained in the group $(\mathrm{GL}(\infty, \mathbb{R}), \mathrm{O}(\infty, \mathbb{R}))$ of the space $H_s^{(n)}$, together with all consequences of that fact (see 4.1, 5.6 and 8.1).

§9. Analytic Continuations

Let G be a real simple Lie group and let G_C be its complexification. Then an irreducible unitary representation ρ of the group G can be extended to G_C only if G is compact (and ρ is hence finite-dimensional). In the opposite case, the representation ρ can be extended, in general, only to a neighbourhood of the identity in G_C and the operators of the representation are unbounded (and their common domain is not invariant). There is, however, one exception; namely, the highest weight representations can be extended to a holomorphic representation of an open semigroup $\Gamma \subset G^\mathbb{C}$; the operators of the representation are unbounded (see Ol'shanskij (1981)).

We shall see that $C^\infty(S^1, K)$ has behaviour similar to that of a compact group. The group Diff$_C$ does not exist but there exists a holomorphic semigroup containing the group Diff (Sections 9.1. and 9.2). Finally, as in the case of finite-dimensional algebras, the unitarizability of representations is not necessary for the integrability of representations from $\mathcal{L}$ to Diff (see 9.3).

9.1. The Partial Complexification of the Group Diff.

9.1.1. The Definition of the Semigroup Γ. Let Diff_a be the group of analytic diffeomorphisms of the circle S^1 realized as the subset $|z| = 1$ of $\mathbb{C}$. Any element $p \in \text{Diff}_a$ has an analytic continuation to an annulus of the form $e^{-\varepsilon} < |z| < e^{\varepsilon}$. Let $A(t)$ denote the map form $\mathbb{C}$ to $\mathbb{C}$ given by the formula $A(t)z = e^{-t}z$. The semigroup Γ consists of all formal (!) triple products of the form

$$p \cdot A(t) \cdot q, \tag{9.1}$$

where $p, q \in \text{Diff}_a$, $t > 0$, $p(1) = 1$. To multiply two elements of the form (9.1), it is sufficient to learn how to convert the product

$$\rho = A(\sigma) \cdot r \cdot A(s), \tag{9.2}$$

where $\sigma > 0, s > 0, r \in \text{Diff}_a$, to the form (9.1).

A) Let s be small enough that r extends to the annulus $e^{-s} \leq |z| \leq e^s$. Then ρ is well-defined as the product of maps in a neighbourhood of the circle S^1. Let K be the domain of annular type between S^1 and $\rho(S^1)$. Let p^{-1} be the canonical conformal map of K onto the annulus $e^{-t} \leq z \leq 1$. Then the map q is found from the equality

$$p \cdot A(t) \cdot q = A(\sigma) \cdot r \cdot A(s)$$

of maps of the circle S^1.

B) Now let s be an arbitrary positive number. Then there exists $\varepsilon = s/n$ sufficiently small such that the product

$$A(\sigma) \cdot r \cdot A(s) = A(\sigma) \cdot r \cdot (A(\varepsilon))^n =$$
$$= ((A(\sigma) \cdot r \cdot A(\varepsilon)) \cdot \ldots \cdot A(\varepsilon)) \cdot A(\varepsilon)$$

can be computed by successive applications of the procedure described in part A.

Theorem 9.1. *The multiplication constructed above is well defined and associative.*

9.1.2. Let Ω be a domain in $\mathbb{R}^n$. We say that a diffeomorphism $p_{t_1,\ldots,t_n}(\varphi)$ depends analytically on parameters $(t_1, \ldots, t_n) \in \Omega$ if the function defined by $P(t_1, \ldots, t_n, \varphi) = P_{t_1,\ldots,t_n}(\varphi)$ is real analytic in the variables $t_1, \ldots, t_n, \varphi$. An element $p \cdot A(s) \cdot q$ depends analytically on parameters $t_1, \ldots, t_n$ if p, q are real analytic in $t_1, \ldots, t_n$.

Theorem 9.2. *Let us suppose that $\gamma_1, \gamma_2 \in \Gamma$ depends analytically on parameters $t_1, \ldots, t_n$. Then $\gamma_1 \gamma_2$ also depends analytically on $t_1, \ldots, t_n$.*

Using Theorem 9.2, it is possible to prove any property of Γ (for example, Theorems 9.1 and 9.3) in a neighbourhood of the identity and then to extend it by real analyticity.

9.1.3. The Complex Structure on Γ. Let $\mathcal{M}$ be the set of all real analytic maps ρ of the circle S^1 into the disk $|z| < 1$ such that:

1) $\rho'(e^{i\varphi}) \neq 0$;
2) the curve $\rho(e^{i\varphi})$ is a positively oriented Jordan curve.

We shall assume that $\mathcal{M}$ is canonically imbedded into Γ (in 9.1.1 A, we have constructed an element of Γ for any $\rho \in \mathcal{M}$). Then the set of maps $p\mathcal{M}$, where $p \in \text{Diff}_a$, forms a complex atlas on Γ. The notion of holomorphic dependence of an element $\gamma_{u_1,\ldots,u_n} \in \Gamma$ on the parameters $u_1, \ldots, u_n$ is defined in an obvious way.

Lemma 9.1. *Suppose that $\gamma_1, \gamma_2 \in \Gamma$ depend holomorphically on parameters $u_1, \ldots, u_n$; then the same is true also for $\gamma_1\gamma_2$.*

9.1.4. Remarks.

1) The covering group of Γ is constructed similarly as was done in 1.2. Theorem 9.3 implies that there exists a central extension of Γ.
2) The semigroup of holomorphic univalent maps of the unit disk (strictly) into itself can be canonically imbedded into Γ.
3) If a product of elements in $\mathcal{M}$ has a sense as the product of maps, then this product coincides with the product in the sense of Γ.

9.2. The Continuation of Representations to the Semigroup Γ.

Theorem 9.3. *Let ρ be a unitary highest weight representation of the group Diff_a and let $d\rho$ be the corresponding representation of $\mathcal{L}$. Then the formula*

$$\hat{\rho}(p \cdot A(t) \cdot q) = \rho(p)\exp\left(-td\rho(L_0)\right)\rho(q)$$

gives a projective representation of Γ by nuclear contractions of the Hilbert space.

9.3. The Integrability of the Nonunitarizable Modules $L(h, c)$.

Theorem 9.4. *(Neretin (1986)) Any representation $L(h, c)$; $h, c \in \mathbb{C}$, of the Virasoro algebra $\mathcal{L}$ (not necessarily unitarizable) can be integrated to a projective representation of the group Diff on a Fréchet spaces.*

Proof. We want to extend the construction of 5.1.3 to any values $\alpha, c \in \mathbb{C}$. Let us consider any representation of the group Diff belonging to the trivial series (see 1.4). Fix $k \in \mathbb{Z}$ and let $H_\alpha = H_\alpha^+ \oplus H_\alpha^-$, where H_α^+ is generated by e_n, $n \geq k$, and H_α^- is generated by e_n, $n < k$. Let us consider the group $\text{GL}(\infty, \mathbb{C}) \times \text{GL}(\infty, \mathbb{C})$ on H_α consisting of all operators preserving both subspaces $H_\alpha^\pm$. The standard technique of Sect.4 shows that the operators $R_{\alpha,s}(q)$, where $q \in$ Diff, belong to the group $(\mathrm{GL}(\infty, \mathbb{C}), \mathrm{GL}(\infty, \mathbb{C}) \times \mathrm{GL}(\infty, \mathbb{C}))$. The imbedding of the group $(\mathrm{U}(2\infty), \mathrm{U}(\infty) \times \mathrm{U}(\infty)) \to (\mathrm{O}(4\infty, \mathbb{R}), \mathrm{U}(2\infty))$ (see 3.6.5) can be analytically continued to the imbedding

$$(\mathrm{GL}(2\infty, \mathbb{C}), \mathrm{GL}(\infty, \mathbb{C}) \times \mathrm{GL}(\infty, \mathbb{C})) \to (\mathrm{O}(4\infty, \mathbb{C}), \mathrm{GL}(2\infty, \mathbb{C})).$$

If we restrict the spinor representation of the group $(\mathrm{O}(4\infty, \mathbb{C}), \mathrm{GL}(2\infty, \mathbb{C}))$ to the group Diff, we get the series $A_{\alpha, s, k}$ of projective representations of Diff. On the level of the Lie algebra $\mathcal{L}$, these modules are basically given by the formulas (5.4). All modules $L(h, c)$ can be found among their submodules. The last point to be checked is that subrepresentations $A_{\alpha, s, k}$ of Diff in the space Λ_0 (see 3.4.3) are in one-to-one correspondence with submodules of $\mathcal{L}$ in the space of finite vectors in Λ_0.

9.4. Representations of Complex Groups $C^\infty(S^1, K)$.

Theorem 9.5. (Goodman and Wallach (1984)) *Any representation $L(\Lambda)$ of the algebra $\widehat{\mathfrak{g}}$ can be integrated to a holomorphic representation of the complex group $C^\infty(S^1, G)$ in a topological vector space.*

Let us now describe an example of such a representation Neretin (1986). Let H_- be the space of functions on S^1 with values in $\mathbb{C}^n$ satisfying the condition $f(\varphi + \pi) = -f(\varphi)$. Let $C^\infty(S^1, \mathrm{SO}(n, \mathbb{C}))$ acts on H_- by the multiplication by functions with values in the orthogonal matrices satisfying the condition $g(\varphi + \pi) = g(\varphi)$. Let $f = (f_1, \ldots, f_n)$. The scalar product in H_- is given by $\int \sum f_j \overline{g_j} d\varphi$. Let the group $\mathrm{O}(2\infty, \mathbb{C})$ consist of operators preserving the symmetric bilinear form $\int \sum f_j \overline{g_j} d\varphi$ and let its subgroup $GL(\infty, \mathbb{C})$ be defined as the group of all operators commuting with the Hilbert transform (8.1).

Theorem 9.6. *The group $C^\infty(S^1, \mathrm{SO}(n, \mathbb{C}))$ is contained in the group $(\mathrm{O}(2\infty, \mathbb{C}), \mathrm{GL}(\infty, \mathbb{C}))$ on the space H_-.*

(The proof is the same as in 8.1.) It is now sufficient to restrict the spin representation of $(\mathrm{O}(2\infty, \mathbb{C}), \mathrm{GL}(\infty, \mathbb{C}))$ to the subgroup $C^\infty(S^1, \mathrm{SO}(n, \mathbb{C}))$. This is, of course, nothing else than the continuation of the construction from 8.1 to the complex group. The representation of $C^\infty(S^1, \mathrm{SO}(n, \mathbb{C}))$ constructed in this way correponds to the basic module (see 8.1).

Appendices

A.9.1. Siegel Domains for Diff$_a$.

Let $H \subset$ Diff$_a$ be the group of rotations. Then an almost complex structure, invariant with respect to Diff$_a$, can evidently be defined on the manifold Diff$_a/H$: the tangent space at the identity is isomorphic to the space of vector fields on S^1 with zero mean, the almost complex structure I on T is the Hilbert transform (see (4.10)), $I^2 = -1$. The existence of a complex structure on Diff$_a/H$ follows from the following fact.

Theorem 9.7. (A.A.Kirillov) *There exists a canonical one-to-one map ρ of the manifold Diff_a/H onto the space Z of all univalent functions holomorphic on the closed disk $|z| \le 1$ which are normalized by the conditions $f(0) = 0, f'(0) = 1$.*

The Construction of ρ. Let $p \in \mathrm{Diff}_a$. Consider the one-dimensional complex manifold M obtained by gluing the domains $U_1 : |z| \le 1$ and $U_2 : |z| \ge 1$ along the boundary in such a way that the point $e^{i\varphi} \in U_1$ is identified with the point $e^{ip(\varphi_1)} \in U_2$. Then M is simply connected; hence there exists a one-to-one holomorphic map f of the manifold M onto the Riemann sphere $\mathbb{C}$. We can assume that $f(0) = 0$, $f(\infty) = \infty$, $f'(0) = 1$. Then the restriction of f to the domain U_1 will be the desired univalent function in Z.

The Construction of ρ^{-1}. Let $f \in Z$. Then $f(e^{i\varphi})$ gives a Jordan curve in the plane, denoted by C. Let μ be a univalent map of the exterior of the curve C onto the exterior of the unit disk $|z| \ge 1$ such that $\mu(\infty) = \infty$. Then $\mu \circ f$ induces a diffeomorphism of the circle $|z| = 1$, which is defined only up to composition with a rotation.

Corollary. There exists a natural action of the group Diff_a on the space Z of univalent functions.

Postscriptum: Category Shtan

The text presented above was written (in Russian) in winter 1986–1987 and something has changed meantime.

A description of the state of the theory in 1986 is given in Sect. 9 but mysterious constructions of this section became simple and clear meanwhile.

The category Shtan. Objects of the category Shtan are nonnegative integers.

The first definition of a morphism. Let m, n be objects of Shtan. A morphism $m \to n$ is a collection (R, r_i^+, r_j^-), where

1°) R is a compact one-dimensional complex manifold with a boundary; the boundary of R consists of $(m + n)$ components,

2°) $r_i^+ : S^1 \to R$ ($1 \le i \le m$); $r_j^- : S^1 \to R$ ($1 \le j \le n$) are fixed analytic parametrizations of the components of the boundary such that the orientations given by the parametrizations r_i^+ agree with (and those given by r_j^- are opposite than) the orientations induced on the components of the boundary by the canonical orientation of the surface R.

Two morphisms (R, r_i^+, r_j^-) and (Q, q_i^+, q_j^-) from m to n are equivalent if there exists a biholomorphic map $\theta : R \to Q$ such that $r_\alpha^\pm(e^{i\varphi}) = q_\alpha^\pm(e^{i\varphi})$.

Now let us define a product of morphisms $(R, r_i^+, r_j^-) : m \to n$ and $(Q, q_j^+, q_\alpha^-) : n \to k$ by pasting together R and Q at points $r_j^-(e^{i\varphi})$ and $q_j^+(e^{i\varphi})$. As a result we get a surface S and the parametrizations $r_i^+(e^{i\varphi})$ and $q_\alpha^-(e^{i\varphi})$ of its boundary components.

The second definition of a morphism. Let m, n be objects of Shtan. Let us define domains $\mathcal{D}_+ = \{z : |z| \le 1\}$ and $\mathcal{D}_- = \{z : |z| \ge 1\}$ in the Riemann sphere $\overline{\mathbb{C}} = \mathbb{C} \cup \infty$. A morphism from m to n is a collection $(S, \sigma_i^+, \sigma_j^-)$ such that

$1^+)$ S is a compact one-dimensional complex manifold without boundary,
$2^+)$ $\sigma_i^+ : \mathcal{D}^+ \to R$ and $\sigma_j^- : \mathcal{D}^- \to R$ are fixed holomorphic univalent maps,
$3^+)$ the intersections

$$\sigma_\alpha^+(\mathcal{D}_+) \cap \sigma_\beta^-(\mathcal{D}_-); \sigma_\alpha^+(\mathcal{D}_+) \cap \sigma_\beta^+(\mathcal{D}_+); \sigma_\alpha^-(\mathcal{D}_-) \cap \sigma_\beta^-(\mathcal{D}_-)$$

are empty.

The two definitions are equivalent. Indeed, let a collection $(S, \sigma_i^+, \sigma_j^-)$: $m \to n$ satisfy the conditions $1^+ - 3^+$. Let us define

$$R = S \setminus \left[\left(\cup_i \sigma_i^+(\mathcal{D}_+^o) \right) \bigcup \left(\cup_j \sigma_j^-(\mathcal{D}_-^o) \right) \right],$$

where $\mathcal{D}_\pm^o$ denotes the interior of $D_\pm$ and $r_\alpha^\pm(e^{i\varphi}) = \sigma_\alpha^\pm(e^{i\varphi})$. Then the collection (R, r_i^+, r_j^-) satisfies the conditions $1^o - 2^o$.

The semigroup Γ. Let us consider a semigroup $\tilde{\Gamma}$ of all morphisms from 1 to 1. Let $\Gamma \subset \tilde{\Gamma}$ be the subsemigroup consisting of all triples (R, r_+, r_-) such that R is an annulus. This semigroup Γ is isomorphic to the semigroup which was constructed in 9.1.

Representations of categories. Let $\mathcal{K}$ be a category. A projective representation (R, ρ) of the category $\mathcal{K}$ is a rule associating a linear space $T(V)$ to each object V of $\mathcal{K}$ and a linear operator $\rho(P) : R(V) \to R(W)$ to each morphism $P \in \text{Mor}_\mathcal{K}(V, W)$ in such way that for any three objects V, W and Y and for any two morphisms $P \in \text{Mor}_\mathcal{K}(V, W), Q \in \text{Mor}_\mathcal{K}(W, Y)$, there is a nonzero complex number $c(Q, P)$ such that

$$\rho(Q)\rho(P) = c(Q, P)\rho(Q, P).$$

As was shown in Nazarov et al. (1989), Neretin (1989b), Neretin (1990), the spin representation and the Weil representation are in fact representations of certain categories $\overline{G\mathcal{D}}$ and $\overline{\text{Sp}}$. The simplest way to construct representations of the semigroup Γ and the category Shtan is to imbed the category Shtan to the categories $\overline{G\mathcal{D}}$ and $\overline{\text{Sp}}$, see Neretin (1989a), Neretin (1989b) (the last construction for Shtan is equivalent to physical construction from Alvarez-Gaume et al. (1988)).

The category Shtan and the categories $\overline{G\mathcal{D}}$ and $\overline{\mathrm{Sp}}$ are partial cases of train constructions, see Neretin (1991).

Conformal field theory. The category Shtan was constructed independently by M. Kontsevich and G. Segal in 1987 – 88. They showed that the conformal field theory is, roughly speaking, representation theory of category Shtan. This statement should not be taken literally but it is closed to reality.

Fusion, tangle categories, etc. In the period 1990 – 94, the main activity was concentrated on the following question. Let $\mathrm{Mor}^g_{\mathrm{Shtan}}(m,n) \subset \mathrm{Mor}_{\mathrm{Shtan}}(m,n)$ be the subset consisting of Riemann surfaces equivalent to the sphere with $g-1$ handles. Let $\Gamma^{(n)}$ be the product $\Gamma \times \Gamma \times \ldots \times \Gamma$ (n copies). It is evident that $\Gamma^{(n)} \subset \mathrm{Mor}_{\mathrm{Shtan}}(n,n)$. Let $\pi_1^+, \ldots, \pi_m^+$, resp. $\pi_1^-, \ldots, \pi_m^-$, be holomorphic representations of the semigroup Γ in spaces $V_1^+, \ldots, V_m^+$, resp. $V_1^-, \ldots, V_m^-$, with the same level $c = \mathrm{const}$. Then there exist multivalued holomorphic functions $\mathcal{Z}$ on $\mathrm{Mor}^g_{\mathrm{Shtan}}(m,n)$ with values in operators $\otimes_j V_j^+ \to \otimes_i V_i^-$ such that

$$\mathcal{Z}\left((R_1^-, \ldots, R_n^-) \cdot Q \cdot (R_1^+, \ldots, R_n^+)\right) = \left[\otimes_i \pi_i(R_i^-)\right] \cdot \mathcal{Z}(Q) \cdot \left[\otimes_j \pi_j(R_j^+)\right],$$

where $Q \in \mathrm{Mor}^g_{\mathrm{Shtan}}(m,n)$, $(R_1^-, \ldots, R_n^-) \in \Gamma^{(n)}$, $(R_1^+, \ldots, R_n^+) \in \Gamma^{(m)}$.

The function $\mathcal{Z}$ can be clearly interpreted as a weak variant of the definition of a representation. The monodromy transformations of the function $\mathcal{Z}$ are linear operators. Consider, for example, the case $g = -1$ (i.e. the Riemann surface is the sphere with $m+n$ holes. The fundamental group of moduli of such surfaces is the (pure) braid group. Hence the monodromy operators of the function $\mathcal{Z}$ form a representation of the braid group.

More careful considerations related to the completion of modules show that the pictures related to different pairs (m,n) can be considered together and that the monodromy operators in fact form a representation of a "tangle category" (the tangle category is a hybrid of Brouer semigroups and braid groups; the groups of authomorphisms of objects of this category are the braid groups).

See Verlinde (1988), Tsuchia, Kanie (1989), Reshetikhin, Turaev (1991), Kazhdan, Lusztig (1993).

Comments on the References

The method of second quantization is described in Berezin (1965), see also Kashiwara and Vergne (1978), Neretin (1986), Segal (1981). For the theory of representations of (G, K)-pairs see Ol'shanskij (1983, 1990). Representations of the Virasoro algebra are discussed in Feigin and Fuks (1982), Goddard and Horsley (1976), Goddard et al. (1986), Goodman and Wallach (1984), Goodman and Wallach (1985),

Kent (1991), Lepowsky et al. (1985), Lepowsky and Wilson (1978), Lepowsky and Wilson (1984), Neretin (1983), Neretin (1986, 1993), Segal (1981); representations of affine Lie algebras in Frenkel (1981), Frenkel and Kac (1980), Garland (1980), Goodman and Wallach (1984), Kac (1983), Kac et al. (1981), Lepowsky et al. (1985), Lepowsky and Wilson (1978), Lepowsky and Wilson (1984), Macdonald (1972), Segal (1981); the theory of characters (not mentioned here) in Kac (1983), Lepowsky et al. (1985), Lepowsky and Wilson (1978); analytic questions can be found in Goodman and Wallach (1984), Goodman and Wallach (1985), Neretin (1983), Neretin (1986), Segal (1981) and physical applications in Frenkel (1981), Garland (1980), Goddard and Horsley (1976), Lepowsky et al. (1985).

For other themes of representation theory see Albeverio et al. (1983), Guichardet (1972), Ismagilov (1971), Kirillov (1974), Vershik et al. (1973), Vershik et al. (1983).

An extended bibliography can be found in Kac (1983), Lepowsky et al. (1985).

References*

Albeverio S., Høgh-Krohn R., Testard D., Vershik A. (1983): Factorial representations of path group. J. Funct. Anal. *51*, No. 3, 115–130, Zbl. 522.22013

Alvarez-Gaume L., Gomes G., Moore D., Vafa C. (1988): Strings in operators formalism. Nucl. Phys. *B303*, No.3, 455-521

Arnol'd V.I., Il'yashenko Yu.S. (1985): Ordinary differential equations. Itogi Nauki Tekh., Ser. Sovrem. Probl. Mat., Fundam. Napravleniya *1*, 7–149. English transl.: Encycl. Math. Sci. *1*, 1–148 (1988), Zbl. 602.58020

Berezin F.A. (1965): The method of second quantization. Moscow, Izdat. Nauka, 235 pp. English transl.: Academic Press, London (1966), Zbl. 131,448

Drinfel'd V.G. (1989): Quasi-Hopf algebras. Algebra i analiz *1*, No.6. English transl.: Leningrad Math.J. *1*, 1419-1457

Feigin B.L., Fuks D.B. (1982): Invariant skew-symmetric differential operators on the line and Verma modules over the Virasoro algebra. Funkts. Anal. Prilozh. *16*, No. 2, 47–63. English transl.: Funct. Anal. Appl. *16*, 114–126 (1982), Zbl. 505.58031

Frenkel I.B. (1981): Two constructions of affine Lie algebra representations and boson – fermion correspondence in quantum field theory. J. Funct. Anal. *44*, 259–327, Zbl. 479.17003

Frenkel I.B., Kac V.G. (1980): Basic representations of affine Lie algebras and dual resonance models. Invent. Math. *62*, No. 1, 23–66, Zbl. 493.17010

Garland H. (1980): The arithmetic theory of loop groups. Publ. Math., Inst. Hautes Etud. Sci. *52*, 5–136, Zbl. 475.17004

Goddard P., Horsley R. (1976): The group theoretic structure of duals vertices. Nucl. Phys. *B111*, No. 2, 272–296

Goddard P., Kent A., Olive D. (1986): Unitary representations of the Virasoro and super Virasoro algebra. Commun. Math. Phys. *103*, No. 1, 105–119, Zbl. 588.17014

---

*For the convenience of the reader, references to reviews in Zentralblatt für Mathematik (Zbl.), compiled using the MATH database, and Jahrbuch über die Fortschritte in der Mathematik (FdM) have, as far as possible, been included in this bibliography.

Goodman R., Wallach N.R. (1984): Structure and unitary cocycle representations of loop groups and the group of diffeomorphisms of the circle. J. Reine Angew. Math. *347*, 69–133, Zbl. 514.22012

Goodman R., Wallach N.R. (1985): Projective unitary positive-energy representations of Diff(S^1). J. Funct. Anal. *63*, No. 3, 299–321, Zbl. 636.22013

Guichardet A. (1972): Symmetric Hilbert Spaces and Related Topics. Lect. Notes Math. *261*, 179 pp., Zbl. 265.43008

Ismagilov R.S. (1971): Unitary representations of the group of diffeomorphisms of the circle. Funkts. Anal. Prilozh. *5*, No. 1, 45–53. English transl.: Funct. Anal. Appl. *5*, 209–216 (1972), Zbl. 235.58006

Kac V.G. (1983): Infinite-dimensional Lie Algebras. Boston, Birkhauser, 251 pp., Zbl. 537.17001

Kac V.G., Kazhdan D.A., Lepowsky I., Wilson H.L. (1981): Realisation of basic representations of Euclidean Lie algebras. Adv. Math. *42*, No. 1, 83–112, Zbl. 476.17003

Kashiwara M., Vergne M. (1978): On the Segal-Shale-Weil representations and harmonic polynomials. Invent. Math. *44*, No. 1, 1–47, Zbl. 375.22009

Kazhdan D., Lusztig G. (1993): Tensor structures arising from affine Lie algebras I, II. J. Amer. Math. Soc. *6*, 905-1001

Kent A. (1991): Singular vectors of Virasoro algebra. Phys Lett. *B273*, 56-62

Khafizov M.U. (1990): A quasiinvariant smooth measure on the diffeomorphism group of a domain. Mat. Zametki *48*, 134-142. English transl.: Math. Notes *48*, 968-974

Kirillov A.A. (1974): Unitary representations of the group of diffeomorphisms and some of its subgroups, Preprint IPM, Moscow, 40 pp. English transl.: Sel. Math. Sov. *1*, No. 4, 351–372 (1981), Zbl. 515.58009

Kuo, Hui-Hsung (1975): Gauss measures in Banach spaces. Lect. Notes in Math. *463*, Springer-Verlag

Lepowsky J. (1985): Calculus of vertex operators. Proc. Nat. Acad. Sci. USA *82*, No.24, 8295-8299

Lepowsky J., Singer J.M., Mandelstam S. (eds.) (1985): Vertex Operators in Mathematics and Physics. New York, Springer-Verlag, 482 pp., Zbl. 549.00013

Lepowsky J., Wilson R.L. (1978): Construction of the affine Lie algebra A_1^1. Commun. Math. Phys. *62*, No. 1, 45–53, Zbl. 388.17006

Lepowsky J., Wilson R.L. (1984): Structure of standard moduls. I. II. Invent. Math. *77*, 199–290, Zbl. 577.17009; Invent. Math. *79*, No. 3, 417–442 (1985), Zbl. 577.17010

Macdonald I.G. (1972): Affine root systems and Dedekind's η-function. Invent. Math. *15*, No. 1, 91–143, Zbl. 244.17005

Nazarov M.L., Neretin Yu.A., Ol'shanskij G.I. (1989): Semigroupes engendre par la representation de Weil du groupe symplectique de dimension infinie. C. R. Acad. Sci. Paris, Ser. I *309*, No. 7, 443–446, Zbl. 709.43007

Neretin Yu.A. (1983): Unitary highest weight representations of the group of diffeomorphisms of the circle. Funkts. Anal. Prilozh. *17*, No. 3, 85–86. English transl.: Funct. Anal. Appl. *17*, 235–237 (1983), Zbl. 574.43005

Neretin Yu.A. (1986): On the spinor representation of O(∞, C). Dokl. Akad. Nauk SSSR *289*, No. 2, 282–285. English transl.: Sov. Math. Dokl. *34*, 71–74 (1987), Zbl. 616.22008

Neretin Yu.A. (1989a): Holomorphic extensions of representations of the group of diffeomorphisms of the circle. Mat. Sb. *180*, No. 5, 634–657. English transl.: Math. USSR, Sb. *67*, No. 1, 75–97 (1990), Zbl. 696.58010

Neretin Yu.A. (1989b): Spinor representation of infinite dimensional orthogonal semigroup and Virasoro algebra. Funkts. Anal. Prilozh. *23*, No. 3, 32–44. English transl.: Funct. Anal. Appl. *23*, No. 3, 196–207 (1989), Zbl. 693.17010

Neretin Yu.A. (1990): On some semigroup of operators in bosonic Fock space. Funkts. Anal. Prilozh. *24*, No. 2, 63–73. English transl.: Funct. Anal. Appl. *24*, No. 2, 135–144 (1990), Zbl. 738.47035

Neretin Yu.A. (1991): Infinite-dimensional groups. Their mantles, trains and representations. In: Topics in P representation theory (ed. A.A.Kirillov), Adv. Sov. Math. *2*, 103-171

Neretin Yu.A. (1993): Some remarks on quasiinvariant actions of the group of diffeomorphisms of the circle and the loop groups. Preprint Scuola Norm. Sup. Pisa, No.15

Ol'shanskij G.I. (1981): Invariant cones in Lie algebras, Lie semigroups and holomorphic discrete series. Funkts. Anal. Prilozh., *15*, No. 4, 53–66. English transl.: Funct. Anal. Appl. *15*, 275–285 (1982), 484.22023

Ol'shanskij G.I. (1983): Unitary representations of infinite-dimensional (G, K)-pairs and the Howe formalism. Dokl. Akad. Nauk SSSR *269*, No. 1, 33–36. English transl.: Sov. Math., Dokl. *27*, 290–294 (1983), Zbl. 532.22019

Ol'shanskij G.I. (1990): Unitary representations of infinite-dimensional pairs (G, K) and formalism of Howe. In: Representation of Lie groups and related topic. Adv. Stud. Contemp. Math. *7*, 269–463, Zbl. 724.22020

Reshetikhin N.Yu., Turaev V.G. (1991): Invariants of 3-manifolds via link polynomials and quantum groups. Inv. Math. *103*, 547-597

Segal G. (1981): Unitary representations of some infinite dimensional groups. Commun. Math. Phys. *80*, No. 3, 301–342, Zbl. 495.22017

Shavgulidze E.T. (1988): On a measure which is quasiinvarinat with respect to action of the diffeomorphism group of a finite-dimensional manifold. Dokl. Akad. Nauk SSSR *303*, No.4, 811-814. English transl.: Sov. Math. Dokl. *38*, 612-629 (1989)

Tsuchia A., Kanie Y. (1988): Vertex operators in conformal field theory on $\mathbb{P}^1$ and monodromy representations of braid group. Adv. Stud. Pure Math. *16*, 297-327

Vershik A.M., Gel'fand I.M., Graev M.I. (1973): Representations of the group $SL(2, \mathbb{R})$, where R is a ring of functions. Usp. Mat. Nauk *28*, No. 5, 83–128. English transl.: Russ. Math. Surv. *28*, No. 5, 87–132 (1973), Zbl. 288.22005

Vershik A.M., Gel'fand I.M., Graev M.I. (1983): Commutative model of representations of the current group $SL_2(\mathbb{R})^X$ connected with a unipotent subgroup. Funkts. Anal. Prilozh. *17*, No. 2, 70–72. English transl.: Funct. Anal. Appl. *17*, 137–139 (1983), Zbl. 536.22008

Zhelobenko D.P., Shtern A.I. (1983): Representations of Lie groups. Moscow, Izdat. Nauka, 360 pp. (Russian), Zbl. 521.22006

Author Index

Subject Index

Encyclopaedia of Mathematical Sciences
Editor-in-Chief: R. V. Gamkrelidze

Dynamical Systems

Volume 1: **D. V. Anosov, V. I. Arnol'd** (Eds.)
Dynamical Systems I
Ordinary Differential Equations and Smooth Dynamical Systems
2nd printing. 1994. IX, 233 pp. 25 figs.
ISBN 3-540-17000-6

Volume 2: **Ya. G. Sinai** (Ed.)
Dynamical Systems II
Ergodic Theory with Applications to Dynamical Systems and Statistical Mechanics
1989. IX, 281 pp. 25 figs.
ISBN 3-540-17001-4

Volume 3: **V. I. Arnol'd** (Ed.)
Dynamical Systems III
Mathematical Aspects of Classical and Celestial Mechanics
1993. 2nd ed. XIV, 291 pp. 81 figs.
ISBN 3-540-57241-4

Volume 4: **V. I. Arnol'd, S. P. Novikov** (Eds.)
Dynamical Systems IV
Symplectic Geometry and its Applications
1989. VII, 283 pp. 62 figs.
ISBN 3-540-17003-0

Volume 5: **V. I. Arnol'd** (Ed.)
Dynamical Systems V
Bifurcation Theory and Catastrophe Theory
1994. Approx. 280 pp. 130 figs.
ISBN 3-540-18173-3

Volume 6: **V. I. Arnol'd** (Ed.)
Dynamical Systems VI
Singularity Theory I
1993. 245 pp. 55 figs.
ISBN 3-540-50583-0

Volume 16: **V. I. Arnol'd, S. P. Novikov** (Eds.)
Dynamical Systems VII
Nonholonomic Dynamical Systems. Integrable Hamiltonian Systems
1994. VII, 341 pp. 9 figs.
ISBN 3-540-18176-8

Partial Differential Equations

Volume 30: **Yu. V. Egorov, M. A. Shubin** (Eds.)
Partial Differential Equations I
Foundations of the Classical Theory
1991. V, 259 pp. 4 figs.
ISBN 3-540-52002-3

Volume 31: **Yu. V. Egorov, M. A. Shubin** (Eds.)
Partial Differential Equations II
Elements of the Modern Theory. Equations with Constant Coefficients
1995. VII, 255 pp.
ISBN 3-540-52001-5

Volume 32: **Yu. V. Egorov, M. A. Shubin** (Eds.)
Partial Differential Equations III
The Cauchy Problem. Qualitative Theory of Partial Differential Equations
1991. VII, 197 pp.
ISBN 3-540-52003-1

Volume 33: **Yu. V. Egorov, M. A. Shubin** (Eds.)
Partial Differential Equations IV
Microlocal Analysis and Hyperbolic Equations
1993. VII, 241 pp. 6 figs.
ISBN 3-540-53363-X

Volume 63: **Yu. V. Egorov, M. A. Shubin** (Eds.)
Partial Differential Equations VI
Elliptic and Parabolic Operators
1994. VII, 325 pp. 5 figs.
ISBN 3-540-54678-2

Springer

B4.10.003

Encyclopaedia of Mathematical Sciences
Editor-in-Chief: R. V. Gamkrelidze

Analysis

Volume 13: **R.V. Gamkrelidze** (Ed.)
Analysis I
Integral Representations and Asymptotic Methods
1989. VII, 238 pp. 3 figs.
ISBN 3-540-17008-1

Volume 14: **R.V. Gamkrelidze** (Ed.)
Analysis II
Convex Analysis and Approximation Theory
1990. VII, 255 pp. 21 figs.
ISBN 3-540-18179-2

Volume 26: **S.M. Nikol'skij** (Ed.)
Analysis III
Spaces of Differentiable Functions
1991. VII, 221 pp. 22 figs.
ISBN 3-540-51866-5

Volume 27: **V.G. Maz'ya, S.M. Nikol'skij** (Eds.)
Analysis IV
Linear and Boundary Integral Equations
1991. VII, 233 pp. 4 figs.
ISBN 3-540-51997-1

Volume 19: **N.K. Nikol'skij** (Ed.)
Functional Analysis I
Linear Functional Analysis
1992. V, 283 pp.
ISBN 3-540-50584-9

Volume 20: **A.L. Onishchik** (Ed.)
Lie Groups and Lie Algebras I
Foundations of Lie Theory. Lie Transformation Groups
1993. VII, 235 pp. 4 tabs.
ISBN 3-540-18697-2

Several Complex Variables

Volume 7: **A.G. Vitushkin** (Ed.)
Several Complex Variables I
Introduction to Complex Analysis
1990. VII, 248 pp.
ISBN 3-540-17004-9

Volume 8: **A.G. Vitushkin, G.M. Khenkin** (Eds.)
Several Complex Variables II
Function Theory in Classical Domains. Complex Potential Theory
1994. VII, 260 pp. 19 figs.
ISBN 3-540-18175-X

Volume 9: **G.M. Khenkin** (Ed.)
Several Complex Variables III
Geometric Function Theory
1989. VII, 261 pp.
ISBN 3-540-17005-7

Volume 10: **S.G. Gindikin, G.M. Khenkin** (Eds.)
Several Complex Variables IV
Algebraic Aspects of Complex Analysis
1990. VII, 251 pp.
ISBN 3-540-18174-1

Volume 54: **G.M. Khenkin** (Ed.)
Several Complex Variables V
Complex Analysis in Partial Differential Equations and Mathematical Physics
1993. VII, 286 pp.
ISBN 3-540-54451-8

Volume 69: **W. Barth, R. Narasimhan** (Eds.)
Several Complex Variables VI
Complex Manifolds
1990. IX, 310 pp. 4 figs.
ISBN 3-540-52788-5

B4.10.003